T0155581

NEUROMETHODS

Series Editor
Wolfgang Walz
University of Saskatchewan,
Saskatoon, SK, Canada

For further volumes:
http://www.springer.com/series/7657

Neuromethods publishes cutting-edge methods and protocols in all areas of neuroscience as well as translational neurological and mental research. Each volume in the series offers tested laboratory protocols, step-by-step methods for reproducible lab experiments and addresses methodological controversies and pitfalls in order to aid neuroscientists in experimentation. *Neuromethods* focuses on traditional and emerging topics with wide-ranging implications to brain function, such as electrophysiology, neuroimaging, behavioral analysis, genomics, neurodegeneration, translational research and clinical trials. *Neuromethods* provides investigators and trainees with highly useful compendiums of key strategies and approaches for successful research in animal and human brain function including translational "bench to bedside" approaches to mental and neurological diseases.

Cognitive Behavioral Therapy in Youth: Tradition and Innovation

Edited by

Robert D. Friedberg

Pacific Graduate School of Psychology, Palo Alto, CA, USA

Brad J. Nakamura

Department of Psychology, University of Hawaii at Manoa, Honolulu, HI, USA

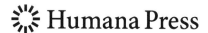 Humana Press

Editors
Robert D. Friedberg
Pacific Graduate School of Psychology
Palo Alto, CA, USA

Brad J. Nakamura
Department of Psychology
University of Hawaii at Manoa
Honolulu, HI, USA

ISSN 0893-2336   ISSN 1940-6045   (electronic)
Neuromethods
ISBN 978-1-0716-0702-2   ISBN 978-1-0716-0700-8   (eBook)
https://doi.org/10.1007/978-1-0716-0700-8

This Humana imprint is published by the registered company Springer Science+Business Media, LLC, part of Springer Nature.
The registered company address is: 1 New York Plaza, New York, NY 10004, U.S.A.

Preface to the Series

Experimental life sciences have two basic foundations: concepts and tools. The *Neuromethods* series focuses on the tools and techniques unique to the investigation of the nervous system and excitable cells. It will not, however, shortchange the concept side of things as care has been taken to integrate these tools within the context of the concepts and questions under investigation. In this way, the series is unique in that it not only collects protocols but also includes theoretical background information and critiques which led to the methods and their development. Thus it gives the reader a better understanding of the origin of the techniques and their potential future development. The *Neuromethods* publishing program strikes a balance between recent and exciting developments like those concerning new animal models of disease, imaging, in vivo methods, and more established techniques, including, for example, immunocytochemistry and electrophysiological technologies. New trainees in neurosciences still need a sound footing in these older methods in order to apply a critical approach to their results.

Under the guidance of its founders, Alan Boulton and Glen Baker, the *Neuromethods* series has been a success since its first volume published through Humana Press in 1985. The series continues to flourish through many changes over the years. It is now published under the umbrella of Springer Protocols. While methods involving brain research have changed a lot since the series started, the publishing environment and technology have changed even more radically. Neuromethods has the distinct layout and style of the Springer Protocols program, designed specifically for readability and ease of reference in a laboratory setting.

The careful application of methods is potentially the most important step in the process of scientific inquiry. In the past, new methodologies led the way in developing new disciplines in the biological and medical sciences. For example, Physiology emerged out of Anatomy in the nineteenth century by harnessing new methods based on the newly discovered phenomenon of electricity. Nowadays, the relationships between disciplines and methods are more complex. Methods are now widely shared between disciplines and research areas. New developments in electronic publishing make it possible for scientists that encounter new methods to quickly find sources of information electronically. The design of individual volumes and chapters in this series takes this new access technology into account. Springer Protocols makes it possible to download single protocols separately. In addition, Springer makes its print-on-demand technology available globally. A print copy can therefore be acquired quickly and for a competitive price anywhere in the world.

Preface

The Value of Discernment

Of course you like it—how can you not like it?! Everyone likes everything nowadays. They like the television and the phonograph and the soda pop and the shampoo and the Cracker Jack. Everything becomes everything else and it's all nice and pretty and likable. Everything is fun in the sun! Where's the discernment? Where's the arbitration that separates what I like from what I respect, what I deem worth, what has . . . listen to me now . . . significance.

Logan (2009)

A quick perusal of websites, blogs, podcasts, social media posts, brochures, and the scholarly literature reveals that cognitive behavioral therapy (CBT) is not a fad but a fashion among psychotherapists. Many practitioners and clinical service systems claim to provide CBT. However, too few practitioners properly deliver the approach at the right dose and direct toward the correct clinical problem [1]. A recent study revealed that practitioners who profess fidelity to a CBT approach do not genuinely implement the procedure [2]. Thus, it is clear that CBT is "popular" and "likeable." However, vulnerable young patients may not be really getting the approach. There is a troubling sequelae to this phenomenon. First and most importantly, good clinical care is truncated. Second, public confidence in authentic CBT can be eroded. Third, the CBT brand is potentially sullied.

These issues gave rise to our motivation to create this text and ask respected CBT scholars and practitioners to contribute chapters that address proper applications of CBT to common clinical presentations. We wanted to create a work that realistically represented both the traditions and innovations that live within the broad CBT realm. It was essential for us that contributors filled their chapters with generalizable principles and procedures. We want readers to be able to transfer material from the page to clinical practice.

Implementing CBT faithfully and with flexibility is the current watchword [3, 4]. Therefore, we invited contributors to plant their innovative procedures in fertile theoretical and empirical ground. Consequently, the chapters provide cutting-edge procedures buttressed by a full-bodied theoretical framework and reinforced by sound science. While there are points of convergence between the ever-growing number of psychotherapy modalities, there are fundamental differences too. Discerning what is and *is not* CBT is essential. Incorporating the theory and research surrounding CBT into the direct clinical care of young patients maintains fidelity. This solid footing enables innovation. We encourage readers to wrap themselves in a cloak of theoretical fidelity as they make the various practices in this book their own.

Our preface began with a passage from the play, Red, by John Logan [5]. His words gave voice to the Mark Rothko character who emphasized the need to discern "liking" from respect, worth, and significance. Many professionals and lay people "like" CBT. AND this is a good thing! BUT, Rothko is right—liking is different from respecting. We want readers to like this book but more importantly, we want you, the reader, to respect the contributors' significant work. Their ideas are indeed worthy, sown in theory, cultivated through science, and ripened by clinical experience.

Palo Alto, CA, USA   *Robert D. Friedberg*
Honolulu, HI, USA   *Brad J. Nakamura*

References

1. Comer JA, Barlow DC (2014) The occasional case against broad dissemination and implementation: retaining a role for specialty care in the delivery of psychological treatment. Am Psychol 69:1–18

2. Creed TA, Wolk CB, Feinberg B et al (2014) Relationship between community therapists' self-report of a cognitive behavioral therapy orientation and observed skills. Adm Policy Ment Health 41:1–8

3. Southam-Gerow MA, Daleiden EL, Chorpita BF et al (2014) MAPping Los Angeles County: taking an evidence-informed model of mental health care to scale. J Clin Child Adolesc Psychol 43:190–200

4. Kendall PC, Gosch E, Furr J et al (2008) Flexibility within fidelity. J Am Acad Child Adolesc Psychiatr 47:987–993

5. Logan J (2009) Red. Oberon Books, London

Acknowledgments

First and always I want to thank the love of my life, Barbara A. Friedberg, who is not only an award winning financial writer but a first-class human being who is my guiding light. My daughter, Rebecca J., whose critical reasoning and sense of humor is something that inspires me and that I aspire to! My heartfelt respect and gratitude goes to my brilliant co-editor Brad J. Nakamura whose keen insight, scholarly acumen, and clinical wisdom brought this text to a higher level. Further, I am humbled by the contributions of our chapter authors. Their work is genuinely astounding and breathtaking. Finally, to our editors Wolfgang Walz and Patrick Marton who invited us to create this text. MUCHO MUCHO THANK YOU!—Bob

I would like to express my heartfelt gratitude for the multitude of conditions that have allowed me to do the work that I love. It is impossible to identify all of these circumstances, but first and foremost thank you to my life partner, Joy N. Nakamura, who has always and continues to sacrifice her time so that I can have the time to do what I want. Thank you to my children, parents, and larger family for the opportunity to be a father, son, and brother. I am also profoundly grateful to my co-editor, Robert D. Friedberg, who has always been a role model for me across a wide range of professional and personal domains. Through this and so many other scientific contributions to our field, he continues to move the needle, helping so many students, youth, and families through his massive body of work. I am supremely thankful for the invitation and the opportunity to work together on this book. Thank you so very much to all of the authors and our editors, Wolfgang Walz and Patrick Marton. Mahalo and aloha!—Brad

Contents

Contributors

BRIAN ALLEN • *Penn State Children's Hospital, Penn State College of Medicine, Hershey, PA, USA*

EVAN ALVAREZ • *Department of Psychology, Rutgers, The State University of New Jersey, Piscataway, NJ, USA*

DARA E. BABINSKI • *Penn State College of Medicine, Hershey, PA, USA; Department of Psychiatry, Penn State Hershey Medical Center, Hershey, PA, USA*

EMILY BADIN • *Department of Clinical Psychology, Graduate School of Applied and Professional Psychology, Rutgers, The State University of New Jersey, Piscataway, NJ, USA*

KELLY N. BANNEYER • *Department of Pediatrics/Psychology Section, Texas Medical Center, Baylor College of Medicine, Houston, TX, USA*

MOLLY BOBEK • *Center on Addiction, New York, NY, USA*

MAYA BOUSTANI • *Department of Psychology, School of Behavioral Health, Loma Linda University, Loma Linda, CA, USA*

MICHELLE L. BUFFIE • *Department of Psychology, University of Maine, Orono, ME, USA*

ESTEBAN V. CARDEMIL • *Frances L. Hiatt School of Psychology, Clark University, Worcester, MA, USA*

MARISSA CASSAR • *Center for the Study and Treatment of Anxious Youth, Palo Alto University, Palo Alto, CA, USA*

BRIAN C. CHU • *Department of Clinical Psychology, Graduate School of Applied and Professional Psychology, Rutgers, The State University of New Jersey, Piscataway, NJ, USA*

BRAD DONOHUE • *University of Nevada – Las Vegas, Las Vegas, NV, USA*

JILL EHRENREICH-MAY • *Department of Psychology, University of Miami, Coral Gables, FL, USA*

JAMAL H. ESSAYLI • *Department of Pediatrics, Division of Adolescent Medicine and Eating Disorders, Penn State Health Milton S. Hershey Medical Center, Penn State College of Medicine, Hershey, PA, USA*

RACHEL FEIN • *Department of Pediatrics/Psychology Section, Texas Medical Center, Baylor College of Medicine, Houston, TX, USA*

KRISTER W. FJERMESTAD • *Department of Psychology, University of Oslo, Norway and Frambu Resource Center for Rare Disorders, Oslo, Norway*

ROBERT D. FRIEDBERG • *Pacific Graduate School of Psychology, Palo Alto, CA, USA*

CHELSIE N. GIAMBRONE • *Department of Psychiatry, Eating Disorder Center for Treatment and Research, School of Medicine, University of California San Diego, San Diego, CA, USA*

ELIZABETH R. HALLIDAY • *Department of Psychology, University of Miami, Coral Gables, FL, USA*

AARON HOGUE • *Center on Addiction, New York, NY, USA*

ANUSHA KAKOLU • *Center for the Study and Treatment of Anxious Youth, Palo Alto University, Palo Alto, CA, USA*

FRANCESCA KASSING • *Trauma and Grief (TAG) Center, Psychology Department, Texas Children's Hospital/Baylor College of Medicine, Houston, TX, USA*

HANNAH R. LAWRENCE • *Department of Psychology, University of Maine, Orono, ME, USA*

JULIETTE M. LIBER • *Department of Psychology, Utrecht University, Utrecht, The Netherlands*

JOHN E. LOCHMAN • *Department of Psychology, The University of Alabama, Tuscaloosa, AL, USA*

ALEXANDRA MACLEAN • *Center on Addiction, New York, NY, USA*

LAILA A. MADNI • *Psychology Department, Antioch University Los Angeles, Los Angeles, CA, USA*

BRYCE D. MCLEOD • *Department of Psychology, Virginia Commonwealth University, Richmond, VA, USA*

BRAD J. NAKAMURA • *Department of Psychology, University of Hawaii at Manoa, Honolulu, HI, USA*

DOUGLAS W. NANGLE • *Department of Psychology, University of Maine, Orono, ME, USA*

JESSICA L. O'LEARY • *Frances L. Hiatt School of Psychology, Clark University, Worcester, MA, USA*

NICOLE P. PORTER • *Center on Addiction, New York, NY, USA*

JENNIFER REGAN • *Hathaway Sycamores Child and Family Services, Pasadena, CA, USA; Los Angeles County Department of Mental Health, Los Angeles, CA, USA*

ELIZABETH RIDEN • *Penn State Children's Hospital, Penn State College of Medicine, Hershey, PA, USA*

DEVON E. ROMERO • *Department of Counseling, University of Texas at San Antonio, San Antonio, TX, USA*

REBECCA A. SCHWARTZ-METTE • *Department of Psychology, University of Maine, Orono, ME, USA*

RANDYE J. SEMPLE • *Department of Psychiatry and Behavioral Sciences, Keck School of Medicine, University of Southern California, Los Angeles, CA, USA*

CHAD E. SHENK • *Penn State Children's Hospital, Penn State College of Medicine, Hershey, PA, USA; Department of Human Development and Family Studies, Penn State University, State College, PA, USA*

CAMEO STANICK • *Hathaway Sycamores Child and Family Services, Pasadena, CA, USA*

ERIC A. STORCH • *Menninger Department of Psychiatry and Behavioral Sciences, Baylor College of Medicine, Houston, TX, USA*

CAITLIN STRAUBEL • *CMS Wellness, LLC, West Boylston, MA, USA*

SARA L. STROMEYER • *Department of Psychology, The University of Alabama, Tuscaloosa, AL, USA*

JASMINE THOMAS • *Center for the Study and Treatment of Anxious Youth, Palo Alto University, Palo Alto, CA, USA*

SANDRA TRAFALIS • *Center for the Study and Treatment of Anxious Youth, Palo Alto University, Palo Alto, CA, USA*

STEPHANIE VIOLANTE • *Department of Psychology, Virginia Commonwealth University, Richmond, VA, USA*

KELLY M. VITOUSEK • *Department of Psychology, University of Hawai'i at Mānoa, Honolulu, HI, USA*

Chapter 1

Overview of CBT Spectrum Approaches

Bryce D. McLeod, Krister W. Fjermestad, Juliette M. Liber, and Stephanie Violante

Abstract

Cognitive behavioral therapy (CBT) is a psychosocial treatment with strong scientific evidence supporting its use with youth for a variety of emotional and behavioral problems. This chapter provides a broad overview of CBT with youth divided into three sections. In the first section, the behavioral and cognitive theories that underlie the CBT approach are described. We also discuss how the two theories have been integrated into theories used to guide CBT. In the second section, a description of specific techniques commonly found in CBT is provided along with a review of the typical modalities employed to deliver CBT. Finally, the third section focuses on the delivery of CBT.

Key words CBT, Youth, Behavior, Cognitive, Evidence-based treatment

1 Introduction

Cognitive behavioral therapy (CBT) is a specific psychosocial treatment that has garnered strong scientific evidence supporting its use with children and adolescents (hereafter called youth) for a variety of emotional and behavioral problems. The CBT model is based on the premise that cognitions, emotions, and behaviors interact to produce and maintain emotional and behavioral problems in youth. CBT has received empirical support for a variety of problems experienced by youth, including depression [1], anxiety [2], post-traumatic stress disorder [3], eating disorders [4], externalizing disorders [5], and substance abuse [6]. CBT is thus commonly used to treat a wide variety of presenting problems for youth. The goal of this chapter is to provide a broad overview of CBT divided into three sections.

The first section focuses on the theories underlying the CBT approach. CBT represents the integration of behavior and cognitive theories. Since the important behavioral (operant, classical conditioning, social learning) and cognitive (Aaron Beck's cognitive theory) theories emerged separately prior to the development of

Robert D. Friedberg and Brad J. Nakamura (eds.), *Cognitive Behavioral Therapy in Youth: Tradition and Innovation*, Neuromethods, vol. 156, https://doi.org/10.1007/978-1-0716-0700-8_1, © Springer Science+Business Media, LLC, part of Springer Nature 2020

the CBT model, we provide a description of these prominent theories and how they are used in CBT. We then conclude the section with a brief description of how these theories were eventually integrated to form a theory designed to guide CBT.

The second section presents information about how CBT works to promote symptom reduction in youth. The behavioral and cognitive techniques commonly found in CBT programs are described. We also review the common modalities (e.g., individual, family-focused) used to deliver CBT to youth and their families.

The final section provides specific information about factors to consider when delivering CBT to youth with emotional and behavioral problems. The section is intended to describe the factors that promote the effectiveness of CBT by helping understand what therapist behaviors may optimize youth clinical outcomes.

2   Theoretical Background

In this section, we detail the theoretical underpinnings of CBT. We discuss the main behavioral and cognitive theories used to develop CBT. We also describe how the cognitive and behavioral theories were integrated to inform the CBT model.

2.1   Behavioral Theory

Behavioral theory is characterized by three types of learning processes: classical conditioning, operant conditioning, and social learning. Each of these theories has been used to develop the therapeutic techniques detailed in the next section.

2.1.1   Classical Conditioning

The theory of classical conditioning, most often associated with Ivan Pavlov [7], posits that learning occurs through repeated pairings between a neutral stimulus and a stimulus that naturally produces a behavior. An unconditioned stimulus (UCS; something that triggers a naturally occurring response) leads to an unconditioned response (UCR; naturally occurring response following the UCS). When the UCS is closely preceded by a conditioned stimulus (CS; an initially neutral stimulus) the CS will eventually produce the same response, termed the conditioned response (CR; the acquired response to the formerly neutral stimulus; [8]).

Classical conditioning is used to explain the development and maintenance of some psychological problems, especially anxiety disorders. For example, if a youth is bitten by a dog (UCS), they are likely to experience a fear response (UCR). The experience of seeing a dog (a previously neutral stimulus) and being bitten by a dog might lead to an association between seeing a dog (CS) and a feeling of fear (CR), potentially leading to the development of a specific phobia of dogs. CBT techniques that target these emotions are based on the principles of classical conditioning. For example, exposure, a therapeutic technique, is based on the principles of classical conditioning theory. To reduce a youth's anxiety about dogs, treatment would involve repeated exposure of the CS (being around dogs) in the absence of the UCS (being bitten by a dog). After extended periods of exposure to the CS in the absence of the UCS, then the CS will no longer produce the CR (fear; i.e., extinction).

2.1.2 Operant Conditioning

The theory of operant conditioning, generally credited to B. F. Skinner [9], emphasizes the role of events in the environment on behaviors. Specifically, it suggests that the likelihood that a behavior will be repeated is a function of events that immediately precede or follow the behavior. In operant theory, behaviors are influenced through positive reinforcement, defined as the addition of a stimulus that leads to an increase in behavior (e.g., giving a youth a toy to reward desired behavior), negative reinforcement defined as the removal of a stimulus that leads to an increase in behavior (e.g., removing a chore to reward desired behavior), positive punishment defined as the addition of a stimulus that leads to a decrease in behavior (e.g., adding a chore to discipline undesired behavior), and negative punishment defined as the removal of a stimulus that leads to a decrease in behavior (e.g., removing a toy to discipline undesired behavior; [10]). The contingencies of reinforcement are a main focus of operant conditioning: antecedents defined as a cue for a youth to engage in a behavior (e.g., caregiver instructions to do homework), behaviors defined as the behavior that is to be altered (e.g., doing homework), and consequences defined as what follows the behavior that makes it more or less likely to happen again (e.g., caregiver verbal praise). Understanding how these three components operate in unison to explain behavior is an important step in understanding the factors that serve to cause (antecedent) and maintain (consequence) a target behavior. This understanding, in turn, is used in CBT to inform the development of interventions that target antecedents or consequences intended to alter the behavior. Operant conditioning thus provides the foundation for many of the assessment and intervention procedures used in CBT designed to increase, or decrease, the frequency of behaviors.

2.1.3 Social Learning

Social learning theory, considered to have developed through the work of Albert Bandura [11], focuses on the influence of social contexts of behavior on the learning process. Its central premise is that behaviors are learned from the environment through observational learning. Specifically, new behaviors, skills, or information is acquired, and existing behaviors are altered, through observing others. While reinforcement and punishment are not a necessary aspect of social learning, social learning theory and operant theory can interact to explain behavior change. Rotter [12] proposed that an individual's behavior is determined not only by observing another, but by their expectation of receiving a reward or punishment. For example, a youth may be more likely to repeat an observed behavior if that behavior was rewarded, and less likely to repeat the behavior if it was punished, because they will then expect the associated reward or punishment in the future. Social learning theory has been applied to youth-focused CBT through the use of therapist modeling techniques employed to teach new skills, educating caregivers about the role modeling can play in influencing youth behaviors, and teaching caregivers to model desired behaviors for their youth [13]. Such interventions will be described in closer detail in the next section.

3   Cognitive Theory

Traditional behavior theory only focused on observable behavior and eschewed a focus on any internal processes. However, some were not convinced that behavior could be explained solely by observable stimuli and turned to the role of internal cognitions on behavior and emotions. The resulting "cognitive revolution" was a paradigm shift in psychology away from a sole reliance on observable behavior to an incorporation of unobservable cognitive processes [14, 15]. The new emphasis on the role of cognitions in psychology prompted the development of Aaron Beck's cognitive theory of depression [16], later extended to other psychological disorders. His theory assumed that cognitive processes become habitual and automatic over time, producing cognitive schemas (i.e., organizational frameworks based on patterns of internal experience), which shape our interpretations of events. These schemas may lead to inaccurate interpretations of innocuous experiences [13]. For example, if a youth holds a schema that they are "weird" or "awkward," they might interpret a group of youth laughing nearby as being directed toward them, ignoring alternative explanations (e.g., somebody told a joke). Beck described these inaccurate interpretations as "illogical thinking processes" [17], and maintained that they underlie emotional disorders.

Beck's cognitive theory suggests that different psychological disorders can be distinguished by their cognitive content. For example, the theory assumes that depression develops and is maintained through negative core beliefs about the self, the world, and the future (i.e., the cognitive triad), which interact with the environment to generate situation-specific negative cognitions [17]. Distinctly, anxiety is more likely to be characterized by cognitions about physical and psychological threat or danger [13]. Although initially conceptualized as relevant to anxiety and depression, Beck's cognitive theory was later applied to a wide range of disorders such as eating disorders [18] and substance abuse [19]. In the 1970s, Beck applied his cognitive theory to the development of cognitive therapy, which eventually resulted in the emergence of CBT.

4 Integration: Cognitive Behavioral Theory

CBT integrates theories that focus on internal (cognitive) and observable (behavioral) processes to explain the development and maintenance of emotional and behavioral problems [20]. Specifically, CBT assumes that our cognitions, behaviors, and emotions are reciprocally linked, and that altering one will result in changes in the others [21]. These reciprocal relationships serve as the central foundation for CBT. CBT integrates strategies that emphasize the influence of cognitive factors and behavioral contingencies, with the goal of altering an individual's cognitions and behaviors in order to reduce negative emotions and encourage positive behaviors.

5 CBT Techniques

Practicing CBT involves a mix of specific behavioral and cognitive techniques that are delivered within a collaborative relationship formed with the youth and their family. In this section, we review specific behavioral and cognitive techniques typically found in CBT programs for youth emotional and behavioral problems. Our coverage is designed to help highlight some of the most common CBT techniques and do not represent a complete list. The section also provides a brief description of the different modalities that have been used to deliver CBT, such as individual and group.

6 Behavioral Techniques

Behavioral techniques are primarily designed to increase the frequency of desired behaviors and decrease the frequency of undesired behaviors. The behavioral techniques are guided by operant, classical conditioning, and social learning theories (see previous section). Some of the most prominent behavioral techniques are described below.

6.1 Relaxation Training

This technique involves teaching a youth a method for self-calming. Relaxation training is found in many CBT programs and can range in type and focus. The simplest form of relaxation is diaphragmatic breathing, which involves training youths how to breath more deeply from the diaphragm. This form of relaxation training has the advantage of being brief and easy to use, which makes it a practical tool for youth in many situations. A somewhat more complicated relaxation strategy is progressive muscle relaxation, which involves teaching a youth to tense and relax different muscle groups. This can be paired with a personalized cue word (e.g., "calm") in cue-controlled relaxation. These more involved strategies help youths recognize tension in their body, so they can identify tension and then engage in relaxation strategies.

6.2   Activity Planning

Activity planning is a common technique used to reduce depression; however, it is also used with youth who may experience chaotic life situations, or perfectionism issues. Activity planning can be done by setting up a regular schedule with a youth. The schedule can include specific activities planned for each day, and a discussion about ways to increase the likelihood that the youth will be able to complete the activity. More specific planning increases the likelihood that an activity will be done. For example, instead of planning "hanging with friends," specifying which friend(s), what to do, where, and how/when to contact the friend increases probability that the activity will be done.

6.3   Exposure

Exposure is a behavioral technique mainly used with youth who experience problems with anxiety. The aim of exposure is to teach youth to approach and cope with a feared stimulus. Exposure is typically comprised of four phases [22]. The first phase is preparation and psychoeducation, in which the youth is provided a rationale for doing exposure. Hierarchy development is the second phase, in which a list of feared stimuli is created. The third phase involves repeated gradual exposure to feared stimuli with postprocessing following each exposure task. Generalization and maintenance is the final phase in which exposures are conducted in a variety of contexts to promote generalization. Through repeated practice facing a feared stimulus, the negative emotions associated with that stimulus are eventually extinguished. Repeated practice allows for habituation toward the feared stimulus. Ratings of anxiety during the exposure tasks help the youth learn that anxiety reduces over time. Exposure can be conducted imaginally (i.e., the youth visualizes the feared stimulus), in role play (i.e., an analog situation that is intended to approximate the feared situation), or in vivo (i.e., exposure to an actual feared stimulus).

6.4   Contingent Reinforcement Plans

These techniques are commonly used to increase or decrease the frequency of behavior and often are used with youth who engage in misbehavior. Contingent reinforcement plans (also called reward plans) can be set up to directly shape a youth's behavior using positive reinforcement (e.g., rewards). Reward plans can be used in-session by the therapist; however, it is more common for the therapist to teach the caregiver or the youth themselves to administer the rewards. A reward plan consists of two components: (a) a list of clearly defined target behaviors that will be rewarded; and (b) a menu of rewards and a schedule for administration. Based on the principles of operant conditioning, reward charts can involve positive reinforcement (i.e., providing a desired consequence, such as a token or money, when the desired behavior is performed) or negative reinforcement (i.e., removing or reducing something undesirable, such as a chore or even a punishment, when the desired behavior is performed). At times, the reward plan can also include negative punishment (also called response cost), as when a desired consequence or situation is removed for an undesired behavior. The simplest example is when a youth has a privilege removed following behavior, such as having screen time reduced or removed for a day for emitting a nondesirable behavior.

7   Cognitive Techniques

Cognitive techniques are designed to address thinking patterns that influence the behavior and emotions of youth. An essential notion guiding cognitive techniques is that cognitions may negatively or positively influence our emotions and/or behaviors. This is particularly so in situations that involve intense or difficult emotions. Cognitive techniques fall into two general categories. First, cognitive techniques can help build cognitive processes designed to help youth cope with difficult situations or emotions (e.g., problem solving) by providing a new way of processing information about the world (e.g., psychoeducation). Second, cognitive techniques can help change unhelpful ways of thinking and counteract cognitive errors (e.g., cognitive restructuring).

7.1   *Problem Solving*

Teaching problem solving skills is a central technique that has cognitive and behavioral components. With problem solving skills, the therapist teaches a youth to cope with difficult situations or emotions by using specific problem solving steps. Problem solving typically involves identifying the problem, listing multiple solutions to the problem, considering the pros and cons of each solution, selecting and using a solution, and then reviewing the results. Youths are often encouraged to reward themselves for performing productive problem solving.

7.2   *Psychoeducation*

This is a commonly employed technique with cognitive components. Psychoeducation provides new information about the symptoms, diagnoses, and treatment process. Providing information about each of these components helps the youth (or their parents) better understand what they are experiencing, normalize their experience, and instill hope regarding the treatment.

7.3 Cognitive Restructuring

This is a core cognitive technique used in CBT programs for a wide variety of presenting problems for youth. This entails attempting to alter dysfunctional cognitions into more functional cognitions. This process can include three stages: (a) identification of current cognitions and the negative feelings and maladaptive behavior that emerge from them along with underlying assumptions and schemas; (b) challenging of the dysfunctional cognitions; and (c) generating alternative cognitions that do not lead to dysfunctional feelings and behaviors, but help the youth cope with certain situations better.

Determining which cognitions lead to improved or impaired feelings is an important first step. It is critical to teach youth that focusing on some cognitions may help control and manage a difficult situation, whereas focusing on other cognitions may escalate and complicate feelings and subsequent coping in the situations that elicit the cognitions. The second and third steps involve challenging dysfunctional cognitions and generating alternate cognitions. In practice, these two steps are often included in the same exercises. Ways of accomplishing these steps include asking for evidence, evaluating how much the youth believes in the cognition(s), and checking for alternative cognitions that modify the initial cognition into a more helpful self-instruction. Besides asking for evidence, lists of questions are often provided, which youth can pick from to challenge their cognitions. Examples of challenging cognitions for anxiety or depression are "Is it really?" "What proof do I have?" "Am I forgetting the positive?" Examples of challenging cognitions for aggression are "What would … (a cool role model) think or do in this situation?" "What would I think about this 5 years from now?" By teaching youth to challenge their cognitions they automatically move forward to the third step which is to help them generate alternative adaptive cognitions. Asking the youth to rate the credibility of their "original cognition" (i.e., anxious, depressed, or aggressive) and the "new cognition," one can explore the power of the "new cognition." This also invites the youth to carefully consider and evaluate the usefulness of the new cognition.

7.4 Self-Monitoring

Monitoring can help increase awareness of feelings, behaviors, and cognitions. When such topics are being monitored, this may increase the client's (and the therapist's) insight of the quality, frequency, and intensity of the topic. Self-monitoring can be administered in many variations. Two common approaches are the use of a diary or brief measures that permit data collection that can be used to build graphs. Monitoring is often a starting point for treatment, and it helps establish the baseline and offers room to discuss realistic goalsetting. Most CBT programs involve self-monitoring activities. For example, CBT programs for depression include mood monitoring and CBT for disruptive behavior problems include monitoring of positive and negative behaviors [23].

8   Techniques Targeting Emotion Regulation

Emotions, including awareness and regulation, are a target of CBT that involves both cognitive and behavioral techniques. A common first step is evaluating emotional competence to determine their basic understanding and awareness of emotions. Emotive education involves building the prerequisites for emotion regulation: emotion recognition or understanding. For young children, emotive education is likely to include information about the difference between cognitions, feelings, and behaviors along with recognition of nonverbal behavior and facial expressions. Education about emotions can also include helping a client understand the physiological basis of their emotions. This may take the form of psychoeducation about the nervous system, and body mapping tasks to identify from where in the body physiological signals derive.

The second component involves grading emotions, which is typically done using various forms of scaling (e.g., 0–8 to rate level of anxiety). Teaching youth to grade their emotions is an important building block on the road to improved regulation. That is, increased awareness of the intensity of emotion is essential to help youth learn how to regulate their emotions. Developing a system to grade emotions in terms of intensity is a useful tool in subsequent evaluations of CBT tasks (e.g., did an exposure reduce the youth's fear?) and is also helpful to address youth expectations. For example, if it turns out that a youth expects to always feel no sadness (a "0") following CBT for depression, psychoeducation about realistic outcome expectations may be required. Finally, and related, developing a benchmarking system for youth's emotional experiences helps prepare them for behavioral tasks. For example, if a youth with agoraphobia rates their anxiety for taking a one-hour bus ride during rush hour with the same intensity as approaching a bus station outside rush hour, further work may be needed to develop a more nuanced fear hierarchy. Learning how to grade emotions is thus an important target in itself as a CBT technique and it is an essential mean for more successful delivery of subsequent CBT tasks.

9   Modalities of Cognitive Behavioral Therapy

CBT can be delivered in many different modalities that involve youth, caregivers, and other family members to varying degrees. Depending on the presenting problems, some modalities may be a better fit than others. Other factors, such as cost or number of available therapists, can also determine which modality is best.

9.1 Individual CBT

In individual CBT, the primary focus of treatment is the youth. Thus, this modality is a one-on-one collaborative and interactive process between the therapist and youth [13]. With this modality, there is a variable level of caregiver involvement. For younger children the level is likely to be higher, as children below age 12 cannot be relied upon for accurate information and are dependent upon their caregivers to get to the treatment location. Moreover, this offers a natural opportunity to involve caregivers in the treatment process (e.g., by explaining homework assignments to both youth and their caregivers).

9.2 Group CBT

In group CBT, treatment is delivered to multiple individuals in a group setting. Several factors make group CBT desirable: cost-effectiveness, opportunities for socialization, normalization of psychopathology, and positive peer influence on the group process [13]. Group CBT can be either youth-, caregiver-, or family-focused: that is, the groups can consist of youth only, caregivers and youth together, or caregivers and youth separately. In youth with antisocial behavior, participation in group treatment may be contraindicated due to the 'peer-contagion effect' or 'deviancy training' [24].

9.3 Family CBT

Family CBT involves regular and intensive involvement of caregivers. Sometimes, other family members, such as siblings, may be involved. There are several variants of this approach. In some versions of family CBT, the caregiver and the youth meet with a therapist at the same time. In other versions, the caregiver and the youth meet individually with the therapist. With family CBT, the caregivers can be regarded as co-clients or co-therapists.

9.4 CBT and New Technologies

Blended CBT refers to the use of internet- or computer-delivered CBT with face-to-face CBT. Ebert et al. [25] explain that using the computer or the internet to provide CBT may overcome some of the limitations of traditional treatment services. Advantages of computer- and internet-based CBT include: (a) availability; (b) anonymity; (c) accessibility at any time and place; (d) flexibility in self-direction and self-pacing; (e) reduced travel time and costs for both clients and therapists; and (f) the appeal of interactivity and visual attractiveness of internet-based programs [25].

Tele-CBT refers to CBT by means of telephone or other communication mode that is not face-to-face. Tele-CBT has been evaluated primarily with adults for a variety of problems, but evaluation of the approach with youth has begun. Tele-CBT is particularly well suited for rural and underserved areas and thus shares some of the advantages as with blended CBT. *Serious games* are defined as alternative education tools that go beyond entertainment with the aim of enabling learning in a digital and interactive fashion (i.e., a game [26]). A *serious game* tends to be brief and implemented for a specific objective and can be used to enhance desired outcomes. *Virtual reality therapy* includes the use of a three-dimensional computer generated environment in which a person can move around and interact as if they were in that world. Due to the reality of that world, the application of techniques such as desensitization and exposure are facilitated. Virtual reality exposure therapy has been supported by meta-analytic findings [27]. For youth, *serious games* and *virtual reality therapy* are promising avenues, and are not restricted to the treatment of anxiety.

10   CBT Delivery

Our final section focuses on general factors that need to be considered when using CBT to address emotional and behavioral problems experienced by youth. The CBT techniques outlined above can be delivered with varying levels of therapist competence, which can influence the effectiveness of CBT for youth with emotional and behavioral problems. Therapist competence is defined as the extent to which the therapist balances technical and relational skills and uses them to facilitate therapeutic change. This is a broad concept that encompasses skillfulness and responsiveness in delivering specific CBT techniques [28]. The past decade has witnessed an increased focus on accountability and the importance of establishing therapist competency in the education and practice of psychology (e.g., [29, 30]) and other areas of mental health [31]. This section therefore focuses on aspects of CBT delivery that can help foster the development of therapist competence.

11   Assessment

Assessment plays a key role in the delivery of CBT and is defined as the process used to collect, interpret, and use clinical information to produce a description of a youth [32]. Most CBT models emphasize the importance of continuous data collection over the course of treatment. Typically, these data are used to help shape and guide the treatment process from intake to termination. When grounded in science and theory, assessment can help make CBT more efficient and effective [33]. For example, symptom and diagnostic measures that help a clinician arrive at the correct diagnosis can inform the development of a case conceptualization and facilitate the subsequent delivery of CBT. Assessment can thus play an important role in guiding CBT over the course of treatment.

Increasingly, the principles of the evidence-based assessment movement have informed the use of outcome monitoring and evaluation in CBT. This method is defined as an approach to clinical evaluation that utilizes science and theory to guide the assessment process [34]. A critical aim of the movement is the development and promotion of guidelines to direct research, structure training, and inform clinical practice [34]. The principles represented in the movement include a scientific approach to assessment, a strong emphasis on the score reliability and validity of tools, and a data-informed approach to clinical practice. While it is beyond the scope of the present chapter to provide a thorough review of evidence-based assessment, readers interested in learning how the principles of this movement can be used to guide the treatment selection and planning process can *see* McLeod et al. [35].

12   Case Conceptualization

Case conceptualization represents an important tool used to guide the delivery of CBT and is defined as a set of hypotheses about the causes, antecedents, and maintaining factors of a youth's presenting problems [36, 37]. Considered an important component of evidence-based practice by the American Psychological Association, the association states that "The purpose of evidence-based practice in psychology is to promote effective psychological practice and enhance public health by applying empirically supported principles of psychological assessment, case conceptualization, therapeutic relationship, and intervention" (p. 273 [38]). A good case conceptualization identifies the factors that serve to cause and maintain a youth's target problem(s) and helps the therapist understand how best to translate findings from the empirical literature for use with a particular youth. Assessment tools used through the course of treatment are critical to generating data used to inform the development and refinement of a case conceptualization [39]. Thus, the two skills–assessment and case conceptualization—ideally are used together in CBT.

13   Cultural Considerations

CBT delivery should be informed by knowledge of ways in which culture and diversity can influence the experience and expression of youth emotional and behavioral problems. Culture is defined as "an integrated pattern of human behavior that includes thought, language, action, and artifacts and depends on man's capacity for learning and transmitting knowledge to succeeding generations" (p. 5 [40]). The failure to take a person's nationality, ethnicity, acculturation level, socioeconomic status, and gender into consideration can lead to poor clinical outcomes. For example, behavior may be labeled as pathological when the behavior is, in fact, accounted for by cultural factors. Therapists thus need to take into consideration culture when determining how to deliver CBT. First, therapists must understand how the expression of symptoms or distress may be influenced by culture. The available evidence suggests that the expression of psychological distress may vary across cultures [41]. This variation may be due to value systems that find different symptoms more, or less, acceptable. Second, therapists must be aware that cultural factors may influence reporting practices. The acceptability of certain symptoms may influence what symptoms are reported as problematic. Understanding how symptom expression and reporting practices may be influenced by culture is an important component of conducting a culturally sensitive assessment. Therapists can use the Diagnostic and Statistical Manual of Mental Disorders Cultural Formulation Interview [42] with youth and their families to understand how cultural and diversity issues may influence the treatment process. Finally, it is important to consider cultural factors in treatment planning and delivery. Some treatment components may need to be modified or replaced altogether in order to be relevant and appropriate for the individual. For example, many cultures do not approach thoughts and feelings as firmly distinct concepts, so general psychoeducation about the relation between thoughts, feelings, and behaviors may need to be adjusted.

14   Developmental Considerations

One of the most distinguishing characteristics of youth is rapid change and this has direct implications for the delivery of CBT. Differentiating between developmentally appropriate behavior and disorder-specific behavior is important. At certain age-periods, normally developing youth may show anxiety for ghosts, show rigidity with regard to rules, oppose limit-setting, cross social boundaries, get absorbed in computer games, or have difficulties with homework planning. Youth with emotional and behavioral problems may show such behaviors with an inappropriate intensity, frequency, or at an inappropriate age. When behaviors exceed what is typical for normal development and lead to clinical impairment, CBT may help, either directed at the youth, their caregivers, or their family. Understanding family dynamics and triggers that precede inappropriate behavior are crucial to target the right 'problem' in treatment. Likewise, many factors are important to take into account to determine whether CBT is appropriate, or for whom CBT is appropriate.

How CBT is delivered should be influenced by knowledge of developmental norms. Developmental factors determine what delivery approaches (e.g., play-based, talk) will provide the most accurate and valid information as well as increase client involvement and motivation. Language development and comprehension directly influence a youth's ability to understand questions, report and reflect upon their experience, and be comfortable in different clinical situations (e.g., play versus interviews). Therapists should therefore take a developmentally informed approach to delivery and select an approach from the perspective of developmental norms.

In general, CBT is offered to youth aged eight or above, though some have questioned the use of CBT with younger youth [43]. Grave and Blissett argued that the cognitive capacities of young children are still developing and that a rational approach, as with CBT, may not be suitable. Notably, a previous meta-analysis on CBT for youth anxiety did not show age effects on treatment outcome [44]. Treatment programs for younger children tend to include adaptations to fit with the developmental level of the participants. Finding a fit to the developmental level of a youth and their family may require adaptations to language, materials, activities, or timing of treatment [45]. If, and how, to involve caregivers may change over the course of a youngster's life from toddler to young adult. For instance, The Incredible Years program stresses the importance of spending positive time together (e.g., participating in youth's play; [46]). Obviously, it would be misjudgment to ask caregivers to involve an adolescent to participate in building a tower. Translating 'shared play' into 'shared activities' may be fitting for younger adolescents, whereas for older adolescents, translation into 'shared time' is most fitting.

15   Relational Skills

The delivery of specific CBT techniques does not take place in a vacuum, but needs to be adapted to youth developmental stage, mood, temperament, and personality. All specific CBT techniques are intertwined with the therapists' relational skills and abilities to create a facilitative therapeutic process. The alliance is commonly defined as the quality of the youth–therapist affective bond and the degree of collaboration between the youth and therapist on therapeutic activities [47]. The alliance has consistently been shown to predict outcomes of treatment with youth [48, 49]. A stronger alliance has also been associated with more consistent youth attendance [50] and higher youth treatment satisfaction ratings [51, 52].

Given the link between a strong alliance and positive clinical outcomes, an important issue for therapists is how the alliance can be enhanced. Broad therapist competencies associated with forming a strong therapeutic relationship in CBT have been identified [45, 53]. These include instillation of hope and optimism for change, and engaging youth with developmentally appropriate activities. The suggestions also include a focus on alliance building early in treatment, using reflective statements, using collaborative language (e.g., *we*, *us*, and *let's*), and tracking and validating feelings. It is important to work with youth and caregivers to achieve mutually agreed upon treatment goals and activities. This is because youth rarely refer themselves to treatment, so they may not initially agree with their caregivers about the goals for treatment, especially in youth with externalizing disorders. It is also recommended that therapists provide a treatment rationale linking the purpose of the treatment activities to the overall treatment goal each time a new treatment activity is introduced [53]. In terms of instilling hope and optimism, therapists can assess for unrealistic treatment expectancies, provide psychoeducation about the expected length and outcome of treatment, give an explanation of the treatment rationale, and set up therapeutic activities that help build youths' self-efficacy. Using these approaches to strengthen the alliance can help maximize youth involvement in treatment, defined as the youths' active participation in the therapeutic activities, including homework compliance.

16   Conclusions

CBT is a potent treatment approach developed through the integration of two separate theoretical traditions, behavioral theory and cognitive theory. CBT has been tested in a number of formats, including individual, family, and group. Evidence supporting CBT has been amassed for a number of problems common in youth, including anxiety disorders, depression, disruptive behavior disorders, and eating disorders.

References

1. Weersing VR, Jeffreys M, Do MCT et al (2017) Evidence base update of psychosocial treatments for child and adolescent depression. J Clin Child Adolesc Psychol 46:11–43

2. Higa-McMillan CK, Francis SE, Rith-Najarian L et al (2016) Evidence base update: 50 years of research on treatment of child and adolescent anxiety. J Clin Child Adolesc Psychol 45:91–113

3. Dorsey S, McLaughlin KA, Kerns SE et al (2017) Evidence base update for psychosocial treatments for children and adolescents exposed to traumatic events. J Clin Child Adolesc Psychol 46:303–330

4. Lock J (2015) An update on evidence-based psychosocial treatments for eating disorders in children and adolescents. J Clin Child Adolesc Psychol 44:707–721

5. Battagliese G, Caccetta M, Luppino OI et al (2015) Cognitive-behavioral therapy for externalizing disorders: a meta-analysis of treatment effectiveness. Behav Res Ther 75:60–71

6. Hogue A, Henderson CE, Ozechowski TJ et al (2014) Evidence base on outpatient behavioral treatments for adolescent substance use: updates and recommendations 2007–2013. J Clin Child Adolesc Psychol 43:695–720

7. Pavlov IP (1927) Conditioned reflexes: an investigation of the physiological activity of the cerebral cortex. Oxford University Press, London

8. Benjamin CL, Puleo CM, Settipani CA et al (2011) History of cognitive-behavioral therapy in youth. Child Adolesc Psychiatr Clin 20:179–189

9. Skinner BF (1938) The behavior of organisms: an experimental analysis. Appleton-Century, Oxford

10. Skinner BF (1969) Contingencies of reinforcement: a theoretical analysis. Meredith, New York, NY

11. Bandura A (1969) Social learning theory of identificatory processes. In: Goslin DA (ed) Handbook of socialization theory and research. McNally & Company, Chicago, IL, pp 213–262

12. Rotter JB (1954) Social learning and clinical psychology. Prentice-Hall, Inc, Englewood Cliffs, NJ

13. Southam-Gerow MA, McLeod BD, Brown RC et al (2011) Cognitive-behavioral therapy for adolescents. In: Brown BB, Prinstein M (eds) Encyclopedia of adolescence. Elsevier, London, pp 100–108

14. Miller GA (2003) The cognitive revolution: a historical perspective. Trends Cogn Sci 7:141–144

15. Neisser U (1967) Cognitive psychology. Appleton-Century-Crofts, New York, NY

16. Beck AT, Rush AJ, Shaw BF et al (1979) Cognitive therapy of depression. Guildford Press, New York, NY

17. Beck AT (1967) Depression: clinical, experimental, and theoretical aspects. Guilford Press, New York, NY

18. Williamson DA, Muller SL, Reas DL et al (1999) Cognitive bias in eating disorders: implications for theory and treatment. Behav Modif 23:556–577

19. Wright FD, Beck AT, Newman CF et al (1993) Cognitive therapy of substance abuse. NIDA Res Monogr 137:123–146

20. Lochman JE, Pardini DA (2008) Cognitive-behavioral therapies. In: Rutter M, Bishop D, Pine D et al (eds) Rutter's child and adolescent psychiatry. Wiley-Blackwell, Hoboken, NJ, pp 1026–1045

21. Gresham FM, Lochman JE (2008) Methodological issues in research using cognitive-behavioral interventions. In: Mayer MJ, Van Acker R, Lochman JE et al (eds) Cognitive behavioral interventions for emotional and behavioral disorders: school-based practice. Guildford Press, New York, NY, pp 58–81

22. Ollendick TH, Hovey LD (2009) Competencies for treating phobic and anxiety disorders in children and adolescents. In: Thomas J, Hersen M (eds) Handbook of clinical psychology competencies. Springer, New York, NY, pp 1219–1244

23. Maalouf FT, Brent DA (2012) Depression and suicidal behavior. In: Szigethy E, Weisz JR, Findling RL (eds) Cognitive-behavioral therapy for youth and adolescents. American Psychiatric Association, Arlington, VA, pp 163–184

24. Gifford-Smith M, Dodge KA, Dishion TJ et al (2005) Peer influence in children and adolescents: crossing the bridge from developmental to intervention science. J Abnorm Child Psychol 33:255–265

25. Ebert DD, Zarski AC, Christensen H et al (2015) Internet and computer-based cognitive behavioral therapy for anxiety and depression in youth: a meta-analysis of randomized controlled outcome trials. PLoS One 10: e0119895

26. Gros B (2017) Game dimensions and pedagogical dimension in serious games. In: Zheng R, Gardner MK (eds) Handbook of research on serious games for educational applications. IGI Global, Hershey, PA, pp 402–417

27. Morina N, Ijntema H, Meyerbröker K et al (2015) Can virtual reality exposure therapy gains be generalized to real-life? A meta-analysis of studies applying behavioral assessments. Behav Res Ther 74:18–24

28. Barber JP, Sharpless B, Klostermann S et al (2007) Assessing intervention competence and its relation to therapy outcome: a selected review derived from the outcome literature. Prof Psychol Res Pract 38:493–500

29. American Psychological Association (2015) Standards of accreditation for health service psychology. American Psychological Association, Washington, DC

30. Baker A (2001) Crossing the quality chasm: a new health system for the 21st century. BMJ 323:1192–1195

31. Unützer J, Schoenbaum M, Druss BG et al (2006) Transforming mental health care at the interface with general medicine: report for the president's commission. Psychiatr Serv 57:37–47

32. Hunsley J (2002) Psychological testing and psychological assessment: a closer examination. Am Psychol 57:139–140

33. McLeod BD, Jensen-Doss A, Ollendick TH (2013) Overview of diagnostic and behavioral assessment. In: McLeod B, Jensen-Doss A, Ollendick TH (eds) Diagnostic and behavioral assessment in children and adolescents: a clinical guide. Guilford Publications, Inc., New York, NY, pp 3–33

34. Hunsley J, Mash EJ (2007) Evidence-based assessment. Annu Rev Clin Psychol 3:29–51

35. McLeod BD, Jensen-Doss A, Ollendick TH (2013) Diagnostic and behavioral assessment in children and adolescents: a clinical guide. Guilford Publications, Inc., New York, NY

36. Eells TD (1997) Psychotherapy case formulation: history and current status. In: Eells TD (ed) Handbook of psychotherapy case formulation. Guilford Press, New York, NY, pp 1–25

37. Nezu AM, Nezu CM, Peacock MA et al (2004) Case formulation in cognitive-behavioral therapy. In: Haynes SN, Heiby EM, Hersen M (eds) Comprehensive handbook of psychological assessment, Behavioral assessment, vol 3. Wiley, New York, NY, pp 402–426

38. APA Presidential Task Force on Evidence-Based Practice (2006) Evidence-based practice in psychology. Am Psychol 61:271–285

39. Christon LM, McLeod BD, Jensen-Doss A (2015) Evidence-based assessment meets evidence-based treatment: an approach to science-informed case conceptualization. Cogn Behav Pract 22:36–48

40. Frisby CL, Reynolds CR (2005) Comprehensive handbook of multicultural school psychology. John Wiley & Sons, Inc, Hoboken, NJ

41. Weisz JR, Sigman M, Weiss B et al (1993) Behavioral and emotional problems among Embu children in Kenya: comparisons with African-American, Caucasian, and Thai children. Child Dev 64:98–109

42. American Psychiatric Association (2013) Diagnostic and statistical manual of mental disorders (DSM-5). American Psychiatric Publishing, Arlington, VA

43. Grave J, Blissett J (2004) Is cognitive behavior therapy developmentally appropriate for young children? A critical review of the evidence. Clin Psychol Rev 24:399–420

44. Bennett K, Manassis K, Walter SD et al (2013) Cognitive behavioral therapy age effects in child and adolescent anxiety: an individual patient data metaanalysis. Depress Anxiety 30:829–841

45. Sburlati ES, Schniering CA, Lyneham HJ et al (2011) A model of therapist competencies for the empirically supported cognitive behavioral treatment of child and adolescent anxiety and depressive disorders. Clin Child Fam Psychol Rev 14:89–109

46. Webster-Stratton C, Reid MJ (2003) The incredible years parents, teachers and children training series: a multifaceted treatment approach for young children with conduct problems. In: Kazdin AE, Weisz JR (eds) Evidence-based psychotherapies for children and adolescents. Guildford Press, New York, NY, pp 224–240

47. Elvins R, Green J (2008) The conceptualization and measurement of therapeutic alliance: an empirical review. Clin Psychol Rev 28:1167–1187

48. Karver MS, De Nadai AS, Monahan M et al (2018) Meta-analysis of the prospective relation between alliance and outcome in child and adolescent psychotherapy. Psychotherapy 55:341–355

49. McLeod BD (2011) The relation of the alliance with outcomes in youth psychotherapy: a meta-analysis. Clin Psychol Rev 31:603–616

50. Shirk SR, Karver MS, Brown R (2011) The alliance in child and adolescent psychotherapy. Psychotherapy 48:17–24

51. Fjermestad KW, Lerner MD, McLeod BD et al (2017) Motivation and treatment credibility predict alliance in cognitive behavioral treatment for youth with anxiety disorders in community clinics. J Clin Psychol 74:793–805

52. Ormhaug SM, Shirk SR, Wentzel-Larsen T (2015) Therapist and client perspectives on the alliance in the treatment of traumatized adolescents. Eur J Psychotraumatol 6:27705

53. Brown RC, Southam-Gerow MA, Mcleod BD et al (2017) The global therapist competence scale for youth psychosocial treatment: development and initial validation. J Clin Psychol 74:649–664

Chapter 2

Cognitive Behavioral Therapy with Depressed Youth

Hannah R. Lawrence, Michelle L. Buffie, Rebecca A. Schwartz-Mette, and Douglas W. Nangle

Abstract

Depression impacts millions of children and adolescents worldwide, causing significant distress for them and their families. Cognitive behavioral therapy (CBT) has been identified as an evidence-based treatment for depressed youth. CBT targets maladaptive patterns of cognitions and behavior in order to improve mood. This chapter provides a primer on core CBT strategies and techniques used when treating depressed youth. To illustrate these techniques, we provide example dialogue between youth and therapists, sample worksheets and handouts, and other resources. Where possible, we suggest helpful adaptations to these techniques for use when working with younger children. Through this text, we hope to provide practitioners with portable knowledge and materials for use in their practice, increasing access to this valuable treatment for depressed youth.

Key words Child and adolescent depression, Psychosocial treatment of child and adolescent depression, Cognitive behavioral therapy, Cognitive interventions for depressed youth, Behavioral interventions for depressed youth, Relaxation strategies for depressed youth

1 Introduction

Youth depression is prevalent and impairing. As the second most common psychological disorder, depression results in substantial distress for millions of children and adolescents [1]. Rates of depression increase from childhood (1–2%) through adolescence (7–8%) with a cumulative incidence of 20% by age 18 [2–4]. Females are at especially high risk starting in adolescence [5] and continuing through adulthood [6]. Many more youth experience subclinical symptoms of depression, which cause significant suffering despite not meeting full diagnostic criteria [7]. Given the widespread and impairing nature of youth depression, effective and timely intervention is critical.

Depression is characterized by depressed or irritable mood, anhedonia, changes in appetite, weight, and/or sleep, psychomotor agitation or retardation, feelings of worthlessness or guilt, a

Robert D. Friedberg and Brad J. Nakamura (eds.), *Cognitive Behavioral Therapy in Youth: Tradition and Innovation*, Neuromethods, vol. 156, https://doi.org/10.1007/978-1-0716-0700-8_2, © Springer Science+Business Media, LLC, part of Springer Nature 2020

Table 1
Symptoms of Depression

Symptoms of Depression
• Depressed or irritable mood
• Anhedonia
• Changes in appetite/weight
• Changes in sleep
• Psychomotor agitation or retardation
• Feelings of worthlessness/guilt
• Reduced ability to think clearly/concentrate
• Low energy
• Suicidal ideation

reduced ability to think or concentrate, low energy, and suicidal ideation (Table 1). A notable difference in diagnosing youth is that irritability can replace depressed mood as a core symptom [8]. Not surprisingly given this symptom constellation, the impact of depression is extensive, affecting social, academic, and family functioning. For youth, depression is associated with delays in cognitive, social, and emotional development, problems with school performance, difficulties in social relationships, and increased risk for additional mental health disorder diagnoses [9–12]. Experiencing depression during childhood or adolescence also increases risk for future psychological and physical health problems [4]. Specifically, earlier onset enhances risk for later depression, a more severe and recurrent course of depression, lower quality of life, and physical illness and mortality [2, 13, 14]. Given the concurrent and prospective functional impairment and negative outcomes associated with youth depression, early identification and treatment are particularly important. One of the more prominent evidence-based treatment options is cognitive behavioral therapy (CBT; [15]). First, we discuss the theoretical and empirical bases for CBT and highlight CBT as an efficacious treatment for depressed youth.

2   Theoretical and Empirical Foundations

According to the Cognitive Model of Depression (Fig. 1; [16]), depressed individuals have negative thoughts about themselves, the world, and the future. Over time, these cognitions lead to engrained beliefs about themselves as helpless, worthless, and/or inadequate. These beliefs lead individuals to feel sad or down and

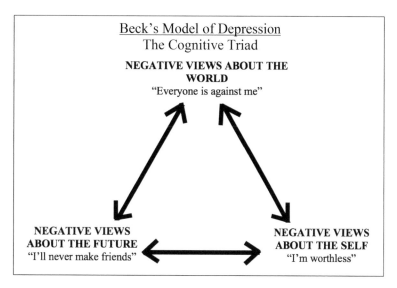

Fig. 1 Cognitive model of depression

behave in unhelpful ways, which, over time, exacerbate depression. CBT works by providing depressed youth with the tools needed to think and behave in more realistic and helpful ways.

Meta-analyses find CBT to be an effective treatment for depressed youth [17]. Treatment effect sizes for CBT with youth are consistently greater than 0.50 [18, 19], with approximately 62% of youth experiencing significant improvement after treatment [20]. CBT is designated a "well-established" treatment for adolescents, indicating that at least two independent research teams have demonstrated that CBT was superior to a placebo or another active treatment [21]. Although no interventions are considered "well-established" for treating depression in children (i.e., youth aged 13 years and under), CBT is considered "possibly efficacious," meaning that in at least one randomized control trial, CBT was superior to a control condition. CBT does appear to be the treatment with the strongest evidence for this age group [21].

What follows is an introduction to the evidence-based core strategies and techniques comprising CBT for youth depression. We provide helpful example dialogue,[1] worksheets for use with youth, diagrams to help illustrate these techniques, and offer guidance on how to select CBT treatment techniques based on youth presentation. Bearing in mind the sharp increase in prevalence and resulting likelihood of encountering adolescents in clinics, we focus our discussion more on treating depressed teens. Working with younger children requires adaptations to accommodate for

---

[1] The examples used are based on the authors' collective clinical experience. The cases and sample dialogue were made up for the chapter and care was taken to ensure that they were not based on particular clients but rather confabulated to illustrate particular points.

Table 2
Typical Course of Treatment

Typical Course of Treatment	
1–2 sessions:	Assessment
1–2 sessions:	Case conceptualization
5–10 sessions:	Intervention/skill building
1–2 sessions:	Relapse prevention

differences in developmental and cognitive levels. Where possible, we offer helpful tips for modifying intervention techniques for younger children. At the end of the chapter, we leave the reader with a list of additional resources and evidence-based treatment manuals.

3 Intervention

A typical course of treatment progresses from assessment, to case conceptualization, to intervention, to relapse prevention (Table 2). First, sessions are spent gathering information and conducting assessment to establish primary area(s) of difficulty and/or diagnosis. Next, the therapist works with youth and/or their families to develop a case conceptualization. Then, treatment begins and youth learn and practice new skills. Finally, once symptoms have diminished, youth and their families are prepared to continue to use these skills beyond their time in therapy and are encouraged to engage in booster sessions or return to therapy should symptoms recur. This chapter emphasizes the case conceptualization and treatment stages of the intervention process.

4 Case Conceptualization

CBT helps depressed youth identify and adjust the maladaptive patterns of cognition and behavior posited to give rise to depressed mood. Depressed youth experience negative cognitions, which lead them to behave in less adaptive ways, which, over time, exacerbates depression. A youth may have the thought *Everyone hates me*, stay at home in bed, and over time, experience persistent depressed mood. CBT equips youth with the tools to interrupt these unhelpful cycles of cognition and behavior.

A strength of the cognitive behavioral approach is its flexibility. Based on the specific cognitions and behaviors a particular youth experiences, different CBT techniques are applied. A strong case

Table 3
Case Conceptualization Resources

Additional Resources for Case Conceptualization
• Beck Institute Cognitive Conceptualization Diagram
– www.beckinstitute.org
• www.psychologytools.com
– CBT worksheets for Case Conceptualization
• Chap. 3, Beck, J. S. (2011). *Cognitive behavior therapy: Basics and beyond* (second ed.). New York, NY: Guilford Press

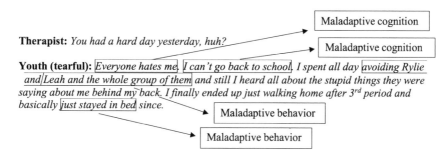

Fig. 2 Identifying unhelpful patterns of cognition and behavior

conceptualization is vital to fully understand which behaviors and cognitions are most problematic and which intervention techniques may be most helpful (see Table 3 for additional resources for case conceptualization). The therapist listens carefully to descriptions of youth distress with an ear towards identifying unhelpful patterns of cognition and behavior and their connections to depressed mood. Consider the dialogue in Fig. 2 below in which a therapist identifies maladaptive cognitions and behaviors as a youth describe their mood.

Over time, the therapist begins to recognize patterns in how depressed mood is connected with these cognitions and behaviors. What thoughts do youth have when they feel sad? What do they do when they feel sad? How do they view and respond to the world around them? Once these patterns are identified, the therapist can begin to form an initial case conceptualization to help guide treatment decisions. Consider the example cognitive behavioral case conceptualization in Fig. 3. Connections between depressed/irritable mood and maladaptive patterns of cognition and behavior are displayed, while precipitating factors that may exacerbate this cycle are also considered.

An important next step is to check in with youth (and/or parents/guardians depending on developmental level) to see how

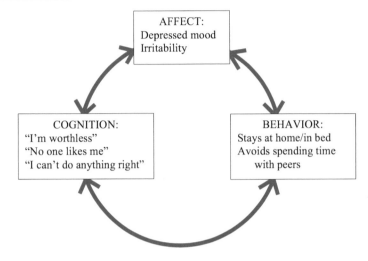

Fig. 3 Example cognitive behavioral case conceptualization

this conceptualization fits with their understanding of youth's distress and adjust the conceptualization accordingly. A good case conceptualization evolves with input from youth and relevant informants (e.g., parents/guardians, teachers) and changes as treatment progresses.

5   Treatment

Once a solid working case conceptualization is formed, treatment can progress. Although the structure of each treatment session will be somewhat individualized and may vary across sessions, it is helpful to consider how a typical session may proceed (Table 4). In general, a session begins by reviewing the prior week's "action plan," or between-session work (not surprisingly, calling this plan "homework" may not be motivating to many youth). Next, the therapist works with youth to identify the most pressing concern for the week. The goal of the therapist is to identify cognitions and behaviors that may be getting in the way. Then, the therapist selects cognitive and/or behavioral interventions that fit best with the specific concerns. Finally, the therapist collaborates with youth to develop a new "action plan" for the following week to practice the new skill(s) they learned in session. Of course, the therapist should attend to the therapeutic relationship throughout each session.

Table 4
Typical Session Structure

Typical Session Structure
1. Establish therapeutic rapport
2. Check in on last week's "action plan"
3. Identify most pressing concern for this week with youth
4. Apply cognitive or behavioral intervention(s) that fit best
5. Collaboratively develop "action plan" for following week

6 Intervention Techniques

The remainder of this chapter is devoted to outlining specific CBT intervention techniques and discussing when they may be used. These techniques were selected for inclusion in this chapter as they are the core components most often included in CBT intervention packages for youth with depression and are the techniques with the most empirical evidence [22]. For each technique, we provide a brief overview, steps to implement the technique, example dialogue between a therapist and youth, and where relevant, handouts, activities, and tips for implementing with different age groups.

6.1 Cognitive Interventions

Cognitive interventions focus on changing negative or unhelpful patterns of thinking. Youth with depression often have negative beliefs about themselves, the world, and the future. When these beliefs are accepted as true, youth feel worse and behave in less adaptive ways. A youth who thinks "*I'm worthless,*" "*No one likes me,*" and "*I'm never going to make friends*" is likely to feel more sad and further isolate themselves. Although research has typically examined cognitive interventions as part of larger treatment packages for youth [22], cognitive change (e.g., decreased negative thoughts) has been found to mediate the relation between CBT treatment and outcome [23], suggesting that cognitive interventions may be a key component of CBT treatments for youth. Cognitive interventions target the cognitive component of the case conceptualization (Fig. 4). By changing patterns of thinking, youth begin to behave in more adaptive ways and, as a result, feel better.

First, youth are introduced to the idea that thoughts and feelings are connected. By learning to "catch" the thoughts connected with their feelings, they can begin to learn to change their thinking patterns.

> **Therapist**: *Let's pretend you're sitting at home, and all of a sudden your phone vibrates. You look over and see that you have a text message. You open the message, and it's REALLY mean. It says that you're an "awful person" or that*

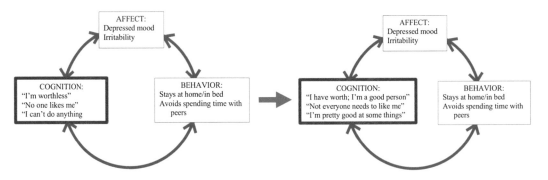

Fig. 4 Targeting cognitions

Table 5
Tips for Identifying Cognitions

Clinician Question: "What do I do if youth have trouble reporting their thoughts?"

It can be challenging (even for adults) to report thoughts/images going through their minds. Here are a few helpful tips:

1. Catch the emotion. Notice a shift in affect during session? In that moment, ask youth what is going through their mind. It is easier to report thoughts in the moment than it is to report thoughts retrospectively.

2. Relive the experience. Have youth close their eyes and imagine the situation in question. Walk through each moment of the situation in detail, asking for thoughts along the way.

3. Ask about imagery. Sometimes youth experience maladaptive cognitions in the form of mental imagery. Along with asking about their thoughts, also ask if they have a picture or movie pop into their minds when they notice they feel sad.

---

you'll "never make anything out of yourself" or that you'll "never feel better." How would that feel?

Youth: *I mean, horrible. Who would do something like that?*

Therapist: *Yeah, of course it would make you feel really bad, right? Now, pretend that, instead, your phone vibrates and you see a really encouraging text message. Maybe it says how hard you're working to feel better or what a kind friend you are. How would that feel then?*

Youth: *Well, better than that first text.*

Therapist: *Right? Well, just like your phone is constantly getting text messages, our brains constantly have thoughts that just pop into our minds. Sometimes these thoughts are negative, unhelpful thoughts that make us feel worse and may not really be completely true. Other times, these thoughts are positive helpful thoughts that are more accurate and help us feel better.*

Youth: *So you're saying that what I think to myself can change how I feel?*

Therapist: *Exactly! Thoughts, feelings, and behavior are all connected. When we feel sad, we have sad thoughts like, "I can't do anything" or "I'll never feel better." Instead of just trying to stop feeling sad, we can change the thoughts we have in our heads—or the text messages we send ourselves—to change how we feel.*

Next, the therapist helps youth identify the thoughts going through their head when they notice themselves feeling sad or depressed (see Table 5 for discussion of a common stuck point).

Table 6
Example Thought Record

Thought Record			
Situation	**Thought**	**Emotion**	**Behavior**
I heard girls talking about me behind my back	*"Everyone hates me"*	*Sad*	*Walked home and went to bed*
I stayed home instead of going to theater rehearsal	*"I don't have any friends"*	*Lonely*	*Didn't answer my phone when my friend called*

The easiest way for youth to "catch" the thoughts going through their heads is to pay attention to when their mood changes. The therapist encourages youth to pay attention to times when they notice that they feel sad or depressed, and to ask themselves, *What was just going through my mind?* Getting in the habit of catching what thoughts go with what feelings makes it easier for youth to identify and then change the cognitions connected with unwanted emotions. The skill of identifying unhelpful cognitions takes a lot of practice. One way to promote practice of this skill is for the therapist to incorporate a thought record as part of youths' weekly action plan (see Table 6).

6.1.1 Identifying Cognitive Distortions

Unfortunately, depression itself leads youth to think in more negative and less helpful ways. In fact, depression is associated with a number of cognitive distortions, or "thinking mistakes." Some common examples include those illustrated in Fig. 5.

One way for youth to begin to understand that their negative cognitions may not be completely true or accurate is for therapists to help them identify the "thinking mistakes" they tend to make.

Youth: *Either I'm going to get perfect score on the SATs, or I'm never going to get into college or make anything out of myself.*

Therapist: *I can see why that thought made you feel so stressed this week! The idea that either you're going to get a perfect score on the SATs or you're never going to make anything out of yourself is an example of what is called black and white thinking. Can you think of why we might call it that?*

Youth: *Because there's no gray?*

Therapist: *Exactly. What might be the gray part in this situation?*

Youth: *I guess I could get a high score that isn't perfect and probably still get into college. And even if I didn't do well this time I could always take it again.*

Therapist: *Well that sounds more realistic, huh? Sometimes when our brains get into patterns of black and white thinking, we can only think of the absolute best case or the absolute worst case whereas the most likely thing to happen is usually somewhere in the middle. Let's keep an eye out for black and white thinking. When you catch yourself thinking in this way, let's work on finding the gray together.*

Fig. 5 Cognitive distortions

Adapting for children: One fun way to help younger children identify some of the "thinking mistakes" they are making is to play a matching game. The therapist can have youth write or draw their thoughts when they feel sad on different slips of paper. Then, youth can match their thoughts to the different thinking mistakes listed above.

There are evidence-based cognitive interventions to help youth think in more accurate and helpful ways. Next, we will review these techniques and provide helpful tips for implementing them with children and adolescents.

6.1.2 Cognitive Restructuring

Cognitive restructuring teaches youth to identify and change unhelpful or negative patterns of thinking. When youth catch themselves thinking in maladaptive ways, cognitive restructuring helps them to restructure those thoughts to be more adaptive. Cognitive restructuring has been found to be effective in increasing the incidence of more adaptive thoughts and, as a result, is associated with decreased depressive symptoms [24].

Once youth feel comfortable identifying the thoughts associated with their depressed mood, the therapist guides them through the process of challenging these thoughts to determine whether they are (a) accurate and (b) helpful (*see* Table 7 for sample questions). There are a number of strategies to help youth question the accuracy and helpfulness of their thoughts. No one strategy will work for all youth or kinds of negative thinking. Instead, a combination of these strategies, applied flexibly depending on the individual youth, will likely be most effective.

Evaluating the evidence. One way to evaluate how accurate cognitions are is to weigh the evidence the thought is true against the evidence that the thought is not true. Because most thoughts have some truth to them, we recommend by starting with evidence *for* the thought. Youth often fail to recognize the evidence their thought might not be true, however, so next we ask for evidence *against* the thought (see Table 8 for common clinician questions).

> **Therapist**: *Let's be detectives for a minute. What evidence do you have that Sam hates you?*
>
> **Youth**: *He didn't invite me over to watch the NBA finals last week with the rest of the guys!*
>
> **Therapist**: *That's right. I remember that really upset you. What else?*
>
> **Youth**: *When I went to the grocery store this weekend with my mom I ran into him and I could tell he didn't even want to say hi. He said "hey" and then basically ran off.*
>
> **Therapist**: *Anything else?*
>
> **Youth**: *I guess that's it.*
>
> **Therapist**: *Now we wouldn't be good detectives if we didn't ask if there is any evidence that your thought isn't true, that Sam doesn't hate you?*
>
> **Youth**: *Well we've been friends for a while.*
>
> **Therapist**: *What tells you that you're friends?*
>
> **Youth**: *He does invite me to hang out pretty much any time there's something cool going on.*
>
> **Therapist**: *So Sam didn't invite you over for the NBA finals but he does invite you to lots of things.*
>
> **Youth**: *Yeah and I'm usually the one he comes to when he has a fight with his parents or needs to vent.*
>
> **Therapist**: *So on one hand, Sam didn't invite you over last weekend and just said a quick "hey" at the grocery store but on the other hand, you and Sam hang out a lot and he seems to really rely on you for support. Let's look at the situation again. Could there be a more accurate and helpful way of thinking about the situation other than Sam hates you?*
>
> **Youth**: *Maybe he just wanted to watch the game with other Cavs fans. He knows I root for the Warriors. I guess he probably doesn't hate me he just felt bad he didn't invite me when he saw me at the grocery store.*

Table 7
Questions to Assist in Challenging Cognitions

Key Questions
"Are there other ways of thinking about this situation?"
"What would your mom/dad/sister/brother/friend think about this situation?"
"What is the worst thing that could happen? What is the best thing that could happen? What is most likely to happen?"
"How helpful is that thought?"

Table 8
Tips for Beginning to Restructure Cognitions

Clinician Question: "What thought should I start with?"
A common question is where to begin with cognitive restructuring. The good news is any unhelpful or inaccurate thought will work! The goal is for the therapist to help youth get used to challenging their thoughts rather than accepting them as true. One tip is to start with a thought that just pops into mind when youth notice they feel sad, like *"Everyone is always mean to me"* (called an "automatic thought" in CBT). Youth may have an easier time restructuring this kind of thought rather than a deeply held belief like *"I'm worthless"* (called a "core belief" in CBT)

Table 9
Tips for Identifying Realistic Expectations

Clinician Question: "What if youth think the best-case scenario is the most realistic scenario?"
Sometimes, when asked for the "best thing that could happen," youth respond with the most realistic outcome. It is important that the best outcome really is the absolute best-case scenario, even if highly unlikely. By doing so, there is more room for youth to identify a realistic outcome that falls somewhere in the middle

Decatastrophizing. At times, youth with depression expect the worst to happen. Decatastrophizing helps them to identify the worst- and best-case scenarios, as well as what is most likely to happen. Through this exercise, youth come to a more accurate appraisal of what is most likely to occur (see Table 9 for common clinician question).

Therapist: *What is the worst thing that could happen in this situation?*
    Youth: *I'll never get better. I'll feel this way forever.*
    Therapist: *That does sounds pretty bad. How about the best thing that could happen?*
    Youth: *I guess that I'll eventually get better?*
    Therapist: *Is that really the absolute BEST thing that could happen? What if we snapped our fingers and you felt 100% happy for 100% of the time from this second on?*
    Youth: *Ha, ok I guess that's better.*
    Therapist: *And so what's the most realistic outcome?*

Youth: *Probably that I won't feel better tomorrow or the next day but that eventually I'll feel happy again.*

Socratic questioning. Socratic questioning involves the therapist asking a series of guided questions that help youth consider and voice why their thoughts may not be true. Therapists need to be aware of the "righting reflex," or the tendency to correct youth or offer unsolicited advice. Although well meaning, youth's response is typically to counter that advice with all the reasons why it may not work or fit with their experience. Instead, prompting youth to articulate for themselves reasons why their cognitions may not be accurate, increases belief in those reasons.

Youth: *I'm a complete loser.*
   Therapist: *What makes you say that?*
   Youth: *No one invited me to homecoming. So yeah, I'm a loser.*
   Therapist: *Tell me more about how that translates to you being a complete loser?*
   Youth: *I'm literally the only one without a date. I must be a loser.*
   Therapist: *In the past have you had dates to the school dances?*
   Youth: *Well yeah. I had a date to junior prom last year at least.*
   Therapist: *Oh ok. And it sounds like you're still going to the homecoming dance?*
   Youth: *Yeah I'm going with a big group of friends.*
   Therapist: *Ah ok and they are all paired up as dates?*
   Youth: *No. . . there are two couples but then like eight of us just going with the group.*
   Therapist: *So it sounds like your reason for feeling like a complete loser is that you don't have a date to this year's homecoming, but when we dig a little deeper you mentioned that a lot of your friends don't have dates either and that in the past you've had dates to other dances?*
   Youth: *Ok yeah I guess I'm not a complete loser.*
   Therapist: *So how might you rephrase your first thought given this new information?*
   Youth: *I guess that I'm bummed I don't have a date, but it will probably still be alright going with friends who don't have dates either.*

Imagery rescripting. Sometimes, youth experience their maladaptive cognitions in the form of mental imagery. They see a mental image or movie in their minds rather than a verbal thought or sentence going through their minds. They may experience an image or movie in their minds of getting rejected by friends, failing an exam, or sitting at home feeling sad. Just as verbal thoughts can be inaccurate and unhelpful, so can mental imagery. Imagery rescripting helps youth "redraw the picture" or "change the ending to the movie." First, the therapist has youth bring to mind the inaccurate/unhelpful mental image or movie. They then have youth describe this mental image or movie in detail. The therapist uses the same cognitive strategies as used to address verbal thoughts above to lead youth through evaluating whether that mental image or movie is accurate and/or helpful. For example, they may ask for the best, worst, and most realistic ending to the movie. Finally, the therapist guides youth through describing a new mental image or a new ending to the movie that reflects a more accurate and helpful

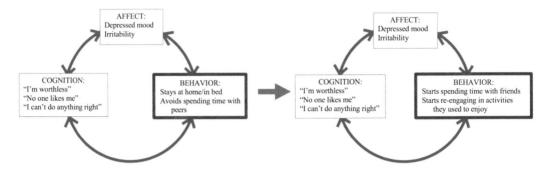

Fig. 6 Targeting behavior

appraisal of the situation. Younger children may also benefit from imagining a parent, guardian, or other caregiver providing them with comfort or help.

> Adapting for children: Younger children may have difficulty with cognitive techniques given their level of cognitive development. Rather than focusing on identifying and challenging their maladaptive cognitions, the therapist can focus on increasing their accurate, helpful cognitions. This can be done by the therapist providing youth with ideas of adaptive cognitions that apply across situations, like "I can do it," "I can try," or "I can be brave." To help increase the incidence of these more adaptive cognitions, the therapist could consider making "Coping Cards" – notecards that youth can keep with them to have these coping statements on hand.

6.2 Behavioral Interventions

Just as cognitive interventions focus on changing maladaptive patterns of thinking, behavioral interventions focus on changing maladaptive patterns of behavior (*see* Fig. 6). Behavioral theories of depression suggest that symptoms are related to a lack of positive reinforcement from the environment [25]. Thus, the primary aim of intervention is to increase engagement in adaptive activities, such as with behavioral activation. Behavioral interventions have been found to be effective in alleviating depression in youth both as a stand-alone treatment [26] and in conjunction with cognitive intervention techniques [15]. These interventions target the behavioral component of the case conceptualization. By changing the way they act, youth also begin to feel and think in more helpful ways.

6.2.1 Behavioral Experiments

One way to help youth test the accuracy of their thoughts is to use the cognitive strategies detailed above. Another way is to set up a behavioral experiment (see Table 10). Behavioral experiments are experiential exercises designed to maximally violate what youth

Table 10
Example Structure of Behavioral Experiment

Steps to Set Up a Behavioral Experiment
1. Specify the problematic thought
2. Design an experiment to test the accuracy of that thought
3. Carry out the experiment
4. Record what was observed
5. Ask, "What did you learn?"

Table 11
Example Behavioral Experiments

Cognition	Example Behavioral Experiment
"I'll just die if I don't get asked to prom"	Go to prom with group of friends
"I can't do anything right"	Make a plan to volunteer at animal shelter
"I just don't have the energy to do anything"	Take a walk around the block
"Getting together with friends won't help"	Plan a fun day out with a friend

expect to happen (see Table 11). Through these experiences, youth gather new information to test the validity of their thoughts.

Youth: *Therapy won't work for me.*
Therapist: *You're wondering whether I can actually help you.*
Youth: *I'm not wondering, I'm sure. Therapy won't work for me.*
Therapist: *How could we test that? How could we know for sure that therapy doesn't work for you?*
Youth: *Huh. I've never really thought about that.*
Therapist: *What do you think- 7 sessions, 8 sessions, 10 sessions?*
Youth: *I guess 8 sessions to be sure?*
Therapist: *Great. And you said you're 100% sure it won't work at all for you. So zero improvement after 8 sessions right?*
Youth: *Yep, none.*
Therapist: *Perfect. And will we know for sure that therapy doesn't work at all for you if you halfway engage in therapy, show up 50% of the time?*
Youth: *No, I guess not. Then we'd wonder whether therapy didn't work or if I didn't give it a good enough shot.*
Therapist: *Ah, I see. That's a great observation!* (writes down plan) *So we'll try therapy for 8 sessions, all in. If after that we see zero improvement we'll know therapy doesn't work and we can try something else. We have those questionnaires you filled out before your first session and we can fill out those same questionnaires after session 8 and see what happens. Are you ready to jump into session number one?*

6.2.2 Behavioral Activation

Symptoms of depression include low energy and loss of interest in activities one used to enjoy. Thus, it is common for youth with depression to limit their engagement in their environment, for

Table 12
Example/Behavioral Activation Plan

Activities I enjoy	Day I plan to do activity	Time I plan to do activity	How I felt before activity (0 = very sad, 10 = very happy)	How I felt after activity (0 = very sad, 10 = very happy)
Practice soccer in backyard	Monday	4 p.m.	2	5
Walk Tyler's dog	Tuesday	8 a.m.	3	4
Go for a hike with Nick	Wednesday	3 p.m.	3	7
Play monopoly with Corey	Thursday	6 p.m.	2	6
Go for a 1-mile run	Friday	3 p.m.	1	5

example, staying in bed or not socializing with friends. Unfortunately, this pattern of behavior limits their chances to experience positive reinforcement, further exacerbating their symptoms of depression. Behavioral activation helps youth make and implement plans to engage in positive activities. By scheduling daily pleasant events, youth learn that engaging in the things they used to enjoy, and with people they are close with, helps them feel better. Behavioral activation, as part of larger treatment packages, has been found to reduce symptoms of depression in youth [27, 28].

Behavioral activation includes (a) collaboratively developing a list of activities youth might enjoy, (b) making a plan to engage in these activities, and (c) monitoring engagement in, and pleasure derived from engagement in, these activities (see Table 12 for an example self-monitoring form for behavioral activation). It is important that the therapist acknowledge that youth likely will not feel like engaging in these activities before doing them. Instead, youth should use behavioral activation to test out whether these activities result in affect change over time.

> **Therapist**: *It's been pretty hard to get out of bed lately, huh?*
> **Youth**: *Yeah. I hate getting up in the morning, and I usually just hang out in my room after school.*
> **Therapist**: *I'm curious, before you started feeling down, what did you used to enjoy?*
> **Youth**: *I don't even know. I was on the soccer team, which was fine I guess. And I'd meet up with friends or hang out with my cousin after practice.*
> **Therapist**: *What did you guys used to do together?*

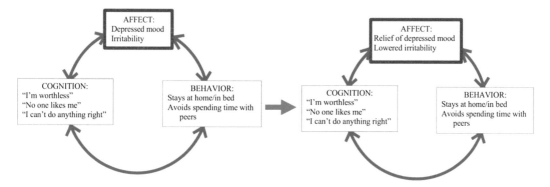

Fig. 7 Targeting affect

Youth: *We used to walk my friend Tyler's dog sometimes. Or we'd go hike the trails behind the school. If the weather was bad we would sometimes get a bunch of snacks and do a board game marathon.*

Therapist: *How about if your friends were busy or out of town?*

Youth: *I actually used to run quite a bit. I was pretty into soccer and liked to stay competitive.*

Therapist: *You have quite the list of hobbies! Playing soccer, running, hiking, and hanging out with friends- going for hikes or playing board games. It sounds like you haven't been doing a lot of those things lately?*

Youth: *Yeah you're right, I haven't been.*

Therapist: *Often, when people feel sad, they stop doing the things they enjoy. This makes a lot of sense. When people feel down they don't feel like doing much. The problem with this is that we then miss out on chances to do things that may make us feel better. Rather than waiting to feel happy to do things you used to enjoy, how would you feel about making a plan to try out some of these activities again and then see if they help?*

Youth: *I guess I could try that. Nick has been bugging me to go on a hike.*

Therapist: *Great! For the rest of today's session let's make a plan for when you'll try out a few of those hobbies this week. You can even rate how you're feeling before and after each activity so that we can problem solve together what sorts of activities help you feel the best. We'll also talk about how to stick to your plan, even if you don't feel like doing these activities in the moment.*

6.3 Relaxation/ Emotion Regulation Techniques

Youth may need additional skills to allow them to regulate their emotions before using other skills. For example, if a youth is crying uncontrollably, it may be unrealistic to expect them to restructure an unhelpful thought or make a plan for behavioral activation. In these cases, relaxation techniques may be particularly helpful. Two such evidence-based techniques include diaphragmatic breathing and progressive muscle relaxation. Here we are directly targeting the affective component of the case conceptualization to help youth regulate enough to engage in other CBT skill work (see Fig. 7).

6.3.1 Diaphragmatic Breathing

Diaphragmatic breathing, a form of deep breathing, teaches youth to contract the diaphragm as they breath, activating the body's relaxation response. Often when youth try to take deep breaths, they end up taking shallow breaths into the tops of their lungs (called "chest" or "shoulder" breathing). Instead, diaphragmatic breathing teaches youth to fill their entire lungs with air. This is key as it contracts the diaphragm (the muscle between the stomach and ribs), which automatically activates our bodies parasympathetic, or relaxation, response. By only filling the tops of their lungs with air, youth miss out on this benefit. Diaphragmatic breathing has been shown to slow down heart rate and lower blood pressure [29], both of which are associated with feelings of relaxation [30]. The key with diaphragmatic breathing is to teach youth to (a) slowly breathe in through their nose and out through their mouth and (b) fill their lungs completely with air.

> **Therapist**: *I can see you're pretty upset right now. I know in the past you've mentioned that sometimes you're chest feels pretty tight and that it's hard to catch your breath when you feel overwhelmed like this. I wonder if we can try a skill together that might help?*
>
> **Youth (through tears)**: *Ok I guess.*
>
> **Therapist**: *This is called diaphragmatic breathing or "belly breathing." It's called that because the point of diaphragmatic breathing is to activate your diaphragm or the muscle that sits right in between your stomach and ribs* (points to her diaphragm). *Normally when we breath, we only breathe air into the top part of our lungs* (demonstrates tense chest/shoulder breathing). *We get some oxygen in, but we don't always feel totally relaxed. By doing diaphragmatic breathing, and breathing deep down into our stomachs, we activate that diaphragm, which turns on our body's relaxation system. Basically it FORCES our body to relax! Let's give it a shot and see if it helps.*
>
> *I'm going to do this with you and walk through each step together. Let's both sit nice and relaxed in our chairs. We're going to take one hand and place it on our chest and our other hand and place it on our stomach. Now the goal as we breathe is to keep the hand on our chest as still as possible but to feel our hand on our stomachs move out as we breath in and in as we breath out. I want you to focus on filling your stomach alllllllll the way up with air and try to keep your chest nice and still. Let's take a slow deep breath in through your nose for 1... 2... 3, hold your breath for 1... 2 and then breathe out slowly out of your mouth for 1... 2... 3. Let's try it one more time breath in through your nose for 1... 2... 3, hold your breath for 1... 2 and then breathe out slowly out of your mouth for 1... 2... 3.*

Adapting for children: A fun way to explain diaphragmatic breathing to younger children is to teach them to "bubble breathe." The underlying skill is the same; youth learn to activate their diaphragm by breathing in through their nose to "fill your bellies alllllllll the way up with air like your belly is a BIG balloon" and then "blow alllllll the air nice and slowly out of the balloon" as they breath out.

6.3.2 Progressive Muscle Relaxation

It can be hard for youth to relax their muscles, especially if they are feeling angry, upset, or nervous. Progressive muscle relaxation forces youths' bodies to relax by first tensing all of their muscles and then releasing that tension. The key is to squeeze/tense/tighten one muscle group, then release tension from that same muscle group. This same sequence is followed for each new muscle group (tension then relaxation) moving up the body from the toes to the face. For each cycle of tension and relaxation, the tension is held for 10–15 s and then relaxed for 10–15 s. The more youth squeeze their muscles during the tension period, the more relaxed they will get during the relaxation period. Movement engages muscles so therapists should make sure that during relaxation periods, youth let their muscles go completely limp and relaxed.

Youth (yelling): *I don't care if my thoughts are true or not! Everyone hates me and that's the way it is! It's never going to get better! I'm going to feel this way forever.*

Therapist: *I hear how upset you are. It really feels like these feelings will never end. It sounds like you need a break from these big feelings for a minute. Then we can come back to this to see if we problem solve together.*

Sometimes when we're feeling sad or overwhelmed or angry, it can be really hard to relax our bodies. Instead we can trick our bodies into relaxing by first tensing all of our muscles and then releasing the tension. The key here is to squeeze all the muscles in one part of our body as much as we possibly can, and then relax those same muscles.

Let's try this together. First, let's get nice and relaxed in our chairs.

Let's start with our feet ... I want you to squeeze alllllllll those muscles in your feet. Scrunch your toes up and hold all those muscles nice and tight ... (hold for 10–15 seconds).

Now relax. Let your feet go nice and limp. Let them hang down in front of you. Pay attention to how it feels to let all of that tension out of your feet ... (relax for 10–15 seconds).

Now let's tense our legs ... Stretch both of your legs out in front of you and point your toes ... squeeze all of those muscles in the top of your legs and the bottom of your legs ... squeeze all those muscles as tight as you can!! (hold for 10–15 seconds).

Now relax ... let your legs fall to the ground. Go limp and relax all the muscles in your legs ... notice how your legs feel ... (relax for 10–15 seconds).

Now let's squeeze all those muscles in our bottom ... squeeze all of those bottom muscles really tight!! (hold for 10–15 seconds).

And relax ... good ... feel your body sink into your chair. (relax for 10–15 seconds).

Now, suck in your stomach muscles and squeeze all those abs... hold those stomach muscles tight.

And relax all the muscles in your stomach ... let your stomach fill back up with air....notice how your muscles feel. (relax for 10–15 seconds).

... Continue with shoulders, arms, hands, and face ...

Lastly you're going to squeeze all the muscles in your whole body ... hold them tight .... squeeze your feet and your legs and your bottom and your abs and your shoulders and your arms and your hands and your face ... (hold for 10–15 seconds).

Now relax your whole body ... imagine you're a rag doll ... let everything go nice and limp. Notice how relaxed you feel ... so calm ... (relax for 10–15 seconds).

When you are ready, you can slowly open your eyes.

Therapist: *How was that?*

Youth: *Good. It was like my body had no choice but to relax.*

Therapist: *Exactly. We can trick our bodies into relaxing even when it seems like we can't seem to calm down. Now let's get back to that idea that you're "going to feel this way forever."*

> Adapting for children: A fun way to explain progressive muscle relaxation to younger children is for the therapist to show them how to tense all their muscles like robots and relax all their muscles like jellyfish. To simplify, younger children can also practice squeezing all of their muscles at once and then relaxing all their muscles at once rather than going muscle group by muscle group.

7 Conclusion

Depression causes considerable impairment, impacting youths' social, academic, and familial lives. It is thus critical that depression be treated effectively, especially early in development when future episodes of depression can be prevented. CBT is one such evidence-based intervention for depression that has been shown to relieve youths' distress [15]. The intervention strategies illustrated in this chapter represent the core components of CBT treatment for youth with depression. In providing example dialogue and practical tips and handouts, we hope that we have brought to life what it might look like to use these skills with youth. Effectively implemented CBT involves far more than rote understanding of CBT strategies, however. In order to derive maximal benefit from CBT, clinicians should select those techniques that best fit with the particular presenting concerns and aim to apply these strategies creatively and in ways that best engage youth. This chapter, therefore, is best considered as a beginning. Like any skill, more learning and practice will strengthen and expand clinicians' abilities to intervene with depressed youth. Tables 13 and 14 list additional resources to help with that growth.

Table 13
Additional Resources for Clinical Practice of CBT with Youth

Additional Resources
Beck Institute for Cognitive Behavior Therapy. CBT for youth and CBT for families of youth workshops and online training. www.beckinstitute.org
Friedberg RD, McClure JM (2015) Clinical practice of cognitive therapy with children and adolescents: the nuts and bolts, 2nd edn. Guilford Publications, New York, NY
Nangle DW, Hansen DJ, Grover RL, Kingery JN, Suveg C (2016). Treating internalizing disorders in children and adolescents: core techniques and strategies. Guilford Publications, New York, NY. www.effectivechildtherapy.org

Table 14
Additional CBT Treatment Manuals for Youth

Evidence-Based Treatment Manuals for Depressed Youth
Fristad MA, Arnold JSG, Leffler JM (2011) Psychotherapy for children with bipolar and depressive disorders. Guilford Press, New York, NY
Stark KD, Simpson J, Schnoebelen S, Hargrave J, Molnar J, Glen R (2007) Treating depressed youth: therapist manual for 'ACTION'. Workbook Publishing, Ardmore, PA
Williams KN, Crandal BR (2015) Modular CBT for children and adolescents with depression: a clinician's guide to individualized treatment. New Harbinger Publications, Oakland, CA

References

1. National Center for Health Statistics (2013) Health, United States, 2013: with special feature on prescription drugs. National Center for Health Statistics, Hyattsville, MD. https://www.ncbi.nlm.nih.gov/pubmed/24967476. Accessed 24 Oct 2018

2. Avenevoli S et al (2008) Epidemiology of depression in children and adolescents. In: JRZ A, Hankin BL (eds) Handbook of depression in children and adolescents. Guilford Press, New York, NY, pp 6–32

3. Lewinsohn PM, Rohde P, Seeley JR (1998) Major depressive disorder in older adolescents: prevalence, risk factors, and clinical implications. Clin Psychol Rev 18:765–794

4. Thapar A et al (2012) Depression in adolescence. Lancet 379:1056–1067

5. Avenevoli S et al (2015) Major depression in the National Comorbidity Survey–adolescent supplement: prevalence, correlates, and treatment. J Am Acad Child Adolesc Psychiatry 54:37–44

6. Piccinelli M, Wilkinson G (2000) Gender differences in depression: critical review. Br J Psychiatry 177:486–492

7. Keenan-Miller D, Hammen CL, Brennan PA (2007) Health outcomes related to early adolescent depression. J Adolesc Health 41:256–262

8. American Psychiatric Association (2013) Diagnostic and statistical manual of mental disorders, 5th edn. American Psychiatric Association, Washington, DC

9. Costello EJ et al (2002) Development and natural history of mood disorders. Biol Psychiatry 52:529–542

10. Kovacs M, Goldston D (1991) Cognitive and social cognitive development of depressed children and adolescents. J Am Acad Child Adolesc Psychiatry 30:388–392

11. Hetrick SE et al (2016) Cognitive behavioural therapy (CBT), third-wave CBT and interpersonal therapy (IPT) based interventions for preventing depression in children and adolescents. Cochrane Database Syst Rev 8: CD003380

12. Nagar S et al (2010) Extent of functional impairment in children and adolescents with depression. Curr Med Res Opin 26:2057–2064

13. Ingram R et al (2014) Depression: social and cognitive aspects. In: Millon T, Blaney PH, Davis RD (eds) Oxford textbook of psychopathology. Oxford University Press, New York, NY, pp 203–226

14. Zisook S et al (2007) Effect of age at onset on the course of major depressive disorder. Am J Psychiatry 164:1539–1546

15. David-Ferdon C, Kaslow NJ (2008) Evidence-based psychosocial treatments for child and adolescent depression. J Clin Child Adolesc Psychol 37:62–104

16. Beck AT (1987) Cognitive models of depression. J Cogn Psychother 1:5–37

17. McCarty CA, Weisz JR, Hamilton JD (2007) Effects of psychotherapy for depression in children and adolescents: what we can (and can't) learn from meta-analysis and component profiling. J Am Acad Child Adolesc Psychiatry 46:879–886

18. Chu BC, Harrison TL (2007) Disorder-specific effects of CBT for anxious and depressed youth: a meta-analysis of candidate mediators of change. Clin Child Fam Psychol Rev 10:352–372

19. Klein JB, Jacobs RH, Reinecke MA (2007) Cognitive-behavioral therapy for adolescent depression: a meta-analytic investigation of changes in effect-size estimates. J Am Acad Child Adolesc Psychiatry 46:1403–1413

20. Weersing VR, Brent DA (2006) Cognitive behavioral therapy for depression in youth. Child Adolesc Psychiatr Clin N Am 15:939–957

21. Weersing VR et al (2017) Evidence base update of psychosocial treatments for child and adolescent depression. J Clin Child Adolesc Psychol 46:11–43

22. Nangle DW et al (2016) Treating internalizing disorders in children and adolescents: Core techniques and strategies. Guilford Publications, New York, NY

23. Kaufman NK et al (2005) Potential mediators of cognitive-behavioral therapy for adolescents with comorbid major depression and conduct disorder. J Consult Clin Psychol 73:38–46

24. Compas BE et al (2001) Coping with stress during childhood and adolescence: problems, progress, and potential in theory and research. Psychol Bull 127:87–127

25. Lewinsohn PM (1974) A behavioural approach to depression. In: Freedman RJ, Katz M (eds) The psychology of depression. Wiley, Oxford, pp 157–174

26. Chu BC et al (2009) An initial description and pilot of group behavioral activation therapy for anxious and depressed youth. Cogn Behav Pract 16:408–419

27. McCauley E et al (2016) The adolescent behavioral activation program: adapting behavioral activation as a treatment for depression in adolescence. J Clin Child Adolesc Psychol 45:291–304

28. Ritschel LA et al (2011) Behavioral activation for depressed teens: a pilot study. Cogn Behav Pract 18:281–299

29. Ma X et al (2017) The effect of diaphragmatic breathing on attention, negative affect and stress in healthy adults. Front Psychol 8:1–12

30. Terathongkum S, Pickler RH (2004) Relationships among heart rate variability, hypertension, and relaxation techniques. J Vasc Nurs 22:78–82

Chapter 3

Cognitive Behavioral Therapy for Child and Adolescent Anxiety: CBT in a Nutshell

Emily Badin, Evan Alvarez, and Brian C. Chu

Abstract

Anxiety disorders and related problems affect a large number of children and adolescents, are impairing, and can affect long-term functioning if left untreated. Cognitive behavioral therapy (CBT) is an efficacious treatment approach that teaches anxious youth multiple skills to help manage anxiety symptoms, challenge unrealistic assumptions, and approach frightening situations. This chapter describes and illustrates CBT strategies and principles in the context of anxious children and adolescents. Worksheet templates and case illustrations are provided to demonstrate how a clinician turns principles into practice using a flexible, principles-based approach. The principles and concrete tools provided herein are designed to help practitioners meet the specific needs of their clients, representing a diversity of clinical profiles and settings.

Key words Child, Adolescent, Anxiety, Evidence-based, Cognitive behavioral therapy (CBT)

1 Introduction

Anxiety disorders reflect a broad category of socio-emotional disorders characterized by intense fear reactions, pervasive worries, avoidant behaviors, and self-critical and threat-related thinking. Concerns that bring youth to treatment include a diverse set of disorders, including generalized anxiety, social anxiety, separation anxiety, panic and agoraphobia, specific phobias, and selective mutism. Anxiety disorders are the most common psychological disorder among children and adolescents (hereafter referred to as youth), affecting nearly 32% of 13- to 18-year-olds, followed by behavior disorders (19.1%), mood disorders (14.3%), and substance use disorders (11.4%) [1]. Epidemiological studies report that over 10% of children meet criteria for some form of anxiety disorder at a given point in their lives [2]. In addition to high prevalence rates, youth anxiety disorders are associated with impairment in social, occupational, academic, and family functioning that may extend into adulthood if left untreated [3]. The degree of comorbidity

Robert D. Friedberg and Brad J. Nakamura (eds.), *Cognitive Behavioral Therapy in Youth: Tradition and Innovation*, Neuromethods, vol. 156, https://doi.org/10.1007/978-1-0716-0700-8_3, © Springer Science+Business Media, LLC, part of Springer Nature 2020

between anxiety and depression is also significant and reflected on dimensional measures [4–6].

The presence of anxiety symptoms in childhood often leads to the avoidance of developmentally expected tasks that may be distressful or anxiety-provoking. For example, youth with social anxiety often display deficits in social skills, decreased life satisfaction, increased family problems, poor relationship quality in adulthood, and increased difficulty adjusting to work-related obligations [7–10]. Additionally, anxious youth are often less likely to explore educational and career opportunities and romantic partnerships [11]. The interplay of symptoms and functional impairment can lead to negative developmental cascades in which the consequences of elevated anxiety symptomatology and impairment compound over time [9, 12, 13].

2   Theoretical and Empirical Foundations

2.1 Evidence Base for Cognitive Behavioral Therapy

A substantial body of research has demonstrated the efficacy of cognitive behavioral therapy (CBT) for treating youth anxiety [14–21]. The largest randomized clinical trial, the Child Anxiety Multimodal Study (CAMS), evaluated the efficacy of individual cognitive behavioral therapy (Coping Cat; CBT), medication (sertraline; SRT), a combination of the two treatments, and a pill placebo among 488 youth aged 7–17 years for treating anxiety [22]. Response rates indicated notable symptom reductions of 80.7% among participants in the combined treatments, 59.7% among those in CBT, and 54.9% in the pill placebo condition.

Other trials support efficacy for individual, family, and group versions of CBT. Kendall and colleagues [19] conducted a randomized clinical trial comparing the efficacy of individual cognitive behavioral therapy (ICBT), family cognitive behavioral therapy (FCBT), and a family-based education/support/attention (FESA) active control for treating youth aged 7–14 years with an anxiety disorder. FCBT and ICBT conditions were superior to FESA in remission of principal anxiety diagnosis (57%, 55%, and 37%). Another RCT conducted by Hudson and colleagues [23] randomized 112 children aged 7–16 years with a principal anxiety disorder diagnosis to either a group cognitive behavioral treatment (CBT) or control group (group support and attention [GSA]). Remission rates at 6-month follow-up were significantly higher for CBT compared to GSA condition (68.6% versus 45.5%). CBT has also been extended to community settings and has shown modest results relative to usual care [24].

More recently, transdiagnostic adaptations of CBT, which are designed to simultaneously target shared features across disorder classes, have shown promising effects for youth presenting for either anxiety or depression [25–27]. Follow-up longitudinal studies of youth who have undergone a course of CBT have shown maintained symptom reduction and global functioning [28] and sustained remission rates ranging from 51% to 85% at 5.5- to 9.3-year follow-up [29, 30].

2.2 Theoretical Overview: Anxiety from the CBT Perspective

The cognitive behavioral model helps conceptualize anxiety in the context of three main components (thoughts, feelings, and actions) [31]. Thoughts, feelings, and actions are interrelated and may influence each other reciprocally. Understanding how a youth is responding to a feared stimulus, a therapist can then fashion interventions to make targeted change. For example, if the child's thought is, "I'm going to fail the exam," then the therapist can use cognitive restructuring to identify the assumptions implicit in that thought and devise a more adaptive coping thought. If the child is experiencing ongoing tension, the therapist can teach progressive relaxation skills to help lower the youth's baseline tension. If the child is demonstrating significant avoidance and escape behaviors (e.g., missing class, procrastinating on homework), then the therapist develops a hierarchy of challenges and coaches the youth to approach those feared situations with confidence and calm.

Underlying the CBT model are two principal traditions: cognitive science and learning theory. Cognitive science has shown that anxious youth tend to demonstrate threat-focused attentional biases, negative automatic thinking, and ingrained negative schema [32, 33]. Specifically, anxious children will selectively focus on threat-related information, assume ambiguous situations as threatening, conclude threat-related attributions, expect negative outcomes, and make behavioral choices prioritizing personal safety over goal achievement or mastery [34]. As these processes interact, anxious youth overemphasize threat in various situations and make decisions based on the perceived threat.

Behavioral science has helped identify particular behavioral habits that keep the youth safe from perceived threats while prolonging long-term impairment. Within the framework of classical conditioning, youth who repeatedly experience intense anxiety in specific situations (e.g., giving a class presentation), will learn over time to associate the experiential stimulus (class presentation) with the cognitive, behavioral, and/or physiological response (developing anxiety symptoms). As this process is repeated, operant conditioning reinforces the child's anxiety to the stimuli, making them more likely to experience higher degrees of anxiety with each exposure. Youth are particularly susceptible to long-term increases in anxiety through avoidant [35] or safety behaviors, which are actions performed to prevent, escape, or minimize feared outcomes or potential associated distress [36, 37].

To intervene, CBT models have been developed to counter these natural avoidance patterns. Avoidance can present a number of problems in effectively processing emotions and learning

[35, 38]. According to the emotion processing model of anxiety, fear associated with stimuli, responses, and meaning result in individuals subsequently perceiving a distorted reality For example, a teen isolates herself after a friend fails to return a text, assuming that further efforts will be rejected). Further, repeated avoidance prevents sufficient activation of fear, which prevents new anti-anxiety information from being learned. In our example, the teen's isolation prevents her from feeling the distress of uncertainty, but it also prevents her from acquiring new learning, such as that her friend is usually receptive. Both the activation of emotions (feeling the uncertainty) and experiencing contradicting information is required for new learning. Similarly, the habituation model of anxiety postulates that prolonged exposure is required to decrease anxiety [39]. In this model, it is assumed that the teen must, first be activated, and then subsequently experience a reduction in subjective distress to learn that the threat is non-threatening.

Alternatively, inhibitory learning theory [40] proposes that the intensity to which aversive or anxiety-provoking stimuli are experienced may be reduced and/or eliminated through expectancy violation. A rise and fall in subjective distress is not required. Instead, the focus is on helping the youth identify what expectations she/he has leading into a situation and how that expectation failed to surface once the threat was approached. This strategy is designed to increase the effectiveness of exposures through increasing the degree to which one's expectations of a particular aversive outcome are violated. The more an expectancy can be violated by experience, the greater the effect of inhibitory learning. As this process is repeated over time, the youth gradually learns experientially that such situations are both approachable and non-threatening.

2.3 Cognitive Behavioral Approach to Youth Anxiety: A Case Illustration

To illustrate a CBT approach, we use the remainder of the chapter to present a case illustration of key CBT conceptualization and intervention strategies. Our case example features a 13-year-old female, TB (fictional initials; the case reflects a compilation of relevant clients), with a principal diagnosis of panic disorder and social anxiety disorder. During the initial sessions, the therapist conducts a comprehensive assessment that permits a diagnostic evaluation and a functional assessment of TB's thoughts, feelings, and behaviors. Upon further assessment, we learn that TB's panic attacks often begin with thoughts (e.g., "I want to escape") followed by emotions and physical sensations (e.g., heart pounding, sweating, shaking, difficulty breathing, chest pain, dizziness, and derealization) and a subsequent action/behavior (e.g., immediately fleeing the situation where the anxiety occurs, typically at school). Differentiating each stage of the panic attack helps demystify the experience for the youth and identify prospective intervention targets.

3   Intervention

3.1   Initial Assessment

Initial diagnostic assessment of a youth's presenting concerns is essential and serves several purposes. First, obtaining a diagnosis informs the therapist's treatment plan in what specific CBT strategies will be applied (e.g., treating panic disorder may include more exposures over a shorter period of time compared to treating social anxiety disorder). Additionally, symptomatic and diagnostic improvement can be more precisely evaluated by the change in the youth's diagnosis when reevaluated after treatment.

We recommend using the ADIS-5 for evaluating youth diagnosis [41]. The ADIS-5 a semi-structured interview consisting of independent but congruent parent and child interviews. The parent and child interviews are conducted individually, allowing the clinician to collect data from each reporter to compose a composite diagnosis. Psychometric studies for the previous version (ADIS-IV-TR) have demonstrated good interviewer reliability (e.g., $k = 0.98$, parent interview; $k = 0.93$, child interview) [42], retest reliability (e.g., $r = 0.76$, parent interview) [43], and sensitivity to treatment effects [18]. Three core principles distinguish assessment with youth from adults: (1) collecting information from multiple informants; (2) using multiple measures to assess various constructs and repeating measure assessment over time; and (3) incorporating developmentally appropriate measures [44, 45]. The ADIS-5 possesses several advantages over other diagnostic measures (e.g., K-SADS and DISC). First, it includes a separate parent interview, which allows the therapist to collect and integrate critical information that would not typically be disclosed by the child. Additionally, the ADIS-5 enables the clinician to assign both a diagnosis and interference rating (e.g., ranking from 0 to 8). Ranking interference within various life domains (e.g., school, home, friendships) helps determine the degree of negative impact within specific contexts of the child's life.

Using short, objective measures can also augment assessment and allow clinicians to capture more detailed aspects of a child's symptom profile. The Screen for Child Anxiety Related Emotional Disorders (SCARED) is a short, psychometrically sound measure that can be used to measure anxiety symptoms periodically (e.g., weekly or monthly) throughout treatment. The SCARED has demonstrated good internal consistency ($\alpha = 0.78$–0.87), test–retest reliability (intraclass correlation coefficients = 0.6–0.9), and discriminative validity both between anxiety and other disorders and within anxiety disorders [46]. It has also been validated for use across cultures [47]. The Multidimensional Anxiety Scale for Children—Child and Parent Forms [48] is another empirically validated measure designed for assessments of four domains of anxiety psychopathology (e.g., physical symptoms, social anxiety, harm avoidance, and separation anxiety). The 39-item form (MASC-39) has been normed on large samples and is one of the most widely used anxiety symptom measures. Additionally, the revised Children's Anxiety and Depression Scale; Parent/Youth (RCADS) is a 47-item scale, rated on a 4-point (Never to Always) scale. Subscales demonstrate good factorial validity (e.g., association with common anxiety and depressive disorders), 1-week test–retest reliability, internal consistency, and good convergent and discriminant validity with other validated internalizing and externalizing symptom measures [49] Additional free valid and reliable measures are also available online [50].

While diagnosis and symptom severity help to classify a youth's concerns into recognizable diagnostic profiles, a functional assessment helps identify observable behavioral patterns that call for intervention. Functional assessments identify antecedent–behavior–consequence (ABC) sequences or behavioral chains as they related to target problem behaviors [51, 52]. An *antecedent* is considered to be any trigger or circumstance (e.g., person, place, object, event, thought, feeling, action) that exhibits discriminating properties. We define the youth's response to the antecedent (e.g., thoughts, feelings, action, physiological response) as their *behavior*. *Consequences* are any outcomes that follow the youth's initial response. Identifying the consequences that follow problematic behaviors help the clinician recognize the reinforcers that maintain those behaviors. Conducting functional assessments gives the client (and other family members) an opportunity to disclose specific details of problematic behavior while gaining insight into the triggers and emotional reactions that precede them [52]. Likewise, a functional assessment acts as an essential assessment tool for helping the client, family, and therapist target which behavioral chains lead to the most concerning problem behaviors.

In TB's case, she reported sweating, difficulty breathing, and feeling like she was losing control. She also reported functional impairment in school such that she would freeze, have difficulty concentrating, and ultimately escape from class or other activities when she has a panic attack. Her repeated attacks led to fear of having future attacks and an avoidance of situations she associated with them, including school. The client's mother also reported these same symptoms and functional impairment. As a result, TB showed adequate signs of panic disorder which is marked by repeated panic attacks and persistent fear of having further attacks. It also comes with substantial avoidance of specific situations like school and evaluative situations with peers. Based on TB's report, a clinician might also consider these comorbid problems (feeling depressed, losing interest in activities, irritability), but these were not considered the clinical foci because TB reported higher interference from her panic attacks.

CBT Case Formulation

Situation (target concern): __Playing in a softball game__

Trigger		
Racing heart shallow breathing		

↓

Feelings		
Scared Anxious		

Thoughts	Behaviors
"I'm going to have a panic attack."	Escape – Ask mom to go home.

Current Outcomes	
Short-Term	Long-Term
Feels better. Heart rate decreases. Breathes normally again.	Don't learn I can handle feelings. Anxiety will likely come back for the next game. Misses hangout with friends after game.
Desired Outcomes	
Short-Term	Long-Term
Feel better and/or tolerate negative feelings.	Learn I can handle feelings Keep playing softball. Don't miss hang out with friends.
Barriers to Desired Outcomes	
Short-Term	Long-Term
Thoughts I can't tolerate negative feelings and that they are dangerous.	Don't give self enough opportunities to practice tolerating negative feelings.

Fig. 1 CBT Case Formulation Worksheet that Emphasizes Short- and Long-term Outcomes

3.2 CBT Case Formulation: Capturing the Client Within a Nutshell

Based on the diagnosis and functional assessment, the clinician begins developing a case formulation. A case formulation (or conceptualization) refers to a working hypothesis of the internal (e.g., physiological arousal or maladaptive beliefs) and external (e.g., environmental reinforcers and punishers) factors that contribute to the development and maintenance of a problem behavior [53–56]. In other words, the clinician tries to capture their client, "within a nutshell." Within CBT, the therapist aims to identify the relevant antecedents (triggers), thoughts, feelings, behaviors, and consequences of the target problem (*see* Fig. 1; CBT Case Formulation Worksheet). When evaluating consequences, the clinician should assess for the short-term and long-term outcomes to better understand the current contingencies. Additionally, the clinician should review the desired outcome (what would the youth prefer to happen instead?) and the barriers to achieving that desired outcome (what is currently getting in the way?). This information can illuminate the best point of intervention (e.g., exposure versus cognitive restructuring).

Although a clinician begins developing a case formulation as soon as he/she meets the client, they are continuously revising it as they gather new information. The case conceptualization should be continually revised as treatment progresses and as the clinician tests out their hypotheses in treatment. This process of routinely integrating new information into conceptualization, and therefore informing treatment planning, has been called the "Assessment-Treatment Dialectic" [57].

Clinicians should always consider culture and context when developing a case conceptualization [45, 58]. Cultural intersectionality is even more salient when working with children and adolescents, since multiple systems (e.g., family, school, and communities) are considered. Several frameworks have been developed to aid clinicians in identifying relevant cultural factors, some created to reflect the influence of specific cultural groups, while others broadly encourage the clinician to consider how culture can affect symptom presentation and treatment [58–60]. For example, a broader system used to assess and consider the role of various cultural influences on a client is the ADDRESSING model by Hays [61]. Hays prompts the consideration of age, disability (developmental or acquired later in life), religion/spiritual orientation, ethnicity/racial identity, socioeconomic status, sexual orientation, indigenous heritage, national origin, and gender during assessment and treatment.

3.3 Case Formulation for TB

TB's anxiety seems to be primarily maintained by avoidance of physiological sensations (e.g., difficulty breathing) and feared situations. Related to her diagnosis of panic disorder, when TB begins to experience sympathetic nervous system arousal, she has an automatic thought ("I'm going to have a panic attack"), which prompts her to escape the current situation instead of tolerating the distress associated with high physiological arousal. TB's avoidance and escape patterns result in short-term relief from the distress associated with these experiences (e.g., she feels safe when she escapes a foreign environment). In the long-term, the avoidance ultimately maintains her anxiety as she fails to provide herself with the opportunity to learn that she can tolerate uncomfortable physiological sensations and may miss out on enjoyable opportunities (e.g., hang out with friends after a softball game) when she escapes a situation.

Furthermore, TB does not have opportunities to encounter evidence that disconfirms her belief that she will have a panic attack. Alternatively, TB aims to reduce physiological arousal in the moment and/or learn to tolerate it (desired short-term outcomes) and reduce interference (e.g., continue playing softball and hang out with friends). When asked about barriers to her desired outcomes, TB and her therapist collaboratively discussed her maladaptive thoughts about her ability to tolerate anxiety and her subsequent avoidance of opportunities to practice tolerating high physiological arousal. TB's social anxiety is maintained through similar mechanisms, specifically avoidance of social interactions. For example, TB avoids initiating and maintaining conversations with her peers because she is fearful of being negatively evaluated or embarrassing herself. Although TB's panic disorder and social anxiety manifest in very different ways, they appear to be maintained through a similar mechanism: avoidance. As a result, treatment should focus on targeting TB's avoidance in order to effectively and efficiently intervene.

3.4 Therapist Posture

In addition to a comprehensive case formulation, a strong therapeutic alliance is crucial for treatment success. Labeled a "VIP" (very important process) by Friedberg and Gorman [62] the alliance is defined as the affective bond between the clinician and client, characterized by trusting rapport and high agreement on treatment goals and tasks [63]. Previous research demonstrates that the client-clinician relationship has been consistently associated with positive treatment outcomes across youth presenting problems [64–66]. Specifically, it has been linked to increased treatment retention, attendance to session, and child and family participation in treatment [67, 68]. Indeed, cognitive behavioral therapy considers the therapeutic relationship as a necessary component of effective treatment [62].

In order to foster a positive and strong therapeutic alliance, therapist posture should be considered. This stance is similar to that of the empathic and encouraging coach. In sports, a coach aims to help a child with their technique, as well as keep them motivated to persist, even with multiple failed attempts. The coach would not be as effective if they motivated through yelling, critical comments, or by withdrawing support when frustrated. Instead, the effective coach demonstrates confidence in the youth, praises the child's willingness to try new suggestions, reinforces the child when they incorporate feedback, and applauds the child's efforts to cope with frustration. In therapy, the clinician takes a similar approach, teaching necessary coping skills while empathizing with the client's current emotional state and encouraging efforts to cope.

Overall, the clinician and client should maintain a collaborative relationship in order to promote growth and adaptive learning. This collaborative style has been found to be most critical during the early phases of treatment [69, 70]. Collaboration can be promoted through the solicitation of constant feedback, allowing the clinician to hone which suggestions are most or least effective to each individual client. Additionally, early interventions that focus on setting mutually agreed upon goals, eliciting commitment, and psychoeducation can promote a collaborative environment, demonstrate understanding, and foster trust between a clinician and client. In order for the clinician to learn about a client's goals or values, time should be dedicated to building rapport, specifically learning about the youth's interests, strengths, and hobbies.

3.5 Psychoeducation

In the early phase of treatment, clinicians should provide ample psychoeducation on the nature of anxiety and how treatment will intend to target anxious symptoms. Interested clinicians are referred to an excellent set of free online FAQ sheets on specific anxiety disorders, produced by the Association of Behavioral and Cognitive Therapies (http://www.abct.org/Information/?m=mInformation&fa=fs_ANXIETY), and to information regarding effective therapies for anxiety, produced by the Society of

Clinical Child and Adolescent Psychology (www. effectivechildtherapy.com). By emphasizing that anxiety, like all emotions, is natural and serves an important function, the clinician is also communicating that experiences of intense emotion, particularly anxiety, is not dangerous. The therapist should further explain that anxiety, although a natural response, may be triggered in unhelpful situations and cause daily functional impairment. Therefore, the goal of treatment is to learn to *manage* anxiety, so it does not interfere with daily life. At the same time, anxiety can be highly distressing and comes with characteristic cognitive, physiological, and behavioral reactions. The therapist can help the youth see anxiety on a continuum of intensity, and that coping skills are needed when the intensity crosses the threshold at which the youth's functioning is at risk. In order to monitor this level of intensity, the therapist can create a subjective scale of the youth's distress called a "feelings thermometer." Collaboratively, the clinician and client can create anchors on the thermometer (0–10 scale) that reference events in the past when the youth experienced that level of distress. For example, when creating TB's thermometer, a "2" referred to her distress after her heart started racing after running in softball, while an "8" referred to the distress she experienced while walking around in a local crowded mall.

After providing a basic overview of the nature of anxiety and the overarching goals of treatment, the clinician should then begin to elicit the idiosyncratic thoughts, feelings, and behaviors that contribute to the child's anxiety. Some youth may require initial affective education in order to identify emotions or physiological sensations. In order to provide this information at a developmentally appropriate level, we recommend using various emotion faces (e.g., individuals portraying happy, sad, angry, or scared faces) and body maps (e.g., body outline detailing common bodily experiences of anxiety) to delineate both the internal and external manifestations of anxiety. Therapists can use feelings charades to assess the child's ability to identify various emotions and feelings and build a feelings dictionary to develop their emotional vocabulary if needed. Furthermore, the therapist can conduct experiments with youth, like robot/rag-doll, to learn the difference between feeling tensed and relaxed and introduce the relaxation coping skill [71].

3.6 Progress Monitoring

Substantial evidence has suggested that therapists who track and review their own therapy progress experience client improvement above and beyond that associated with a specific treatment [72]. As such, we recommend that therapists monitor youth progress continuously throughout treatment to track changes in a youth's symptoms and impairment. This information helps the clinician determine if youth are responding to treatment and if the treatment plan should be updated. The following are several brief validated measures that have demonstrated good psychometric properties [50]. The Multidimensional Anxiety Scale for Children; Youth/Parent (MASC) [48] has a 10-item short form that is designed for repeated assessments of anxiety symptoms. The revised Children's Anxiety and Depression Scale; Parent/Youth (RCADS) [49] is a 47-item scale whose items correspond closely to DSM-IV-TR symptoms of anxiety and depression, but its length might dictate that therapists implement several sessions apart. Finally, the Center for Epidemiologic Studies-Depression Scale; Youth/Parent (CES-D) [73] is a parent and child report form that includes 20 items designed to assess depressed mood, feelings of worthlessness/guilt, sense of helplessness/hopelessness, psychomotor retardation, loss of appetite, and sleep disturbance [50].

In addition to anxiety symptoms, functional impairment should also be assessed. Functional impairment describes the impact that psychopathology has on a youth's ability to complete routine and age-appropriate acts in school, family, peer relationships, and work/extracurricular activities [74]. Measures that focus on youth impairment in critical life domains include the Strengths and Difficulties Questionnaire [75, 76] and the Brief Impairment Scale (BIS) [77].

Recent evidence suggests that idiographic "top problem" or target goals measures can also provide useful information for continuous progress monitoring [57, 78]. As one example of this, a therapist could help generate a list of the youth and caregivers' top three problems for the youth by brainstorming things that are getting in the way for the child or that represent goals she/he wishes to achieve. It is important that these goals are reasonably specific and concrete. A goal of "eliminating anxiety" is too diffuse to manage realistically. Rather, the therapist helps the youth identify specific forms of anxiety (e.g., "freezing up when I talk to kids in class;" "obsessing over homework for hours"). The therapist can then help the youth/caregivers assign a trackable distress rating (say, on a 0–10 scale). Alternatively, they could rate "disruption," "avoidance," or "impairment." Once this preliminary set of problems and distress ratings are derived, the therapist can reassess weekly or periodically to monitor success on these targeted goals that are meaningful for the family.

3.7 Exposures

Exposures are behavioral experiments that promote approach behaviors. When a child or adolescent approaches a difficult situation, he/she is given the opportunity to practice skills learned in session, learn that they can manage the distress associated with that situation, possibly habituate to the anxiety (although this is not necessary), and encounter disconfirming evidence that will challenge maladaptive thoughts and/or beliefs [37, 40]. The first step in designing an exposure is to brainstorm relevant challenges that align with the goals of treatment. These challenges should activate some level of fear or distress, which should be rated by the client and ranked in order of difficulty. When working with younger clients, first categorizing the situations from easy, medium, to hard may provide an important first step in ranking each situation individually. The challenges on the hierarchy should vary in order to reflect increased difficulty and promote generalization across contexts. Exposures are then designed to be conducted in session, with the therapist coaching the youth through the challenge and practicing the skills they have learned throughout therapy. Exposures can be conducted in vivo, meaning "in life," or imaginally. Imaginal exposures involve the client recalling or envisioning themselves in the challenging situations. These are typically used when a situation listed cannot be replicated in session. During both in vivo and imaginal exposures, the therapist acts as a coach and cheerleader.

Examples include a youth giving a speech in front of similar-aged peers to practice speaking in front of a group for an extended period of time. The youth might use this scenario to practice challenging negative automatic thoughts about peer judgment, taking deep breaths to release muscle tension, and practice performing even in the face of emotional discomfort. Another example could be if the youth engages in an interoceptive exposure, like hyperventilating, for approximately 30 s, their expectancy of dying or fainting due to a breathing disturbance will be violated and they will learn disconfirming evidence that physiological arousal will not lead to their feared outcome. Lastly, if a child or an adolescent is having difficulty separating from a caregiver at bedtime, an exposure that includes his/her caregiver only lying in bed with him/her for a predetermined amount of time (e.g., 5 min) will lead to an expectancy violation that the youth can handle the distress associated with the separation and disconfirming evidence that no harm will come to themselves or his/her caregiver at night. This separation experience may also lead to feelings of autonomy and mastery, since the child or teen may not have ever been able to fall or stay asleep without his/her caregiver. In order to enhance learning, the therapist should not engage in strategies to reduce expectancies (e.g., cognitive restructuring or relaxation exercises) to maximize expectancy violation [40].

Case Example: Prior to conducting an exposure with TB, the therapist worked to create a fear hierarchy that listed her feared situations in order of least to most distressing. Some of the situations on TB's hierarchy included initiating conversations (rating: 3/10), joining conversations (5/10), and giving class presentations (8/10). Next, the therapist asked questions to determine which contexts presented easier or harder challenges. For example, TB found that it was easier to initiate conversations with friends compared to acquaintances. Additionally, she preferred to speak via text instead of face-to-face. After making these situations more specific, TB re-rated these situations in order of difficulty, which she then placed on her hierarchy. In order to develop a hierarchy, the clinician can manipulate duration, setting, number of individuals, and topic for any given situation. These strategies will create diversity among associated distress ratings by the client.

The goal of establishing a fear hierarchy is to develop a wide array of scenarios that could help the youth practice their skills in difficult situations. After completing a fear hierarchy, the clinician is ready to begin exposures. Although traditionally exposures were conducted in a hierarchical manner, from least to most distress elicited, inhibitory learning models suggest that it may be more effective to choose exposure tasks less methodically in order to activate emotions and promote learning [79, 80]. More importantly than where you start on the hierarchy is the degree to which the youth experiences emotional activation. Thus, the key role of the therapist is to ensure that the exposure is realistic and evocative, even while they are helping to coach the youth through this challenging situation. The "art" of exposure therapy is finding that "optimal" zone where you create an exposure that is challenging enough to fully activate the youth emotionally but is also manageable enough such that the youth is willing to engage it. If the youth refuses or escapes prematurely, opportunities to learn new behaviors or to counter negative expectations are truncated. In addition, the therapist will want to ensure that the youth focuses her/his entire attention on the task or challenge at hand. If the therapist sees the youth distracted or engaging in safety behavior, the therapist intervenes to return the youth's attention to the aspects of the exposure that will activate targeted fears.

Once the clinician and therapist have created a challenge hierarchy, they can choose a situation and try it out! Clinicians can view the "In vivo Exposure form" for a step-by-step guide (*see* Fig. 2). First, the therapist and child need to choose a situation from the hierarchy that activates the youth (aiming for an optimal activation of about a 7 on a 10-point scale) and seems feasible for the youth so they engage in the practice. The second step is to describe the exposure situation, which should be as realistic as possible and planned collaboratively with the child or adolescent in order to foster a sense of control. For example, TB chose the location in the mall that she wanted to practice in (food court) and the duration that she hyperventilated for (30 s). The goal of this was to challenge the unproven expectancy that she would faint and be embarrassed if she exposed herself to an open setting. Next, the therapist should outline achievable behavioral goals with the client in order to provide objective feedback after the exposure is complete. These goals should be concrete, observable, and realistic, in order to set the youth up for success. While TB's goal was to last 30 s, the clinician also suggested making eye contact with five

In vivo Exposure Form

1. **Situation: (What's the situation?)**

   Stand in the center of the mall's food court and hyperventilate for 30 seconds.

2. **What are you expecting to happen?**

   I'm going to faint in the food court and embarrass myself in front of everyone.

3. **Achievable Behavioral Goals: (What do you want to accomplish?)**

Goal	Accomplished?
a. *Just start hyperventilating instead of delaying.*	
b. *Last 30 seconds.*	✔
c. *Make eye contact with 5 people who pass by.*	✔

4. **Rewards**

Reward	Earned?
Earn 30 minutes of additional time to Facetime best friend.	✔

5. **Feelings:**   **Distress Rating: __5__**
   Scared, racing heart, butterflies in stomach.

   ## Time to Begin!

   Distress Rating: __8__
   At 15 seconds

   Distress Rating: __3__
   At 30 seconds (right after stopping)

6. **Coping Skills: (Which coping skills did you use?)**

   Coping thoughts: "These feelings are safe. I can handle this."

7. **How did it go? What actually happened?**

   I didn't faint and nobody noticed I was panicking. Also, I learned that I could handle hyperventilating for 30 seconds.

Fig. 2 Exposure Worksheet that Emphasizes Feared Outcomes

individuals who passed by because TB frequently avoided eye contact while completing interoceptive exposures. After establishing behaviorally specific goals, the therapist and client can discuss an external reward that can be given shortly after the end of the exposure. The next section provides a greater description of effective rewards for kids and teens in a therapeutic context. After identifying a reward, the therapist will solicit the youth's distress rating in order to evaluate anticipatory anxiety.

Now it is time to do the exposure! During the exposure, the therapist will monitor distress ratings intermittently and coach the client to use coping strategies while in the anxiety-provoking situation. Exposures will vary in duration, depending on the target; however, a therapist should aim to complete multiple exposures in one session if the practice is short (e.g., 3–4 min). In the case of TB, distress ratings were solicited every 15 s. After the exposure is complete, the clinician and client should review the results and the clinician can provide corrective and supportive feedback. Post-exposure processing should include asking about what the child learned during the practice in order to gauge expectancy violation and any successful coping experiences or barriers to implementing coping skills. The clinician should remember to always praise the kid or teen's *effort* and willingness to engage in a difficult situation and challenge themselves. Additionally, the clinician can provide a "take home message" to consolidate learning. Overall, it is important to emphasize that approaching is better than avoiding and that coping skills can help manage distress.

3.8   Rewards

External rewards are essential to foster motivation and engagement in child anxiety treatment. Anxious children and teens tend to set perfectionistic standards for self-reward, often being overly critical of their performance and/or minimize their successes. One goal of treatment is to shift the teen's focus to their efforts rather than the outcome. However, you can imagine that a perfectionistic teen will be difficult to convince. One way to counter this is to communicate that you cannot succeed unless you try. As Wayne Gretzky once said, "You miss 100% of the shots you don't take." Additionally, rewards powerfully reinforce the youth's efforts during the exposure, which increases their motivation to engage in a difficult task. When brainstorming rewards, the clinician should focus on incentives that can be provided daily and are renewable so the child or adolescent can earn it every day they engage in the behavior. The best rewards are those that are available, feasible, and meaningful to the youth. The incentives should also be provided immediately after the behavior occurs, which strengthens its reinforcing value. Here, TB valued spending time with family members, and so they were motivating. They also were inexhaustible; that is, TB could earn 30 min of family time one night and it could still be rewarding the following night. Rewards that can be "renewed" on a daily basis are some of the most sustainable. Other examples include access to desirables like cell phones, the car, and family privileges (picking dinners; family activities).

3.9 Problem Solving   The goal of problem solving is to help youth gain control over challenging situations and become more flexible in how they approach problems. We employ the problem solving STEPS acronym first employed by Weisz and colleagues [81] to operationalize the process and include: (S) Say what the problem is; (T) Think of solutions; (E) Examine each solution (List pros and cons for each solution); (P) Pick one solution and; try it (S) See if it worked. In the first step, the youth specifically describes the problem they are facing (S: Say what the problem is). Clients frequently identify broader goals (e.g., "feeling anxious," "getting scared at school") that need to be refined further by identifying the specific context (e.g., "not knowing how to talk with kids at school") and desired outcomes (e.g., "having a conversation before class"). Although these translations do not guarantee solving all anxious feelings, they offer the youth targets that may help them move closer to achieving their primary goal ("feeling less anxious").

For the second step, the youth practices brainstorming as many solutions as possible to solve the problem (T: Think of solutions). It is important that the therapist encourage the youth not to interrupt the process with criticisms of the solutions, as youth often dismiss them because of anxiety about choosing the "right" one. For example, if a child or adolescent identifies "difficulty completing homework," problem solving solutions could be to "ask my parents for help," or "post flyers around my neighborhood for a tutor." Even if the youth were to offer a truly controversial solution (e.g., "burn down the school so I don't have to do homework"), the therapist simply writes it down and waits for the next step to evaluate the pros and cons of all offered solutions. The main objective is to help the youth get unstuck from their usual entrenched patterns by seeing that many options exist, and the youth has choice in how they proceed.

After generating ideas, the therapist and youth review the pros and cons of each solution (E: Examine each solution). Instead of pushing their own opinion on a favored solution, the therapist should objectively assist the youth in thinking out the realistic consequences (both positive or negative) for each one. If one of the solutions to preparing for a test is "studying with a friend," the therapist simply helps identify pros ("working as a group," "being in a different environment") and cons ("getting more distracted," "having conversation with friend"). For the controversial solution of "burning down the school," the therapist helps the youth evaluate all the possible consequences of enacting such a plan. If the youth and therapist take the process seriously, the solutions that reflect the best outcomes for the youth overall will come to the top. Of course, if one is working with a youth with a specific history of delinquent behaviors, a similar approach can be taken, but the therapist will want to keenly remind the youth of ways in which past decisions have resulted in undesirable consequences.

Once pros and cons have been listed, the youth can now select a solution to try (P: Pick one solution and try it). It is critical that the youth understand that the first solution might not work. Specifically, the main idea is that many solutions can be generated to solve a particular problem, even if they are not "perfect." The youth is then encouraged to select one solution and evaluate the outcome (S: See if it worked). Next, the therapist and youth determine how successful the solution was in solving the problem and what changes could have produced an even better outcome. Finally, both therapist and youth decide whether to modify the solution and try again or select a different solution from the list.

Returning to our case example, TB first stated (S) a concrete problem she wanted to solve (e.g., preparing a presentation for class). Both the therapist and TB collaboratively picked this goal because it was specific and achievable. Next, TB thought (T) of multiple solutions to the problem with the therapist encouraging all possibilities nonjudgmentally (e.g., prepare the night before, ask for an extension to have more time, rehearse with a friend, ask for an alternative assignment so that I do not have to present). After generating various solutions, the therapist and TB evaluated (E) each potential solution to determine their likelihood in solving the problem. Although TB was initially interested in asking for a different assignment so she would not have to present, both she and the therapist determined that this would be a less effective solution; substituting the presentation was not guaranteed and would likely put TB at a greater disadvantage when she would need to present in the future. TB ultimately picked (P) the solution of rehearsing with a friend given that this option seemed to be the most effective. Finally, TB applied the solution to see how it worked (S) for helping prepare for the presentation. She noted that practicing was helpful for learning the material but also anxiety provoking because she had never rehearsed with a friend before.

3.10 Cognitive Restructuring

Cognitive restructuring consists of techniques designed to help the youth increase awareness of the impact of unrealistic negative thoughts on emotions and learn that they can modify those thought patterns. These techniques can be broken down into several core skills, including thought tracking, identifying thinking traps, and generating more realistic and flexible coping thoughts.

Initially, the therapist helps the youth observe the relationship between their thoughts, emotions, and the events that precede them. For younger children, thoughts can be described as "things you say in your head like thought bubbles in comic books." For teens and adolescents, this process can be explained as their inner dialogue or private thoughts. Negative thoughts often precede sad feelings, threat-related thoughts lead to anxious feelings, and optimistic thinking precedes hopeful feelings. Therefore, the first major skill in cognitive restructuring is teaching the youth to identify those thoughts in different situations and how they impact mood.

Next, the therapist helps the youth to identify unrealistic assumptions and label thinking traps (*see* Fig. 3; Common Thinking Traps worksheet). Distorted thinking is typically expressed as high levels of self-criticism, negative outlook, and helplessness when confronted with threatening situations. Alternatively, they reflect assumptions about the likelihood of risk or the catastrophic nature of the outcomes. After identifying automatic thoughts, the therapist helps the youth realistically examine the evidence for each assumption. If the youth can offer evidence to support the more realistic conclusion, then cognitive restructuring may not be warranted. If the youth cannot provide evidence for her conclusion, then she is likely jumping to conclusions and "filling in the blanks" from partial information. The youth is likely falling into one of several thinking traps in these circumstances. At this point, the therapist helps the youth label the thought with a relevant thinking trap. For example, if the youth says, "I'm never going to make any friends because nobody likes me," they would be falling into the two thinking traps of Fortune Telling (predicting the future) and Mind Reading (assuming we know what others are thinking).

Next, the therapist teaches the youth to develop more realistic and adaptive coping thoughts. Adaptive coping thoughts directly target specific thinking traps the youth is engaging in. It is important, however, that coping thoughts avoid being overly positive. Encouraging unrealistic "positive thinking" is as equally unhelpful and may set the youth up for disappointment. Instead, the objective is to encourage the youth to use a more balanced, flexible perspective to help them push forward without ruminating on false negatives (*see* Fig. 4; Coping Thoughts Tracker). For example, TB reported experiencing a panic attack that caused her to vomit while on a camping trip with peers. Although her anxiety symptoms were unpleasant in the moment and she felt embarrassed, TB was able to develop the flexible coping thought "I may experience bad anxiety, but I can live through the worst of it and still do other things." Adopting a coping thought may not protect the youth from all possible negative outcomes, but it keeps the teen from dwelling on negatives that have not happened yet.

3.11 Working with Parents and Caregivers

Parents and caregivers have been found to play a role in the development, maintenance, and subsequent course of treatment of youth anxiety disorders [82]. The various caregiver factors that have been shown to negatively impact the maintenance of anxiety and course of treatment broadly fall under the categories of parental overcontrol and rejection. Parental overcontrol is broadly defined as excessive regulation by parents of their child's activities, instruction on a child or teen's cognitions or affect, and hindering their child's burgeoning independence [82]. One highly cited

Common Thinking Traps

Mind Reading: You assume you know what other people think without sufficient evidence of their thoughts.
Example: "She thinks I'm weird."

Fortune-telling: You assume negative events in the future without enough evidence.
Example: " "I have no chance making the soccer team." "I'm going to fail this test."

Catastrophizing (Doomsday Predictor): You take a disappointing, neutral, or ambiguous event and see it in the worst possible light.
Example: "Now I'm going to fail this class because I got this 'C' on my test."

What if's (Tell me, tell me): You keep asking a series of questions, or the same question, because you are never satisfied with the answer. No answers seem to reassure you, no matter how many times you ask.
> *Example:* "What if they give a pop quiz tomorrow?" "What if I don't study enough and get a bad grade?"

Discounting the Positives (Nothing special): You minimize the positives of a situation or minimize your contributions. You claim the positive things you do are trivial or you disregard positive things that may have happened.
Example: "Everybody was invited to that party. I'm not special."

Looking for the Negative (Walking with Blinders On): You focus on the negative outcomes or the negative aspects in a situation.
> *Example:* "I only got a 90% on the test instead of a 100%."

Overgeneralizing: You see a global pattern of negatives after seeing something once or twice.
Example: After stuttering in a class presentation, you think "I always screw up presentations. I can never present without sounding stupid."

All-or-nothing Thinking: You view events or people in all-or-nothing (right-or-wrong) thinking.
Example: "If I don't score this goal, I'm horrible at soccer."

Should statements: You see events in terms of how things should be, rather than simply focusing on how things are.
Example: "I should ace all my exams."

Emotional Reasoning: Allowing emotions to dictate your actions.
> *Example:* When you feel overwhelmed and you think "I can't do this," because you *feel* the task is insurmountable without evidence supporting that conclusion.

Fig. 3 Common Thinking Traps for Anxious Youth

Coping Thoughts Tracker. Coping thoughts that can get you out of your thinking trap!					
Situation	Thought	Thinking Trap	Anxiety (0-10)	Helpful/Coping Thought	Anxiety (0-10)
Giving class presentations	"Everybody is going to see I'm embarrassed and think I'm weird"	Mind reading	8	I can't know for certain what people are thinking	2

Fig. 4 Coping Thoughts Tracker that Emphasizes Thinking Traps and Coping Thoughts

dimension of parental overcontrol is parental intrusiveness, which refers to caregivers who will perform tasks for their children, even if their child can do it independently, leading to impairment in self-efficacy. Therefore, caregivers are not likely to acknowledge or foster their child's autonomy. As a result, children with anxiety disorders may not believe in their ability to tolerate or effectively manage a novel situation, without parental assistance [82]. Alternatively, parental rejection is often described as a caregiver's coldness, disapproval, and unresponsiveness toward their child [82]. Findings suggest that rejecting caregivers can prompt children to have greater difficulty regulating negative emotions and increase anxiety sensitivity [82].

Clinicians are encouraged to work directly with parents in order to reduce parental rejection and overcontrol, and subsequently promote warmth and autonomy-granting. Additionally, caregiver involvement can be leveraged to promote skill generalization and reinforce approach behaviors. For example, caregivers can coach their child through exposures at home, which can lead to more successful approach experiences. We often teach parents effective communication strategies in order to reduce overcontrol and rejection and improve the child's mastery of new skills. One of these strategies is to provide choices to children and teens, especially when they are indecisive, in order to promote autonomy. Additionally, it is important to work with parents to tolerate watching their child struggle and occasionally fail instead of rescuing their child from an anxious situation, thereby perpetuating the avoidance. Many parents will require you to brainstorm coping skills with them in order to tolerate seeing their child in distress, including counting backward from 10, singing a song in their head, or developing their own coping thoughts.

Lastly, in order to most effectively communicate with their child during an emotional or anxiety-provoking experience, we teach parents to "empathize and encourage." This strategy refers to empathizing with a child or teen's current emotional state (e.g., "I can see this is really hard for you …") and then encouraging them to make efforts to cope (e.g., "I know you can stick through this. Why don't we try using a coping thought?"). We recommend using a statement with both validation and change elements, because even well-meaning parents can unintentionally lean too heavily on validation or commands to cope. Additionally, caregivers who tend to be rejecting (e.g., low warmth) may benefit from explicit instruction on validation strategies. For example, in TB's case, her panic attacks occurred so frequently that it often evoked frustration and impatience from her mother. Her mother would often couple a dismissive, "Why do you make such a big deal over this?" with an accommodation that rescued TB (e.g., running into the mall to pick something up instead of sending TB). Instead, we taught the mother to actively listen and reflect TB's fear (empathize), "I see how scared it makes you to go into the mall by yourself," and then to advocate for approach solutions (encourage), "… and I know you'll be able to take it one step at a time." This reorientation from an "either-or" (if one is scared, one should not do it) to a "both-and" (one can be scared and do it) perspective helps send a message of validation and approach over time.

3.12   Out-of-Session Practice (Homework)

Homework or "out of session practice" should be assigned at the end of every session in order to continue honing skills and promote generalization to diverse settings. If the kid or teen only practices skills in session, they are unlikely to use them in their daily lives. Each homework assignment should be related to the content reviewed or learned during that day's session. For example, a clinician can assign an exposure outside of session (e.g., raise your hand two times in science class three times this week) or daily use of a new skill in order to promote acquisition. Clinicians can also assign self-monitoring forms during early phases of treatment to aid assessment or monitor outcomes. However, it is important that the clinician assigns fewer, simpler tasks that the youth is likely to complete. A simple task should be one that the kid or teen can recall by memory, if needed. Prior to the client leaving session, it is important to confirm that they understand the assignment and what is expected of them to set them up for success.

TB was assigned to practice one relaxation strategy (e.g., deep breathing or progressive muscle relaxation) every day for 5 min each day and rate how she felt prior to the practice (0–10) and after the practice (0–10). Additionally, after an exposure session where TB initiated a conversation with an individual in the clinic, she was assigned to initiate a conversation with a peer every day for 1 week and track her anxiety before the conversation, afterward, and describe the outcome in order to assess for expectancy violations.

In order to communicate and reinforce its importance, the clinician should review the homework at the beginning of the following session. If the youth completes their homework (even partially), the clinician can provide heavy praise and even an external reward (or points toward a reward). This clinician response will further reinforce the youth's homework completion throughout treatment. If the child or adolescent did not complete their homework, it is important that the clinician assess any barriers and collaboratively generate solutions in order to ensure completion over the following weeks.

3.13 TB in a Nutshell

Over the course of treatment, TB's therapist focused on the above interventions with the aim of addressing her panic disorder and social anxiety, which interfered in a variety of domains including her academic achievement and family and peer relationships. The therapist monitored progress by assessing her anxiety symptoms using the SCARED each session and noted a steady improvement (*see* Fig. 5). While steady improvement is not required (for alternative trajectories, *see* [82], here it indicated consistent uptake of skills and lessons across therapy. The most effective interventions for TB in facilitating her decreased anxiety symptoms were psychoeducation, in order to address thoughts that high physiological arousal is inherently dangerous, and exposure, to target TB's avoidance of daily tasks and situations which led to functional impairment. TB's engagement in approach behaviors allowed her to (a) learn that she can tolerate negative feelings, (b) encounter disconfirming evidence that high physiological arousal is dangerous, and (c) allow her to experience "pleasant surprises" (e.g., meet new friends) which were positively reinforcing. As a result, TB experienced a reduction of panic attacks and resumption of a number of important daily activities (e.g., attend all school days during the last two months of school and make honor roll, form new friendships, and try out for a competitive softball team).

3.14 Conducting CBT in the Real World

3.14.1 Comorbidity

Treating youth with anxiety often implies managing one or more other anxiety and depressive disorders concurrently. Anxiety and depression are highly comorbid [6] with rates ranging from 10% to 15% among youth who meet criteria for a principal anxiety disorder. The most common anxiety and depressive disorders in youth are Separation Anxiety Disorder, Social Anxiety Disorder, Generalized Anxiety Disorder, Specific Phobia, Major Depressive Disorder, and Persistent Depressive Disorder [82]. Various combinations of these

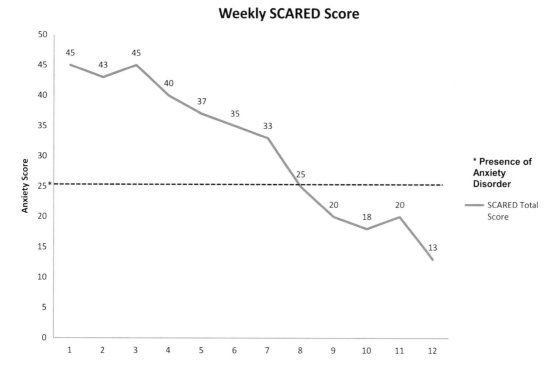

Fig. 5 Sample Progress Monitoring Chart Using the Screen for Child Anxiety Related Emotional Disorders (SCARED)

diagnoses often cluster together strongly within families and across generations, making treating them simultaneously the rule rather than the exception.

Oftentimes, comorbid diagnostic presentations create additional clinical interference that compete with the youth's ability to engage in treatment [83, 84]. As a result, motivation is negatively impacted, and the therapist should plan accordingly to build or rebuild rapport. Playing games or activities with the youth can be an effective method for reengaging the child and strengthening the therapeutic relationship. Physical activities (e.g., tossing a ball, shooting hoops, playing catch, jumping jacks) are particularly helpful for increasing a child's energy level and likelihood to engage in treatment-related tasks. Cultivating parent motivation is also critical, given parents' active role in treatment (confirming whether child completed homework, dropping off child to session, reporting interfamilial conflict, etc.). Nock and Photos [85] developed the Parent Motivation Inventory (PMI) to capture levels of parent motivation to engage in treatment. Results showed that higher scores on parental motivation predicted lower perceived treatment barriers and increased number of treatment sessions attended.

Frequently, children and families may be limited in their ability or willingness to engage with in treatment, which requires the therapist to implement case management strategies. Case management broadly refers to the process of assessing, planning, coordinating and evaluating various external services required to meet a person's mental health needs. Within the context of treatment, these typically include communicating with families to determine what components of treatment might be challenging to complete, finding alternative meeting times due to repeated schedule conflicts, addressing family concerns as they relate to child's treatment, negotiating treatment fees, and obtaining access to reliable transportation.

3.14.2 Clinician Barriers When Conducting Exposures

Although exposure therapies for anxiety disorders have a strong evidence base supporting their use, few clinicians use exposures in practice [86, 87]. Some clinicians harbor negative beliefs about exposure therapy, suggesting that purposefully eliciting a client's anxiety is insensitive, possibly damaging to clients, and may lead to increased rates of drop-out [88]. Despite the evidence against these myths, clinicians are still ambivalent about the utility of exposure therapy. Additionally, some therapists find it difficult to implement exposure due to an experience of secondary distress [89]. When the clinician has an urge to avoid their own anxiety by "rescuing" their client, the exposure becomes ineffective. Therefore, each clinician should consider their emotional tolerance when planning to implement exposure. If tolerance is low, then utilizing coping skills (e.g., coping thoughts and deep breathing) and supervision and consultation can improve the experience and effectiveness of exposure therapy.

Additionally, clinicians may shy away from conducting exposure due to issues of liability when leaving the office for an exposure. In order to best simulate a real-world experience, clinicians are encouraged to leave the office with their clients in order to practice skill-use in diverse settings. As a result, some clinicians worry about breaking confidentiality or crossing proper boundaries by conducting therapy outside of the traditional setting. It is important to remember that crossing a boundary does not equate to violating the boundary. If leaving the office and crossing that boundary is done in the service of implementing effective treatment and informed consent is provided by the client, then out of office exposure can be done ethically. Additionally, many (if not all) exposures do not require the therapist to identify themselves as such when outside of the office. Some planning ahead of extra-therapy exposures can help. For example, planning with a youth/caregiver what a therapist should say should a well-intentioned bystander inquires about their activities or relationship. As long as clinical content is discussed privately and the client has given their informed consent to engage in this therapeutic activity (as they should for all activities in therapy), no liability or ethical issues should arise [88].

3.15   Focus on Diversity

Shaping treatment to the culture and context of a youth and family is essential, especially when working with youth and families who are connected through multiple systems (e.g., school, community, family) where cultures overlap [45]. Families of diverse cultural, racial and sexual minority backgrounds often experience significantly greater difficulty accessing mental health services [90]. The most common barriers include limited financing to pay for treatment, cultural stigma toward mental health treatment and limited access to resources [91]. Although evidence-based interventions cannot always be immediately tailored to a youth family's cultural background, therapists are still encouraged to discuss the topic in treatment. Hays [61] lists ten strategies that fostering "cultural competency" that modify components of treatment to be more culturally salient. Some examples of these include: identifying cultural strengths and supports, validating client's self-reported experiences of oppression, emphasizing collaboration over confrontation, avoiding challenging core cultural beliefs and developing homework assignments that are culturally congruent.

3.16   Implementing Evidence-Based Practices with Flexibility

Leveraging balance between adhering to manual guidelines versus adapting principles to better meet the needs of a family refers to maintaining "flexibility within fidelity" [45, 92]. Preliminary research has shown that applying this framework has been helpful in increasing youth engagement [93]. Similarly, Mazzucchelli and Sanders [94] note a distinction between "low risk" and "high risk" adaptations. Low-risk adaptations might be choosing to offer more sessions, increase/decrease session length to be more consistent within a particular setting (e.g., school vs. outpatient), or to provide sessions within various settings to better target treatment goals (e.g., offsite exposures). High-risk modifications would include arbitrarily selecting specific strategies to use or not use or discarding the planned agenda. These techniques are especially helpful when the presenting problem or context varies, which is more often the rule rather than the exception. Collectively, evidence suggests that thoughtfully planned therapist flexibility can be beneficial under specific circumstances. In so doing, therapists can "breathe life into manuals" by tailoring manualized protocols to individual youth concerns [95].

3.17   Collaborating with Allied Professionals: Working Within the Youth Ecosystem

Unlike adults, youth interface within a variety of outside environmental systems (e.g., school, pediatrics, extracurricular activities) that include allied professionals, or specialists whom have regular contact with and provide support to the child. Therapists will often need to liaise with teachers, counselors, nurses, and pediatricians. Obtaining information from them when needed is helpful for the clinician to inform their conceptualization and when the child may need additional supports or accommodations in environments outside of session. In this way, the therapist can assume various roles, whether it be as a consultant (e.g., providing relevant clinical information to the school), advocate (e.g., pushing for supports to ensure child is protected from bullying) or obtaining ancillary clinical information to inform individual treatment. Before establishing initial contact with any allied professional, the therapist should obtain explicit documented consent from the family.

Since youth spend most of their time at school, many of their presenting clinical concerns often generalize to educational settings. Likewise, children are often referred for treatment due to problems that occur exclusively within school settings. In both situations, the therapist can help a youth struggling with anxiety receive the appropriate amount of school accommodations (e.g., supports that help the youth meet their educational abilities but that do not free them from all challenges). Explicitly communicating treatment plans to teachers and other school personnel greatly facilitates the therapist's ability to implement treatment effectively (e.g., informing teachers that child will be conducting a social anxiety exposure requiring her to give an oral presentation). Teachers and other school professionals also offer a valuable source of supplementary clinical data that would not otherwise be available in regular treatment sessions. School refusal is a common problem among anxious and depressed youth that often necessitates ongoing dialogue between the therapist and teachers [96–98]. Doing so helps with treatment planning by allowing the child to receive the necessary accommodations to maximize the intervention's effectiveness.

Youth who present with more acute impairment will frequently augment psychotherapy treatment with medication. Obtaining this information requires the therapist to obtain a release of information from the parent and also from the treating medical professional (e.g., psychiatrist or pediatrician). The therapist should both aware of any current medication the youth is currently taking and if medications are titrated down or switched. Specific symptoms may arise, remit or worsen as a function of the prescribed medication. Further, communication with the youth's medical health professional helps the therapist differentiate physical health concerns that might influence the youth's presentation or functioning in treatment. If a youth's symptoms and functioning continue to remain poor or deteriorate, the therapist may decide to encourage the family to seek outside medical consultation. Alternatively, if a youth has shown marked functional improvement, titrating medication to a lower dose can be broached by the therapist with the family.

When reaching out to collaborating medical professionals, using simple, non-jargon language helps facilitate and expedite the exchange of information. Focusing on the clinical minutia may result in miscommunication, frustration, and inadvertently direct attention away from the most salient clinical issues. Just as importantly, therapists should arrive to the discussion prepared with specific questions. Given that most mental health professionals' time is limited, keeping the dialogue structured will help ensure another follow-up call is not needed. Working among larger interdisciplinary teams may require more ongoing contact and collaboration depending on the youth's clinical needs. Specifically, establishing a consensus on the youth's conceptualization ensures that the treatment team's interventions are focused and do not conflict with one another.

4 Conclusion/Discussion

Problems in anxiety affect a large number of children and adolescents, are impairing, and can affect long-term functioning if left untreated. Cognitive behavioral therapy is an efficacious treatment approach that teaches anxious multiple skills to help manage anxiety symptoms, challenge unrealistic assumptions, and approach frightening situations. This chapter reviewed a number of strategies that can be used in a flexible principles-based way to address complex comorbidity and be implemented across diverse practice settings with youth from different backgrounds. The principles and concrete tools provided herein are designed to help practitioners meet the specific needs of their clients in their individual settings.

References

1. Merikangas KR et al (2010) Lifetime prevalence of mental disorders in US adolescents: results from the National Comorbidity Survey Replication–Adolescent Supplement (NCS-A). J Am Acad Child Adolesc Psychiatry 49:980–989

2. Copeland WE et al (2014) Longitudinal patterns of anxiety from childhood to adulthood: the Great Smoky Mountains Study. J Am Acad Child Adolesc Psychiatry 53:21–33

3. Swan AJ, Kendall PC (2016) Fear and missing out: youth anxiety and functional outcomes. Clin Psychol Sci Pract 23:417–435

4. Cole DA et al (1998) A longitudinal look at the relation between depression and anxiety in children and adolescents. J Consult Clin Psychol 66:451–460

5. Stark KD, Laurent J (2001) Joint factor analysis of the children's depression inventory and the revised children's manifest anxiety scale. J Clin Child Psychol 30:552–567

6. Kessler R et al (2003) The epidemiology of major depressive disorder: results from the National Comorbidity Survey Replication (NCS-R). JAMA 289:3095–3105. https://doi.org/10.1001/jama.289.23.3095

7. Benjamin CL et al (2013) Anxiety and related outcomes in young adults 7 to 19 years after receiving treatment for child anxiety. J Consult Clin Psychol 81:865–876

8. Essau CA et al (2014) Anxiety disorders in adolescents and psychosocial outcomes at age 30. J Affect Disord 163:125–132

9. Ginsburg GS et al (2018) Results from the child/adolescent anxiety multimodal extended long-term study (CAMELS): primary anxiety outcomes. J Am Acad Child Adolesc Psychol 57:471–480

10. Scharfstein L et al (2011) Children with generalized anxiety disorder do not have peer problems, just fewer friends. Child Psychiatry Hum Dev 42:712–723

11. Hughes AA, Hedtke KA, Kendall PC (2008) Family functioning in families of children with anxiety disorders. J Fam Psychol 22:325–328

12. Masten AS, Cicchetti D (2010) Developmental cascades. Dev Psychopathol 22:491–495

13. Flannery-Schroeder E et al (2004) Comorbid externalizing disorders and child anxiety treatment outcomes. Behav 21:14–25

14. Barrett PM, Dadds MR, Rapee RM (1996) Family treatment of childhood anxiety: a controlled trial. J Consult Clin Psychol 64:333–342

15. Beidel DC, Turner SM, Morris TL (2000) Behavioral treatment of childhood social phobia. J Consult Clin Psychol 68:1072–1080

16. Beidas RS et al (2015) Predictors of community therapists' use of therapy techniques in a large public mental health system. JAMA Pediatr 169:374–382

17. Kendall PC (1994) Treating anxiety disorders in children: results of a randomized clinical trial. J Consult Clin Psychol 62:100–110

18. Kendall PC et al (1997) Therapy for youths with anxiety disorders: a second randomized clinical trial. J Consult Clin Psychol 65:366–380. https://doi.org/10.1037/0022-006X.65.3.366

19. Kendall PC et al (2008) Cognitive-behavioral therapy for anxiety disordered youth: a randomized clinical trial evaluating child and family modalities. J Consult Clin Psychol 76:282–297

20. Martinsen KD et al (2019) Prevention of anxiety and depression in school children: effectiveness of the transdiagnostic EMOTION program. J Consult Clin Psychol 87:212–219

21. Silk JS et al (2018) A randomized clinical trial comparing individual cognitive behavioral therapy and child-centered therapy for child anxiety disorders. J Clin Child Adolesc Psychol 47:542–554

22. Walkup JT et al (2008) Cognitive behavioral therapy, sertraline, or a combination in childhood anxiety. N Engl J Med 359:2753–2766

23. Hudson JL et al (2009) Cognitive-behavioral treatment versus an active control for children and adolescents with anxiety disorders: a randomized trial. J Am Acad Child Adolesc Psychiatry 48(5):533–544

24. Southam-Gerow MA et al (2010) Does cognitive behavioral therapy for youth anxiety outperform usual care in community clinics? An initial effectiveness test. J Am Acad Child Adolesc Psychiatry 49:1043–1052

25. Chu BC et al (2016) Transdiagnostic group behavioral activation and exposure therapy for youth anxiety and depression: initial randomized controlled trial. Behav Res Ther 76:65–75

26. Ehrenreich JT et al (2009) Development of a unified protocol for the treatment of emotional disorders in youth. Child Fam Behav Ther 31:20–37

27. Ehrenreich-May J et al (2017) An initial waitlist-controlled trial of the unified protocol for the treatment of emotional disorders in adolescents. J Anxiety Disord 46:46–55

28. Beidel DC, Turner SM, Young BJ (2006) Social effectiveness therapy for children: five years later. Behav Ther 37:416–425

29. Barrett PM et al (2001) Cognitive–behavioral treatment of anxiety disorders in children: long-term (6-year) follow-up. J Consult Clin Psychol 69:135–141

30. Kendall PC et al (2004) Child anxiety treatment: outcomes in adolescence and impact on substance use and depression at 7.4-year follow-up. J Consult Clin Psychol 72:276–287

31. Kendall PC et al (1992) Anxiety disorders in youth: cognitive-behavioral interventions. Allyn and Bacon, Boston, MA

32. Barrett PM et al (1996) Family enhancement of cognitive style in anxious and aggressive children. J Abnorm Child Psychol 24:187–203

33. Bell-Dolan DJ (1995) Social cue interpretation of anxious children. J Clin Child Psychol 24:2–10

34. Puliafico AC, Kendall PC (2006) Threat-related attentional bias in anxious youth: a review. Clin Child Fam Psychol Rev 9:162–180

35. Chu BC, Skriner LC, Staples AM (2013) Behavioral avoidance across anxiety, depression, impulse, and conduct problems. In: Ehrenreich-May J, Chu B (eds) Transdiagnostic treatments for children and adolescents: principles and practice. Guilford Press, New York, NY, pp 84–110

36. Blakey SM, Abramowitz JS (2016) The effects of safety behaviors during exposure therapy for anxiety: critical analysis from an inhibitory learning perspective. Clin Psychol Rev 49:1–15

37. Hoffman LJ, Chu BC (2019) When is seeking safety functional? Taking a pragmatic approach to distinguishing coping from safety. Cogn Behav Pract 26:176–185

38. Harvey AG, Watkins E (2004) Cognitive behavioral processes across psychological disorders: a transdiagnostic approach to research and treatment. Oxford University Press, New York, NY

39. Foa EB, Huppert JD, Cahill SP (2006) Emotional processing theory: an update. In: Rothbaum BO (ed) Pathological anxiety: emotional processing in etiology and treatment. The Guilford Press, New York, NY, pp 3–24

40. Craske MG et al (2014) Maximizing exposure therapy: an inhibitory learning approach. Behav Res Ther 58:10–23

41. Silverman WK, Albano AM (2017) Anxiety disorders interview schedule for children

(ADIS-5) child and parent interviews. Unpublished Manuscript, New York, NY

42. Silverman WK, Nelles WB (1988) The anxiety disorders interview schedule for children. J Am Acad Child Adolesc Psychiatry 27:772–778

43. Silverman WK, Eisen AR (1992) Age differences in the reliability of parent and child reports of child anxious symptomatology using a structured interview. J Am Acad Child Adolesc Psychiatry 31:117–124

44. Chu BC (2008) Child and adolescent research methods in clinical psychology. In: McKay D (ed) Handbook of research methods in abnormal and clinical psychology. Sage Publications, Thousand Oaks, CA, pp 405–426

45. Chu BC (2019) Evidence-based therapist flexibility: making treatments work for clients. In: Prinstein MJ, Youngstrom EA, Mash EJ, Barkley RA (eds) Treatment of disorders in childhood and adolescence, 4th edn. Guilford Publications, New York, NY, pp 27–46

46. Birmaher B et al (1999) Psychometric properties of the Screen for Child Anxiety Related Emotional Disorders (SCARED): a replication study. J Am Acad Child Adolesc Psychiatry 38:1230–1236

47. Hale WW et al (2011) A meta-analysis of the cross-cultural psychometric properties of the Screen for Child Anxiety Related Emotional Disorders (SCARED). J Child Psychol Psychiatry 52:80–90

48. March JS et al (1997) The Multidimensional Anxiety Scale for Children (MASC): factor structure, reliability, and validity. J Am Acad Child Adolesc Psychiatry 36:554–565

49. Chorpita BF et al (2000) Assessment of symptoms of DSM-IV anxiety and depression in children: a revised child anxiety and depression scale. Behav Res Ther 38:835–855

50. Beidas RS et al (2015) Free, brief, and validated: standardized instruments for low-resource mental health settings. Cogn Behav Pract 22:5–19

51. Kazdin AE (2001) Bridging the enormous gaps of theory with therapy research and practice. J Clin Child Psychol 30:59–66

52. Rizvi SL, Ritschel LA (2014) Mastering the art of chain analysis in dialectical behavior therapy. Cogn Behav Pract 21:335–349

53. McLeod BD et al (2013) Evidence-based assessment and case formulation for childhood anxiety disorders. In: Essau CA, Ollendick TH (eds) The Wiley-Blackwell handbook of the treatment of childhood and adolescent anxiety. John Wiley & Sons, Ltd, Chichester, pp 177–206. https://doi.org/10.1002/9781118315088.ch9

54. Persons JB (1989) Cognitive therapy in practice: a case formulation approach. WW Norton, New York, NY

55. Christon LM, McLeod BD, Jensen-Doss A (2015) Evidence-based assessment meets evidence-based treatment: an approach to science-informed case conceptualization. Cog Behav Pract 22:36–48

56. Persons JB (2006) Case formulation–driven psychotherapy. Clin Psychol Sci Pract 13:167–170

57. Weisz JR et al (2011) Youth top problems: using idiographic, consumer-guided assessment to identify treatment needs and to track change during psychotherapy. J Consult Clin Psychol 79:369–380

58. American Psychological Association (2017) Multicultural guidelines: an ecological approach to context, identity, and intersectionality. Retrieved from http://www.apa.org/about/policy/multicultural-guidelines.pdf

59. Bernal G, Bonilla J, Bellido C (1995) Ecological validity and cultural sensitivity for outcome research: issues for the cultural adaptation and development of psychosocial treatments with Hispanics. J Abnorm Child Psychol 23:67–82

60. Hwang WC (2006) The psychotherapy adaptation and modification framework: application to Asian Americans. Am Psychol 61:702

61. Hays PA (2001) Addressing cultural complexities in practice: a framework for clinicians and counselors. American Psychological Association, Washington, DC

62. Friedberg RD, Gorman AA (2007) Integrating psychotherapeutic processes with cognitive behavioral procedures. J Contemp Psychother 37:185–193

63. Chu BC et al (2005) Alliance, technology, and outcome in the treatment of anxious youth. Cog Behav Pract 11:44–55

64. McLeod J (2011) Qualitative research in counselling and psychotherapy, 2nd edn. Sage, London

65. Karver M et al (2008) Relationship processes in youth psychotherapy: measuring alliance, alliance-building behaviors, and client involvement. J Emot Behav Disord 16:15–28

66. Karver MS et al (2006) Meta-analysis of therapeutic relationship variables in youth and family therapy: the evidence for different relationship variables in the child and adolescent treatment outcome literature. Clin Psychol Rev 26:50–65

67. Hawley KM, Weisz JR (2005) Youth versus parent working alliance in usual clinical care: distinctive associations with retention,

satisfaction, and treatment outcome. J Clin Child Adolesc Psychol 34:117–128

68. McLeod BD et al (2014) The relationship between alliance and client involvement in CBT for child anxiety disorders. J Clin Child Adolesc Psychol 43:735–741

69. Creed TA, Kendall PC (2005) Therapist alliance-building behavior within a cognitive-behavioral treatment for anxiety in youth. J Cosult Clin Psychol 73:498–505

70. Podell JL et al (2013) Therapist factors and outcomes in CBT for anxiety in youth. Prof Psychol Res Pract 44:89–98

71. Kendall PC, Hedtke KA (2006) Cognitive-behavioral therapy for anxious children: therapist manual. Workbook Publishing, Ardmore, PA

72. Bickman L et al (2011) Effects of routine feedback to clinicians on mental health outcomes of youths: results of a randomized trial. Psychiatr Serv 62:1423–1429

73. Radloff LS (1977) The CES-D scale: a self-report depression scale for research in the general population. Appl Psychol Meas 1:385–401

74. Rapee RM (2012) Family factors in the development and management of anxiety disorders. Clin Child Fam Psychol Rev 15:69–80

75. Goodman R (1997) The strengths and difficulties questionnaire: a research note. J Child Psychol Psychiatry 38:581–586

76. Goodman R, Meltzer H, Bailey V (1998) The strengths and difficulties questionnaire: a pilot study on the validity of the self-report version. Eur Child Adolesc Psychiatry 7:125–130

77. Bird HR et al (2005) The Brief Impairment Scale (BIS): a multidimensional scale of functional impairment for children and adolescents. J Am Acad Chil Adolesc Psychiatry 44:699–707

78. Wyszynski CM (2018) Measuring progress: a comparison of three outcome measures as classifiers of treatment response and diagnostic remission throughout CBT for youth anxiety. Doctoral dissertation, Rutgers University-School of Graduate Studies

79. Craske MG et al (2008) Optimizing inhibitory learning during exposure therapy. Behav Res Ther 46:5–27

80. Knowles K, Olatunji B (2019) Enhancing inhibitory learning: the utility of variability in exposure. Cogn Behav Pract 26:186–200. https://doi.org/10.1016/j.cbpra.2017.12.001

81. Weisz JR et al (2005) Therapist's manual PASCET: primary and secondary control enhancement training program, 3rd edn. University of California, Los Angeles, CA

82. Chorpita BF, Barlow DH (1998) The development of anxiety: the role of control in the early environment. Psychol Bull 124:3–21

83. Chu BC et al (2012) Calibrating for comorbidity: clinical decision-making in youth depression and anxiety. Cogn Behav Pract 19:5–16

84. Hudson JL, Krain AL, Kendall PC (2001) Expanding horizons: adapting manual-based treatments for anxious children with comorbid diagnoses. Cogn Behav Pract 8:338–346

85. Nock MK, Photos V (2006) Parent motivation to participate in treatment: assessment and prediction of subsequent participation. J Chil Fam Stud 15:333–346

86. Chu B et al (2015) Sustained implementation of cognitive-behavioral therapy for youth anxiety and depression: long-term effects of structured training and consultation on therapist practice in the field. Prof Psychol Res Pract 46:70–79. https://doi.org/10.1037/a0038000

87. Harned MS et al (2011) Overcoming barriers to disseminating exposure therapies for anxiety disorders: a pilot randomized controlled trial of training methods. J Anxiety Diord 25:155–163

88. Olatunji BO, Deacon BJ, Abramowitz JS (2009) The cruelest cure? Ethical issues in the implementation of exposure-based treatments. Cogn Behav Pract 16:172–180

89. Castro F, Marx BP (2007) Exposure therapy with adult survivors of childhood sexual abuse. In: Handbook of exposure therapies. Academic, Cambridge, MA, pp 153–167

90. Kazdin AE (2017) Addressing the treatment gap: a key challenge for extending evidence-based psychosocial interventions. Behav Res Ther 88:7–18

91. Corrigan PW, Druss BG, Perlick DA (2014) The impact of mental illness stigma on seeking and participating in mental health care. Psychol Sci Public Interest 15:37–70

92. Kendall PC, Beidas RS (2007) Smoothing the trail for dissemination of evidence-based practices for youth: flexibility within fidelity. Prof Psychol Res Pract 38:13–20

93. Chu BC, Kendall PC (2009) Therapist responsiveness to child engagement: flexibility within manual-based CBT for anxious youth. J Clin Psychol 65(7):736–754

94. Mazzucchelli TG, Sanders MR (2010) Facilitating practitioner flexibility within an empirically supported intervention: lessons from a system of parenting support. Clin Psychol Sci Pract 17:238–252

95. Kendall PC et al (1998) Breathing life into a manual: flexibility and creativity with manual-

based treatments. Cogn Behav Pract 5:177–198

96. Chu BC et al (2019) Developing an online early detection system for school attendance problems: results from a research-community partnership. Cogn Behav Pract 26:35–45

97. Kearney CA, Graczyk P (2014) A response to intervention model to promote school attendance and decrease school absenteeism. Child Youth Care Forum 43(1):1–25

98. Kearney CA (2008) School absenteeism and school refusal behavior in youth: a contemporary review. Clin Psychol Rev 28:451–471

Chapter 4

Cognitive Behavioral Therapy for Pediatric OCD

Dara E. Babinski

Abstract

Cognitive behavioral therapy (CBT) is a first-line treatment for pediatric obsessive-compulsive disorder (OCD) that is associated with large reductions in OCD symptoms as well as improvements in functional outcomes. Over the past decades, a number of different treatment manuals have been developed to address pediatric OCD, each demonstrating efficacy for addressing this type of concern. Although these interventions vary in scope and format, they also share many of common elements that can be applied flexibly to allow youth to gain control of their OCD and improve their overall functioning. The goal of this chapter is to highlight the core components of CBT for youth with OCD. Specifically, thorough assessment, psychoeducation, cognitive restructuring, exposure, and response prevention strategies are reviewed. In addition, family involvement strategies and maintenance programming are also covered. Examples of therapeutic dialogue and useful therapy worksheets are also provided to facilitate successful treatment delivery.

Key words Obsessive-compulsive disorder, OCD, Cognitive behavioral therapy, CBT, Youth

1 Introduction

Obsessive-compulsive disorder (OCD) is a neurobehavioral disorder that affects approximately 1–2% of children and involves recurrent thoughts (i.e., obsessions), including excessive worry about aggressive or catastrophic events, contamination, becoming sick, and dying, and/or ritualistic behaviors (i.e., compulsions), including checking, counting, washing, ordering/arranging, confessing/asking, and hoarding. These obsessions and compulsions are difficult to control and cause clinically significant impairment at home, at school, and with peers [1]. OCD often begins in childhood [2], and is associated with increased risk for additional difficulties including other anxiety disorders and depression [3]. Left untreated, OCD may follow a chronic course into adulthood [4]. Thus, treatment for pediatric OCD is critical.

Cognitive behavioral therapy (CBT) is a first line treatment for children with OCD, either alone or in combination with medication [5–7]. Although medication is often the most available

Robert D. Friedberg and Brad J. Nakamura (eds.), *Cognitive Behavioral Therapy in Youth: Tradition and Innovation*, Neuromethods, vol. 156, https://doi.org/10.1007/978-1-0716-0700-8_4, © Springer Science+Business Media, LLC, part of Springer Nature 2020

treatment, CBT is considered the treatment of choice for OCD [5, 8]. Youth and their parents often prefer non-medication over medication treatment options [9], as CBT provides children with skills to actively manage worry. Large improvements in symptoms and functioning have been reported among many youth with OCD in CBT, which are often maintained even after treatment has ended [5–7]. Over the past decades, various CBT manuals have been developed to target pediatric OCD [6, 7]. While these treatments vary in duration and format (i.e., group versus individual versus family-based), there is also notable overlap in the core modules included in these treatments. That is, CBT for pediatric OCD involves exposure and response prevention (EX/RP) to provide practice in facing fears and resisting ritualistic behavior. Most manuals also address cognitive distortions and negative thinking patterns that often reinforce worry or impede motivation in treatment. Furthermore, parental involvement in CBT for pediatric OCD is also often critical. Rather than a one-size-fits-all approach to CBT for pediatric OCD, a flexible approach that considers the specific nature of the child's symptoms, developmental level, family context, and other important factors is necessary. The current chapter outlines the core components of CBT and offers case examples and dialogue to guide the development and treatment of OCD in youth.

2   Theoretical and Empirical Foundations

The development of OCD involves classical conditioning, in which an unexpected frightening experience, the unconditioned stimulus, is paired with a neutral stimulus, or the conditioned stimulus, and is followed by operant conditioning, in which the newly developed fear is further reinforced when it is followed by a satisfying consequence such as avoidance or a ritualistic behavior [10]. For example, an obsessive fear of sickness may develop when 8-year-old Thomas becomes sick after eating chicken. It is unlikely that eating chicken actually led to becoming sick, as no one else who had eaten the chicken that day became ill. In this case, Thomas begins to associate eating chicken with becoming sick. Thus, the unconditioned stimulus is becoming sick and the conditioned stimulus is eating chicken. In addition to worrying about having to each chicken at future family dinners, this fear becomes more generalized. Thomas stops eating chicken altogether, he develops a fear that other foods may contain chicken, and has a difficult time eating in the lunch room at school as he is on high alert looking at the lunches of his classmates, that may contain chicken, and present additional risk for sickness. Operant conditioning serves to further reinforce this fear. That is, the fear of eating chicken becomes more frequent and impairing when followed by a satisfying consequence.

In this case, when Thomas avoids chicken, he experiences relief from the worry of becoming sick, and this temporary relief in turn reinforces the desire to engage in additional obsessions and compulsions focused on avoiding eating chicken. He may also develop rituals such as reciting a prayer in his head when he sees or smells chicken as a means to provide relief from the obsessive fear of eating chicken.

CBT aims to break this cycle of obsessions and compulsions through fear extinction [10]. Fear extinction involves repeated exposure to the conditioned stimulus in the absence of experiencing the unconditioned stimulus. In the above case, for example, the fear response to the conditioned stimulus (i.e., eating chicken) declines through repeated exposure to the food in the absence of the feared stimulus (i.e., sickness) and/or engagement in safety behaviors, such as avoidance, or compulsions. Through repeated exposure to the conditioned stimulus of chicken, a new association between the conditioned stimulus (i.e., chicken) and not experiencing the unconditioned stimulus (i.e., sickness) is learned and the fear response is extinguished.

3   Intervention

3.1   Assessment

Comprehensive evaluation of OCD symptoms, severity, and related functional impairments is critical to developing a meaningful treatment plan. There are a number of validated assessments for pediatric OCD. For example, the Anxiety Disorders Interview Schedule for DSM-IV: Child and Parent Version (ADIS-C/P) [11] and the Schedule for Affective Disorders and Schizophrenia for School-Age Children—Present and Lifetime version (K-SADS-PL) [12] are diagnostic interviews commonly used to asses OCD in children. Additionally, the Children's Yale-Brown Obsessive Compulsive Scale (CY-BOCS) [13] is frequently used as a clinician-rated, semi-structured inventory of pediatric OCD symptoms and severity, although there is also emerging evidence of its use as a self-report and parent-reported measure, as well [14, 15]. Administering a measure of OCD severity such as the CY-BOCS at the initial assessment is also useful as it can be used as a follow-up assessment to track OCD severity throughout the course of treatment.

Therapists should use some time during the initial assessment to describe and define obsessions and compulsions as children and families may have varying levels of knowledge about OCD that could affect their responses. It is important to collect information from both the child and parents about the child's symptoms and functioning, as well as goals for treatment. Doing so will provide the therapist with an opportunity to gauge the child's insight and family's understanding of OCD. In addition, conducting individual assessment with the child may help to build rapport and trust with the therapist, while also potentially eliciting obsessions and fears that the child may not be comfortable discussing in front of his or her parents (e.g., sexual obsessions). As OCD often co-occurs with other disorders, including depression, generalized anxiety disorder, attention-deficit/hyperactivity disorder, and conduct problems, adequate assessment of comorbid psychopathology is also relevant. Consideration of whether treatment for co-occurring problems is necessary prior to initiating CBT for OCD is also important.

Based on the information collected from the initial assessment, the therapist and child can begin to work together to develop a fear hierarchy. The fear hierarchy is a comprehensive list of the child's fears, obsessions, and compulsions ranked in order of severity, often using Subjective Units of Distress (SUDS) [16], in order to rate worry from 0 (*no fear*) to 10 (*extreme anxiety*). The therapist should ensure that the child is rating fears as accurately as possible. For younger children or those with limited insight, it may be helpful to rely on a simpler rating system, such as rating anxiety using a stoplight analogy (i.e., green = I feel little to no anxiety, yellow = I feel some anxiety, but I can generally persist through it; and red = Anxiety is too high). The fear hierarchy should be a working document that is referred to at every subsequent therapy session. The child should be instructed to update the list frequently, as additional fears are identified or as the severity ratings of the fears change over time. An example fear hierarchy is depicted in Fig. 1.

3.2 Psychoeducation

Although some psychoeducation about OCD and CBT may informally occur as part of the initial assessment, it is also important to devote time in the initial sessions to provide youth and their families with more information about OCD and the course of CBT. Importantly, psychoeducation has been shown to result in improvements in functioning, and may instill a sense of hope in the child and family that enhances engagement in CBT [17].

It is important to set clear expectations about the nature of treatment, including discussion of how sessions will be organized, the duration of treatment sessions, and who will attend. Given that youth may be hesitant to engage in treatment, it may also be useful to spend time with the child to set up a contract related to expectations about engagement in treatment, completion of therapy homework, and regular attendance. The dialogue below describes an excerpt of how to begin to have this conversation with youth and their families.

Therapist: *I'm going to be your coach in treatment. We are both going to be working together to fight your OCD and meet your treatment goals. In order to fight your OCD, you will need lots of practice. This is like any sport or activity, where the more effort you put into practice, the bigger the results and rewards.*

Child: *That makes sense.*

Therapist: *In order to practice fighting OCD, we first have to spend some time learning about what OCD is. We need to figure out how OCD is getting in the way of your day-to-day life, and then we will talk about strategies to manage*

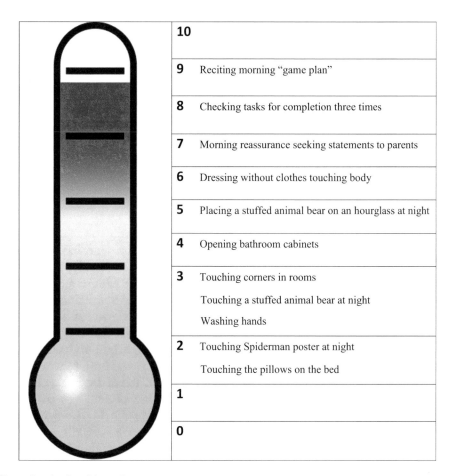

10	
9	Reciting morning "game plan"
8	Checking tasks for completion three times
7	Morning reassurance seeking statements to parents
6	Dressing without clothes touching body
5	Placing a stuffed animal bear on an hourglass at night
4	Opening bathroom cabinets
3	Touching corners in rooms Touching a stuffed animal bear at night Washing hands
2	Touching Spiderman poster at night Touching the pillows on the bed
1	
0	

Fig. 1 Example of a fear hierarchy

OCD symptoms. Every time you come here for therapy, we will set up our sessions in the same way. First, we will briefly check in on your week. I may ask you and mom to tell me generally about how the week went, and if there was anything particularly noteworthy that happened that week, but this should only take a few minutes. We will have a set agenda for every session and have a lot more to discuss. Every week, you will get therapy homework. This will help you continue to practice the skills you are learning here in therapy at home and at school. The more you practice in "the real world" the better you will get at fighting OCD. The rest of the session we will spend discussing different strategies to manage OCD, and I always want to spend some time practicing the skills. At the end, we will plan what you will complete for homework and check in with mom briefly so she knows what you are working on this week.

Child: *That sounds like a lot.*

Therapist: *This is a lot of work, and I can already tell you are committed to doing the hard work to get rid of OCD in your life. As your coach, I will do my part in guiding you to use certain strategies, but there may be times this gets challenging. Even though fighting OCD means facing your fears, I will never force you to do anything in session. I will also never ask you to complete a practice or activity that I would not do myself. So if we decide that you are going to touch a toilet seat one day to fight your fear of germs, then I can be right there touching the toilet too!*

Child: *Hehe.*
Therapist: *Do you have any questions about how we will run our sessions?*
Child: *How long will I have to come here for therapy?*
Therapist: *That's a good question, but there is not a definite answer. Everyone is different and there are some people who may be ready to end treatment after a few months, although there are others who spend much longer in treatment. Regardless of how long you are in treatment, most people experience improvements in their functioning after at least eight to twelve sessions of weekly therapy. I will ask you and your parents to complete some brief questionnaires when you come to therapy, so we can track how helpful therapy has been so far in managing OCD. If we see that OCD symptoms are improving, that will be excellent news that therapy is helpful for you, but if we do not see large improvements in functioning after a month or so, that is also useful information that lets us know we need to try additional treatment strategies to fight back OCD.*

Beyond orienting the child and family to CBT, it is also critical that the child and family learn about the neurobiological origins of OCD. Attention to the neurobiological underpinnings of OCD emphasizes that OCD is not the child's fault or a result of personal shortcomings. Additionally, OCD does not define their identity or character, but instead, OCD may be viewed as a "brain hiccup," [10] that can be controlled by the child through the skills taught during CBT. In some cases, it may be helpful to have the child pick a name to label OCD. This may allow you both to work together to fight the common enemy together. Relatedly, it may be helpful to have the child engage in a number of activities to continue to actively distance himself/herself from OCD. A diagram depicting all of the ways that OCD is interfering with the child's life may provide a useful visual aid to prompt motivation for treatment. Such a diagram may also be important to use to track progress throughout CBT. A journal may also be a useful tool to have children actively describe how OCD negatively impacts their life on a daily basis [10]. Children can write in this journal during the week for homework so that they are continuing to actively identify OCD, and in doing so, the journal may also serve as a useful treatment monitoring tool.

3.3 Cognitive Techniques

It is often the case that youth engage in negative self-talk that can impede treatment outcome and decrease motivation for treatment. This includes statements, such as "I can't do this," "Why am I even trying? This is never going to get any better," and "What's wrong with me?!" While a positive attitude is certainly useful in CBT for OCD, it is more important that children engage in self-talk that is realistic and/or constructive. When therapists identify that children are engaging in negative self-talk, attention to alerting the child to the negative self-talk, and the negative impact of such self-talk on their mood and behavior may likely be useful. The child may benefit from adopting a more positive or realistic mantra to use throughout treatment, such as, "I can do anything I decide to do if I work hard" or "My OCD does not define me."

Additionally, youth often overestimate the likelihood of obsessive fears and their own responsibility in the occurrence of feared events. Thus, challenging and clarifying these maladaptive thoughts through cognitive restructuring may also be useful. To combat these maladaptive thinking patterns therapists may ask, "How likely is it that the event will occur?" and "What's the worst thing that could happen if the event occurred?" to highlight the improbability and potentially catastrophic nature of the child's irrational thinking. Certainly, the relevance of cognitive techniques in treatment for pediatric OCD depends much on the child and the nature of the OCD symptoms. For example, cognitive techniques may be particularly helpful for older youth and/or youth with greater insight into the nature of their OCD symptoms, although may provide minimal benefit for youth with poor insight. The dialogue described below provides an example of a therapist challenging a child's cognitive distortions related to OCD.

> Therapist: *You told me you need to recite your to-do list three times in the morning so that you have a good day at school.*
> Child: *Yes, that is my "game plan." I need to say that in order to make sure nothing bad happens at school. When I do this, nothing embarrassing will happen to me and I can get though my day.*
> Therapist: *What sorts of embarrassing things are you worried may happen at school if you don't say the game plan?*
> Child: *I don't want to let my soccer team down during our games this week. I may make a bad pass or let the other team score a goal.*
> Therapist: *What is the likelihood that you will let down your team if you don't say your game plan in the morning this week?*
> Child: *I don't know . . . I have always said the game plan for as long as I have been on this team, so it is hard to tell. When I first started coming here, I would have answered very high, close to 95%, but I know that is probably my OCD.*
> Therapist: *Ok, so OCD makes it seem really likely that you will let your team down if you don't say the game plan, but what would you say?*
> Child: *I remember when I was in fourth grade in a different soccer league and didn't say the plan. I still played well. I had also qualified for the traveling team then. I probably made some mistakes playing then, but I do not really remember anything terrible, and all my teammates were probably doing the same. Now, I would estimate the likelihood of messing up because I didn't say the game plan was maybe 5% but probably closer to zero.*
> Therapist: *That is pretty low. It sounds like saying the game plan may have very little to do with how you play in soccer. It sounds like OCD may have been making it seem more likely. What if you did make a bad play? What if you missed the winning shot and let your team down? What would happen then?*
> Child: *My coach always tells us "there's no 'I' in team." We are always talking about how it is not one player on the team that makes or breaks the game. I could remind myself of that. On the traveling team, we have lots of games, so it is not really expected that we win all of our games. Plus, there are always substitutes that can play if one of us is having a difficult game. If I made a bad play, I would remind myself that happens to everyone from time to time, and I would use that as an opportunity to continue to practice and improve my game instead of just worrying about how I will play.*
> Therapist: *So if you did make a bad play, how likely is it because you did not say your game plan?*
> Child: *Not at all likely. Maybe I was just having an off day or the other team was really good. I can keep showing up to practice and giving my best, and get better each day.*

3.4 Exposure and Ritual Prevention (EX/RP)

EX/RP is arguably the most important part of CBT for pediatric OCD. Compared to other components of OCD treatment, EX/RP is associated with the largest effects [5, 7]. In this treatment phase, youth actively face their fears, and through repeated exposure to their fears, habituation occurs and anxiety decreases. Youth are exposed to the fears listed on the fear hierarchy, beginning with fears rated as mild to moderate. As anxiety decreases in these early exposures, and as youth become more confident in their skills to complete exposure exercises, more challenging exposures exercises can be conducted. Exposure to feared stimuli is critical, as it serves to weaken the established association between the unconditioned stimulus and conditioned stimulus (i.e., the obsession), while a new association is formed, and the child learns the once feared stimulus does not invoke the feared outcome. However, it is also critical that during these exposure exercises youth do not engage in compulsions, rituals, or other safety behaviors that may attenuate the strength of the exposure. For example, when challenging a fear of germs by rubbing door handles on the clinic bathroom, it will be important that the therapist and child agree on a reasonable time when the child may eventually wash his or her hands. Washing them immediately following the exposure or even at the end of the treatment session may weaken the strength of the exposure.

While EX/RP procedures will have likely been discussed during the initial sessions when discussing psychoeducation, it will be important to review and thoroughly explain the structure of exposure exercises. During this time, it may also be useful to update the fear hierarchy. Ground rules for the sessions should be discussed, including how exposures will be selected, and that youth will be expected to complete the entire exposure to allow sufficient opportunity for anxiety to dissipate. It will be important to describe that during exposure, anxiety is expected to increase and then decrease over time, and ratings will be used to help the child and therapist communicate about how challenging an exercise is, and whether the anxiety has sufficiently dissipated. As the fear hierarchy has already been developed prior to engaging in exposure, the child will already have some insight into rating the severity of different worries. While the 0–10 SUDS is often used to gauge the difficulty of different exposures, the therapist should monitor to ensure that the child is accurately reporting their anxiety. For example, if a child reports relatively low anxiety (i.e., SUDS = 2), but is crying about the upcoming exposure and is refusing to participate, it will be important to further discuss accurate ratings of anxiety. It is likely this exposure may have a much greater SUDS rating. Ideally, a strong therapeutic alliance will have been formed between the therapist and child, so that they can work together to select exposure exercises that will sufficiently challenge anxiety, while at the same time not providing an experience that is too easy or too difficult. Therapeutic alliance is also critical in order to ensure that safety behaviors are not unknowingly being done.

To ensure the success of the exposure exercise, it is also critical that the therapist discuss response prevention (RP). RP requires that youth do not engage in rituals that may alleviate anxiety so that the cycle of obsessions and compulsions is not further reinforced. Ritualistic compulsions associated with OCD often require individuals to engage in behaviors in the same way or engage in rituals more frequently over time, and when they do not engage in these exact rituals, anxiety increases. Thus, while engaging in rituals may temporarily reduce anxiety, over time, obsessive worry is reinforced and increases. Attention to shortening or delaying these rituals or doing these rituals differently helps to challenge the irrational cognition that engaging in the ritual leads to a specific outcome or decrease in anxiety. Examples of EX/RP practices are listed in Fig. 2.

EX/RP exercises can be imaginal or in vivo. Imaginal exposures provide an opportunity for youth to practice EX/RP in their imagination. It may be particularly useful to practice imaginal exposures

Fear	Example EX/RP	SUDS before	SUDS after
Fear of contamination			
	Touch the toilets	8	3
	Do not use hand sanitizer all day	8	3
	Do not use hand sanitizer for 30 minutes	6	4
	Eat a snack without washing hands	5	2
Fear of becoming sick			
	Listening to an audio clip of someone vomiting	9	3
	Shaking people's hands	7	2
	Not going to the nurse's office for the week	5	1
Fear of dying			
	Walking through a hospital	7	2
	Spending time with grandfather in the hospital	7	1
	Only asking mother twice per day about grandfather's health	5	2

Fig. 2 Example EX/RP practices

during the early stages of EX/RP in order to gradually acclimate the child to the structure and process of EX/RP. Imaginal exposure may be most suitable for addressing fears that are relatively mild, as in vivo exposures offer a more real-time experience, where the child is more actively exposed to the feared stimulus and has the opportunity to fully habituate to the fear. Imaginal exposures are also useful in situations where it is not feasible to expose the child to a feared stimulus in session. For example, it may be helpful to set up an imaginal exposure for a child who is unable to visit her grandfather with cancer in the hospital due to fear of contracting an illness. In the therapy session, imaginal exposures could be planned to have the child imagine visiting the grandfather in the hospital, how she would feel touching him, and how she would not wash her hands after the interaction. Below is an example dialogue of an imaginal exposure practice conducted early in CBT for a child with contamination fears.

> Therapist: *One of the first ways we can practice an exposure is by imagining what it would be like for you to face your fear of germs. This is what we call an imaginal exposure.*
> Child: *Even thinking about touching germs makes me feel gross. I don't really even like saying the word "germs" because it makes me think about my OCD even more.*
> Therapist: *For a lot of people, even thinking or talking about their OCD gives them anxiety, so the more we can have you face those thoughts rather than avoid them, the more power you will have over your OCD. On a scale from 0 (no distress at all) to 10 (terrible worry) how would you feel if you were in math class and your friend sitting next to you sneezed on you?*
> Child: *That would be difficult for me, probably a 7 out of 10.*
> Therapist: *I want you to try really hard to imagine this is happening right now. You are in the middle of working on a math worksheet and your friend sneezed on you. Focus on your worry. How does your body feel?*
> Child: *My heart is racing. I just want to get up and run to the bathroom to wash off my arm. Maybe I can change my shirt.*
> Therapist: *What types of thoughts are you having?*
> Child: *I don't want to get sick now. I am worried I will get sick. Am I going to get sick and throw up? This is gross.*
> Therapist: *You have the hang of this! I want you to try to continue to stay in this moment. Keep focusing on your worry about getting sick. Imagine what you would do next in this situation.*
> Child: *OCD is telling me to run to the bathroom and scrub myself clean.*
> Therapist: *That does sound like something OCD would say to you. What are you going to do to fight back?*
> Child: *I am going to try to stay in my seat and resist going to the bathroom to wash off.*
> Therapist: *Good. Keep focusing on those thoughts.*
> Child: *I am going to get sick. I am going to get sick. I am going to get sick.*
> Therapist: *Now, let's pretend it is 15 minutes later. How are you feeling now? What is your 0 to 10 worry rating?*
> Child: *It went down. Maybe a 4 out of 10. I am not crawling in my own skin anymore and I think I can focus enough to work on my math.*

The therapist has a critical role in setting up relevant and useful EX/RP exercises. In selecting initial exercises, it will be important that therapists develop exercises that can be successfully conducted in the therapeutic setting. There are a number of programs and research studies that are able to deliver EX/RP over the course of sessions that are 90 minutes or longer. However, other settings may allow only for 60 min sessions, so it is imperative that as much as possible there is sufficient opportunity for anxiety to decrease in the duration of the session. Clinicians should rely on the child's self-report of distress in the exposure as well as behavioral observations to determine whether anxiety has sufficiently decreased. In situations in which is it not feasible to remain in session until anxiety has been sufficiently reduced, it will be important to talk with the family to try to problem solve how to continue the exposure to completion. For instance, a 60 min therapy session may not allow sufficient time to allow for anxiety to dissipate after touching a public sink and not using hand sanitizer. It may be critical to work with the family to agree upon a time after the session, perhaps two hours later prior to the family eating dinner when the child may be allowed to wash her hands.

3.5 Family Involvement

Parents and other family members are essential in promoting positive CBT outcomes [7, 18]. Family members provide valuable information in the early assessment phase of treatment, and throughout treatment they can continue to provide important information about the symptoms they continue to see, the progress they observe, and other concerns that the child may not discuss in treatment. Therefore, it is critical to regularly check in with family members to assess ongoing progress, and there are a number of ways this can be done. Family members may be provided with rating scales, such as the CY-BOCS [13] to assess OCD severity and impairment while the child is actively engaged in the session. Additionally, some time, perhaps a few minutes, at the beginning and/or end of the treatment session may be devoted to eliciting formal or informal feedback about the child's progress.

Beyond their critical role as informants, family members are often also critical in promoting and maintaining treatment progress. When family members are aware of the specific skills discussed in treatment, they may be better able to remind their child to apply the skills in anxiety provoking situations. Additionally, when relevant EX/RP practice is not feasible in the clinic setting, parents and other family members may help facilitate EX/RP exercises at home or in other relevant settings. Family members may also help to remind their child to complete therapy homework, and provide positive reinforcement or rewards for maintaining engagement in treatment. Careful consideration should be given to how family members will be involved in the treatment. Too little family involvement may be problematic, particularly for younger children or those youth who have difficulty applying CBT skills on their own. On the other hand, too much family involvement may also impede treatment progress, as it is important to promote child autonomy in CBT to ensure that youth can apply CBT skills in situations where their parents and family are not there to actively scaffold and support their use of CBT skills (e.g., with friends, at school). Providing the family with a session checklist each week that briefly describes the skills and homework for the week can be a useful tool for promoting communication between therapist and family members. Additionally, it may be useful to devote a few minutes at the beginning or end of the CBT session to review the session and weekly homework together with the child and family.

Despite the essential role that family members have in promoting positive CBT outcomes, it is also important to consider that family members may be engaging in activities, often unknowingly, that reinforce and maintain obsessions and compulsions. In an effort to help their child alleviate anxiety, family members may actually accommodate OCD symptoms in maladaptive ways. For example, are parents providing objects needed for rituals? Are they allowing their child to follow a certain routine to minimize anxiety? Are they answering their child's reassurance seeking questions repeatedly? Are they decreasing child responsibility or limiting attempts at discipline? A parent may watch their child engaging in a morning ritual before school that involves dressing and eating in a certain way. Over time, this ritual comes to require more time and the parent begins to wake the child earlier to allow for the extended time engaging in the ritual. While it is clear the parent's goal is to help and support the child getting to the bus on time and reducing anxiety before school, the extended opportunity to engage in the ritual only serves to strengthen the OCD symptoms. Therefore, discussing ways that family members may actually be accommodating their child's OCD symptoms is important. It is often the case that parents self-identify how they may be accommodating their child's OCD early on in treatment. For example, when providing psychoeducation about the development and maintenance of OCD, parents may disclose behaviors they had been doing that may have reinforced their child's anxiety. As a result, it may be useful for the parents and the therapist to begin to discuss family accommodation of OCD. A more formal discussion of accommodation may be useful after several sessions of CBT with the child, when the therapist has obtained more information about child symptoms, engagement in treatment, and family dynamics.

Therapist: *Thank you for coming in. Johnny and I wanted to review his therapy homework with you this week.*

Mother: *I am happy to help. I know Johnny is really working hard in here. He seems to have a renewed sense of hope about being able to fight OCD.*

Therapist: *I agree. He is taking a lot of ownership in working hard in and out of session. Johnny, why don't you tell your mother the practice we came up with for you to complete for homework?*

Child: *We talked about how I will put the bottles of hand sanitizer from my book bag away. I will not use any hand sanitizer at school and I won't go to the bathroom at school to wash my hands.*

Mother: *I bought these hand sanitizer keyrings that I put on all my children's book bags. I like for them to be able to wash their hands before they have lunch and after gym class. I use them, too. I guess I may have modeled some of the washing behavior.*

Therapist: *It is not uncommon that parents' best intentions actually turn out to help OCD. OCD can be very tricky. Although it makes a lot of sense to want to wash your hands at certain times of day, in order to practice fighting OCD we sometimes need to do things differently.*

Mother: *I want to help as much as I can. I will plan to take the hand sanitizer out of my car, too, so Johnny knows we are all working on this and we are all supporting him. What else should I do?*

Therapist: *Johnny is doing a great job facing the fears he has listed on his hierarchy, so we want to make sure we continue to fight OCD at his pace. Johnny, what do you think?*

Child: *I think not using hand sanitizer this week will be pretty challenging. I have been relying on it so much over the past two years. I do want to continue to cut out other times OCD makes me wash too much, like cutting down my time in the shower, but I think that may be too much for one week.*

Mother: *That sounds great. What can I do to help this week?*

Therapist: *The most important thing you can do is what you are doing now. Continuing to support and encourage Johnny as he fights OCD. It is so important to work together so we can ensure he is maximizing his time outside of therapy to fight OCD.*

Child: *If I need help, I will let you know. I want to try to do this on my own.*

Mother: *That sounds great. I will let you do your thing, and be here if you need me.*

Therapist: *That sounds like an excellent plan. Keep up the good work!*

3.6 Booster Sessions and Maintenance of Treatment Gains

Existing manuals for pediatric OCD provide CBT in a time-limited format often ranging from approximately 10 to 22 sessions [7, 10]. While many youth are ready to complete treatment at this time, others are not, and extension of treatment may be beneficial. It is important to determine whether treatment goals have been met and whether OCD symptoms are adequately managed. In most cases, it may be that subthreshold symptoms of OCD continue to persist. However, if children and their families report feeling they are able to manage these symptoms well, and that they are not causing clinically significant levels of impairment, termination of treatment may be appropriate. In addition to eliciting informal feedback about treatment progress and the need for additional treatment from children and their family members, reliance on additional assessments is also encouraged. For example, well validated measures such as the CY-BOCS [13] and the NIMH Global Obsessive-Compulsive Scale (NIMH GOC) [19] may aid in clinical decision making about termination. Furthermore, if significant items still remain on the fear hierarchy, additional treatment may be necessary.

If it is determined that additional treatment is needed, it will be important to then determine what type of additional treatment is necessary. For some youth, additional practice with EX/RP may be necessary. For others, additional family difficulties that are impeding treatment may need to be addressed. Still for others, there may be a desire for additional treatment despite clinically significant improvement in functioning. In these cases, it is important to consider whether additional treatment will result in continued improvement, or whether continuing treatment may actually reinforce thoughts and behaviors associated with OCD. That is, is continued contact with the therapist accommodating a need for reassurance? Alternatively, does continued treatment potentially undermine the child and family's autonomy to manage OCD symptoms?

While the answers to these questions are not always clear, it may be useful to consider booster sessions to provide additional treatment, albeit in a scaled back, less intensive and/or less frequent format even if youth have experienced clinically significant improvement. The risk for relapse of OCD symptoms is high [20], and booster sessions may provide a means for youth to gain more confidence in their ability to manage symptoms in the long-term, while discussing important topics such as when symptoms require attention or more intensive therapeutic intervention. The majority of CBT skills should already have been learned up to this point in treatment. Thus, the focus of booster sessions may be placed on reinforcing skills and further promoting autonomy and independence in applying CBT strategies to manage OCD symptoms. Opportunities to allow the child to be more active in the planning of treatment sessions and homework assignments should be encouraged. For example, it may be useful to have youth set the agenda for the treatment session and identify additional EX/RP exposures that would be useful to complete. It may also be beneficial to work with the child to create a daily or weekly report card with goals tracking management of OCD. Figure 3 shows an example daily report card that lists behaviors the child has identified as goals for the week in between treatment sessions. Of note, these daily goals address OCD-specific behaviors (e.g., reassurance seeking), but also included more generalized goals to improve functioning (e.g., calling a friend).

Ultimately, once OCD symptoms and impairment are well managed, treatment goals are met, and the child and family report confidence in being able to manage and control OCD, it is time to terminate treatment. Reaching the end of treatment is an important milestone that should be celebrated. Children and their families have devoted significant time and effort into attending treatment, learning new skills, and facing their fear and discomfort, and are now at an important crossroads where they can enjoy the benefits of their hard work. Although the benefits of living a life in which they have managed OCD is likely rewarding in and of itself, therapists can also provide certifications of CBT completion and devote some time to acknowledging and rewarding child and family efforts.

George's OCD Daily Report Card

	Monday	Tuesday	Wednesday	Thursday	Friday	Saturday	Sunday
1. Brush teeth no more than 2x a day	(Y) N	(Y) N	(Y) N	(Y) N	(Y) N	(Y) N	(Y) N
2. No more than 2 reassurance seeking questions in math class	(Y) N	Y (N)	Y (N)	(Y) N	(Y) N	(Y) N	(Y) N
3. No instances of recopying homework	(Y) N	Y (N)	Y (N)	(Y) N	(Y) N	(Y) N	(Y) N

OTHER

1. Attend science club   (Y) N

2. Call a friend this weekend   (Y) N

Total Number of Yeses ____19____   Total Number of Nos ____4____   Percentage of Yeses ____82.6%____

***Goal for the week = At least 75% Yeses

***Reward if goal met: Skate park with a friend

Fig. 3 Example daily report card to improve long-term maintenance of treatment

4 Discussion/Conclusions

Pediatric OCD can be a debilitating disorder that is associated with substantial impairment across academic, behavioral, and social functioning [1]. Emergence of OCD in childhood may signal risk for recurrent and persistent difficulty, as well as portend high risk for additional psychopathology, including depression and other anxiety disorders across the life span [3, 4]. Thus, treatment addressing pediatric OCD is of critical importance. Fortunately, substantial advancement in treatment for pediatric OCD has accumulated over the last decades [5–7], and CBT has emerged as a first line treatment for pediatric OCD associated with large reductions in symptoms and similarly large improvements in functioning for many children. In fact, for many youth, large improvements in functioning are demonstrated within weeks of treatment, and these improvements are maintained even after treatment has been completed.

Results from the Pediatric OCD Treatment Study (POTS), the largest randomized controlled trial of children with OCD, provide further evidence of the efficacy of CBT. However, the POTS also calls attention to a substantial portion of youth who do not experience remission of OCD at the end of CBT. Specifically, the POTS compared the relative efficacy of CBT, sertraline, and their combination over 12 weeks [21] and showed that all three active treatments were more efficacious than placebo; combined treatment was associated with the highest remission rate, although it was not significantly higher than that of CBT alone. However, 46.4% of children did not experience a remission of symptoms. While this seemingly high rate of CBT nonresponders may be due to complications in delivering treatment with integrity and fidelity in a multisite design [22], other single site studies have still indicated a considerable proportion of children who continue to experience symptomatology with CBT [7]. Poor treatment outcome has been related to long-term consequences for children with pediatric OCD [4]; thus, it is important to understand which children are less likely to benefit from CBT so that adequate interventions can be designed.

Notably, randomized controlled trials of CBT rely on manualized treatment protocols that deliver a one-size-fits-all approach to treating pediatric OCD. This is a necessary approach to examining the efficacy of CBT. However, treatment outcome may be optimized for youth when treatment is personalized. There are core components of CBT for OCD that may be differentially important in the course of treatment depending on the child's presenting symptoms, impairment, level of insight, developmental age, motivation for treatment, and family support, among other important factors. Thorough assessment and psychoeducation are critical components needed to initiate treatment. Additionally, EX/RP practice is perhaps the most potent component of treatment [5]. The amount of attention provided to cognitive restructuring, family involvement, and maintenance planning will vary to a larger degree, and determining how to incorporate these components will involve sound clinical judgement. Certainly all these components can be addressed in treatment, although the extent to which these components are emphasized will vary.

CBT for OCD can be a very effective treatment, and given the active nature of EX/RP sessions, it is possible to see clear benefits in anxiety reduction even in the course of one 60 min session. These visually apparent results (e.g., the child can now touch a toilet with little observable apprehension) are important signals that treatment is beneficial. Other assessments such as rating scales and child and parent open-ended feedback may also be helpful in clinical decision-making about the extent to which certain treatment components are necessary. It is not always the case that a certain treatment skill or practice will be useful for a particular child. In addition, some obsessions and compulsions may be relatively easier to address than others. For example, there may be plenty of opportunities to practice EX/RP related to contamination fears, although addressing other more abstract obsessions, such as sexual obsessions and religious preoccupations may present substantially greater challenge. Consultation with other clinicians about addressing these concerns is critical. However, some trial and error in treatment is also likely and may be helpful in modeling to the child that you are working together as a team to figure out the best way to manage OCD.

Ultimately, CBT for pediatric OCD is a potent treatment that can be applied flexibly to improve youth functioning. There are a number of excellent resources for clinicians and families to consult for up-to-date research and treatment guidelines for pediatric, from the Society of Clinical Child and Adolescent Psychology (www. effectivechildtherapy.org), the International OCD Foundation (kids.iocdf.org), and the Association for Behavioral and Cognitive Therapies (www.abct.org). Despite the substantial support for the efficacy of CBT for pediatric OCD, additional large-scale studies are needed to further demonstrate the broad range benefits of OCD. Important questions remain that can be answered by these studies (e.g., which treatment components work best for which youth with OCD? For which youth is CBT alone versus CBT plus medication treatment most beneficial?) Additionally, other questions about the neural underpinnings of pediatric OCD may also serve to identify clues that will help optimize CBT for youth with OCD. At the same time, clinicians who are implementing OCD in their own practices may provide important insight into optimizing CBT outcomes for youth with OCD, and integration of these clinical and research efforts has great potential in continuing to improve CBT outcomes for youth with OCD and their families.

References

1. Geller DA, Biederman J, Faraone S et al (2001) Developmental aspects of obsessive compulsive disorder: findings in children, adolescents, and adults. J Nerv Ment Dis 189:471–477

2. Walitza S, Wendland JR, Gruenblatt E et al (2010) Genetics of early-onset obsessive–compulsive disorder. Eur Child Adolesc Psychiatry 19:227–235

3. Pine DS, Cohen P, Gurley D et al (1998) The risk for early-adulthood anxiety and depressive disorders in adolescents with anxiety and depressive disorders. Arch Gen Psychiatry 55:56–64

4. Stewart SE, Geller DA, Jenike M et al (2004) Long-term outcome of pediatric obsessive-compulsive disorder: a meta-analysis and qualitative review of the literature. Acta Psychiatr Scand 110:4–13

5. Abramowitz JS, Whiteside SP, Deacon BJ (2005) The effectiveness of treatment for pediatric obsessive-compulsive disorder: a meta-analysis. Behav Ther 36:55–63

6. Barrett PM, Farrell L, Pina AA et al (2008) Evidence-based psychosocial treatments for child and adolescent obsessive–compulsive disorder. J Clin Child Adolesc Psychol 37:131–155

7. Freeman J, Garcia A, Frank H et al (2014) Evidence base update for psychosocial treatments for pediatric obsessive-compulsive disorder. J Clin Child Adolesc Psychol 43:7–26

8. March J, Frances A, Carpenter D et al (1997) Expert consensus guidelines on obsessive-compulsive disorder. J Clin Psychiatry 58 (Suppl 4):2–72

9. McHugh RK, Whitton SW, Peckham AD et al (2013) Patient preference for psychological vs. pharmacological treatment of psychiatric disorders: a meta-analytic review. J Clin Psychiatry 74:595–602

10. March JS, Mulle K (1997) OCD in children and adolescents. Guilford Press, New York, NY

11. Silverman WK, Albano AM (1996) The anxiety disorders interview schedule for DSM-IV—child and parent versions. Graywinds Publications, San Antonio, TX

12. Kaufman J, Birmaher B, Brent D et al (1997) Schedule for affective disorders and schizophrenia for school-age children-present and lifetime version (K-SADS-PL): initial reliability and validity data. J Am Acad Child Adolesc Psychiatry 36:980–988

13. Scahill L, Riddle MA, McSwiggin-Hardin M et al (1997) Children's Yale-Brown Obsessive

Compulsive Scale: reliability and validity. J Am Acad Child Adolesc Psychiatry 36:844–852

14. Gallant J, Storch EA, Merlo LJ et al (2008) Convergent and discriminant validity of the children's Yale-Brown Obsessive Compulsive Scale symptom checklist. J Anxiety Disord 22:1369–1376

15. Storch EA, Murphy TK, Adkins JW et al (2006) The children's Yale-Brown Obsessive-Compulsive Scale: psychometric properties of child- and parent-report formats. J Anxiety Disord 20:1055–1070

16. Wolpe J (1969) The practice of behavior therapy. Pergamon Press, New York, NY

17. Babinski DE, Pelham WE, Waxmonsky JG (2014) Cognitive-behavioral therapy for pediatric obsessive-compulsive disorder complicated by stigma: a case study. Clin Case Stud 13:95–110

18. Merlo LJ, Lehmkuhl HD, Geffken GR et al (2009) Decreased family accommodation associated with improved therapy outcome in pediatric obsessive–compulsive disorder. J Consult Clin Psychol 77:355–360

19. Goodman WK, Price LH (1992) Assessment of severity and change in obsessive compulsive disorder. Psychiatr Clin North Am 15:861–869

20. Hiss H, Foa EB, Kozak MJ (1994) Relapse prevention program for treatment of obsessive-compulsive disorder. J Consult Clin Psychol 62:801–808

21. March JS, Foa E, Gammon P et al (2004) Cognitive-behavior therapy, sertraline, and their combination for children and adolescents with obsessive-compulsive disorder: the Pediatric OCD Treatment Study (POTS) randomized controlled trial. JAMA 292:1969–1976

22. Watson HJ, Rees CS (2008) Meta-analysis of randomized, controlled treatment trials for pediatric obsessive-compulsive disorder. J Child Psychol Psychiatry 49:489–498

Chapter 5

Trauma-Focused Cognitive Behavioral Therapy (TF-CBT)

Brian Allen, Elizabeth Riden, and Chad E. Shenk

Abstract

Trauma-focused cognitive behavioral therapy (TF-CBT) is currently considered by multiple sources to be the "gold standard" treatment for children and adolescents experiencing posttraumatic stress and associated symptomatology subsequent to trauma exposure. Dozens of randomized controlled trials demonstrate the superiority of TF-CBT over rapport-focused treatment for youth experiencing sexual abuse, domestic violence, natural disasters, and other forms of trauma exposure. This chapter discusses the theoretical and empirical basis of TF-CBT, and provides an overview of the specific components of TF-CBT, as well as practical examples of technique implementation. Current standards for training and supervision are reviewed.

Key words Trauma, Child abuse, Treatment, Cognitive behavioral therapy, Exposure therapy, Posttraumatic stress

1 Introduction

Trauma-focused cognitive behavioral therapy (TF-CBT; [1]) is widely regarded by both clinicians and researchers as a front-line intervention for the treatment of posttraumatic stress among children and adolescents [2, 3]. TF-CBT is a component-based protocol initially developed with preschoolers who experienced sexual abuse [4] and later tested with older children and adolescents who experienced diverse forms of trauma exposure [5–7]. In addition, the positive effects attributed to TF-CBT are documented cross-culturally, and varied cultural beliefs and practices have been successfully integrated into the protocol [8–10].

2 Theoretical and Empirical Foundations

From a behavioral perspective, TF-CBT incorporates Mowrer's [11] two-factor model in conceptualizing the development and maintenance of posttraumatic stress symptomatology. The first

Robert D. Friedberg and Brad J. Nakamura (eds.), *Cognitive Behavioral Therapy in Youth: Tradition and Innovation*, Neuromethods, vol. 156, https://doi.org/10.1007/978-1-0716-0700-8_5, © Springer Science+Business Media, LLC, part of Springer Nature 2020

factor, classical conditioning, provides the mechanism by which both memories of a traumatic event and previously innocuous stimuli (e.g., clothing, music, names), through pairing with the actual traumatic experience, are capable of eliciting psychological and physiological distress. One is then motivated to avoid these reminders to prevent the onset of such distress, resulting in any number of maladaptive behaviors, such as refusing to discuss the trauma or its effects, emotional numbing, and substance use. When successful, these efforts at avoidance reduce the tension and stress experienced by the individual, thus negatively reinforcing the avoidance efforts through operant conditioning processes, the second factor of Mowrer's framework.

Two primary and important treatment implications derived from this understanding of posttraumatic stress are integrated into TF-CBT. First, exposure procedures will be necessary to extinguish both the classically and operantly conditioned fear response associated with any number of reminders. A hallmark feature of TF-CBT is that exposure begins in the first session by talking about the index trauma(s), and continues during each session regardless of the specific component being implemented. In addition, a number of sessions are dedicated specifically to the implementation of focused exposure exercises through the development of a factual narrative account of the person's trauma. Second, avoidance is likely to be a strong force in many cases and the clinician must not inadvertently reinforce this process. For instance, if a clinician is in the midst of exposure exercises with an adolescent who becomes standoffish in session, the clinician redirecting the discussion to the adolescent's preferred topics, such as friends or activities, may reinforce the adolescent's avoidance behavior making future exposure sessions more difficult.

From a cognitive perspective, TF-CBT incorporates a number of tenets from classical cognitive theory [12] as well as later revisions [13–15]. In short, one may make any number of attributions regarding the causes, meaning, or impact of the traumatic event. For instance, a girl who was sexually abused by her father may blame herself for not doing more to prevent the abuse, interpret the abuse as meaning that she is forever changed in a negative way, and then take responsibility for the dissolution of her parents' marriage after disclosure. These maladaptive thoughts may lead to a number of emotional and/or behavioral concerns, as well as themselves serve as reminders of the traumatic events, which can then prompt significant distress for the individual. In addition, to guard against experiencing repeated trauma, one's interpretation of threat cues may be significantly altered to view harmless circumstances as signaling potential danger, resulting in hypervigilance, fearful perceptions of the world, and significant anxiety (e.g., separation anxiety). TF-CBT places a premium on identifying maladaptive thoughts throughout the treatment process and implementing focused cognitive restructuring techniques.

From a developmental perspective, TF-CBT recognizes unique considerations that must occur when working with children and teenagers. First, it is not uncommon that youth who experience trauma have underdeveloped emotion regulation, behavior regulation, and interpersonal skills, especially youth with chronic traumatic experiences. It may be difficult for some children to participate in exposure-based exercises until they learn more effective coping skills. As such, teaching a number of affective modulation and relaxation skills is one of the initial tasks of TF-CBT. Also in recognition of this developmental aspect, exposure exercises in TF-CBT are considered gradual and sequential as they unfold over a number of sessions, while utilizing previously learned skills to cope with emotional responses that may occur. Second, caregivers play a crucial role in a child's development and are an important support when a child is faced with distress. During TF-CBT, the child's caregiver is actively involved in each session and provides support in numerous ways, including learning the same skills being taught to the child and encouraging their use outside of session, supporting the child throughout the exposure process, and implementing behavior management skills to prompt the development of improved behavior regulation. Lastly, a child's level of cognitive development is considered during the specification of exercises. For instance, when treating a preschool-age child, the caregiver may be much more involved and the implemented techniques may rely more on the behavioral mechanisms of change; however, adolescents may exhibit entrenched maladaptive cognitions that require significantly more focus on cognitive restructuring techniques.

To date, over 20 randomized controlled trials have tested the efficacy and effectiveness of TF-CBT. Replicated results demonstrate the superiority of TF-CBT over non-directive rapport-focused treatments [5, 16]. suggesting that the mechanisms of effect for TF-CBT are not solely due to generic therapeutic factors, such as therapeutic rapport or the passage of time. Having convincingly established the beneficial effect of the intervention, a number of potential mediators and moderators of TF-CBT treatment response have been investigated. In a dismantling study, Deblinger and colleagues [17] found that including the focused exposure exercises inherent in the construction of a trauma narrative yielded greater effect for children who displayed higher levels of posttraumatic stress avoidance than excluding the narrative. As one might expect, when the narrative was replaced in treatment with additional sessions focused on the development of caregiver behavior management skills, children exhibited greater improvements in behavioral problems. One interesting randomized controlled trial in Norway found that (a) TF-CBT was superior to community-based rapport-focused treatment for ameliorating posttraumatic stress of youth, (b) a better quality of therapeutic rapport improved TF-CBT outcomes, and (c) improvement in

Component

P:	Psychoeducation
	Parenting Skills
R:	Relaxation Skills
A:	Affective Expression and Modulation Skills
C:	Cognitive Coping and Processing Skills
T:	Trauma Narrative & Cognitive Processing
I:	In Vivo Exposure
C:	Conjoint Parent-Child Sessions
E:	Enhancing Safety and Developmental Trajectory

Fig. 1 The PRACTICE acronym, TF-CBT components

maladaptive cognitions were more apparent in the TF-CBT condition and these improved thoughts predicted improvements in post-traumatic stress and depressive symptoms [18, 19]. In addition, sustained effects for TF-CBT have been documented up to two years post-treatment [20].

3 Intervention

TF-CBT incorporates a number of components that are summarized by the acronym PRACTICE (*see* Fig. 1). Each of these components is briefly discussed below.

4 Psychoeducation

The first component of TF-CBT is psychoeducation, which typically begins during a clinician's first interaction with a child and caregiver and continues throughout the course of treatment. The primary goals of psychoeducation are to provide general information related to trauma, specifically the type of trauma experienced, and to help normalize the physiological and/or psychological responses the child and caregiver may be experiencing. This process not only allows the child and caregiver to understand the rationale for treatment, but also may combat any negative feelings, such as shame or guilt, that the child and/or caregiver have related to their experiences. During psychoeducation, the clinician works closely with the child and caregiver to identify relevant trauma reminders and posttraumatic stress symptoms, and to identify correspondingly helpful materials and information that may be reviewed in session.

Psychoeducation should be delivered in a developmentally sensitive manner. With young children, it is helpful to use books, games and worksheets as these activities are interactive and children often respond better to such an approach than the clinician simply reading or having the child read an information sheet. In the case of a 4 year-old girl exposed to domestic violence, the clinician may use a book, such as *A Terrible Thing Happened* [21], to review common posttraumatic stress symptoms in a non-threatening way. This book discusses a "terrible thing" that occurred to a raccoon and his behaviors and feelings afterward. Young children are typically able to identify and verbalize their own symptoms that are similar to those experienced by the raccoon, such as having nightmares and feeling sad most of the time. The acquisition of such information can be reinforced through the use of other activities, such as making a matching game where the child earns a point for every trauma reminder she is able to accurately identify.

With adolescents, interactive tools are a preferred method of providing psychoeducation as well, but the clinician may expect the adolescent being able to engage in more of a discussion. For example, in the case of a 15 year-old boy who was physically abused by his father, the clinician may utilize handouts from the National Child Traumatic Stress Network website (www.nctsn.org) to teach information about trauma types and the specific impact of physical abuse. The adolescent may be expected to ask a number of questions regarding the causes or effects of such abuse and opportunities may arise to take note of maladaptive cognitions evident through the discussion. Regardless of the child's age, the initial sessions of psychoeducation begin the exposure process as the clinician should both define and use appropriate descriptive language (e.g., sexual abuse, hurricane) and not avoid such topics because of the youth's posttraumatic avoidance.

5 Parenting Skills

Experiencing a traumatic event can be extremely disruptive to the life of a child and their family, and caregivers often struggle to function as a parent. It is not uncommon for a clinician to find that a caregiver has become permissive with a child and lax in providing appropriate rules and consistency. In the parenting skills component, a clinician works directly with the caregiver to identify problematic behavior patterns and then develop an effective behavior management plan. In addition, parenting skills may be used to reestablish appropriate boundaries and improve parent–child communication. A positive parenting approach is a key factor in TF-CBT. Caregivers are instructed to use praise, selective attention, and contingency reinforcement programs to promote positive behaviors. Depending upon the situation, other forms of parent training may also be appropriate, such as teaching reflective listening or empathy skills when a parent shows difficulty in providing appropriate emotional support to the child.

6 Relaxation Skills

Children suffering with symptoms of posttraumatic stress may demonstrate an exaggerated startle response, higher resting heart rate, difficulty sleeping, hypervigilance, restlessness and irritability, anxiety, and/or anger/rage reactions. These symptoms may be intermittent or chronic, depending on the child and complexity of the situation, but almost always interfere with a child's ability to function normally at home and school. Relaxation skills are taught early in treatment in order to give the child and caregiver tools to manage everyday stressors as well as when they are exposed to trauma reminders. It is important for the clinician to educate the child and caregiver around normal somatic reactions to trauma reminders and explain how bodies sometimes react to stress. Normalizing these responses and reactions can validate these experiences for both the child and caregiver and improve the likelihood of implementing the relaxation skills taught in session.

Similar to other CBT interventions, relaxation training may take many forms. Some of the more commonly taught skills include controlled breathing, progressive muscle relaxation, and positive imagery. However, other techniques are often promoted as relaxation and may be applicable, such as mindfulness, yoga, and scheduling relaxing activities at home. It is typically advisable to ask parents and children the types of activities they currently utilize that are relaxing. Relying on such methods may be particularly helpful to increase their frequency or connect them specifically to moments of distress. Similar to the previous components, interactive tools are a preferred way of teaching relaxation to both children and adolescents. Using games, stories, and activities creates a fun and often enjoyable atmosphere for the child to learn and practice the skills. For example, with children of all ages, a clinician may use bubbles or balloons to teach controlled breathing while educating the child on the importance of inhaling and exhaling slowly. With this exercise, the child is able to practice along with the clinician and see how it is much easier to blow up a balloon or create more bubbles when taking slow, controlled breaths.

7   Affective Expression and Modulation

After a traumatic event, children and caregivers often experience intense feelings they may not understand or have the vocabulary to express. Young children may even feel confused or overwhelmed by the intensity of their emotions. In this component, the clinician works with the child and caregiver to identify, express, and learn to regulate a wide range of feelings and emotions. The clinician typically begins by having the child identify common, everyday feelings such as happiness or sadness and gradually moves toward exploring more intense feelings such as anger or guilt. In the parallel parenting sessions, the clinician will encourage the caregivers to actively listen to their children and praise any efforts to open up and verbally share emotions. It is important for both the child and caregiver to have a safe therapeutic atmosphere to express and process the difficult emotions that may follow a significant trauma. Any number of interactive games and activities are available to teach this skill. Multiple resources describing such activities and providing clinical examples of their use are available ([1, 22]; https://tfcbt.org/category/resources/).

In addition to labeling of feelings, it is important for children to learn how to rate the intensity of feelings so as to understand the moments when active coping skills may be required. This is usually accomplished by teaching the child to use a Subjective Units of Distress (SUD) scale, usually a rating of 1–10. The clinician works with the child to identify different experiences that fall along a continuum of intensity for a given emotion. For instance, receiving a piece of candy will likely result in the feeling of happiness, but at a low intensity, perhaps a 1 or 2. Going with the family on a vacation to a desired destination, such as a theme park, will also result in happiness, but at a much higher intensity, say an 8 or 9. Using various examples, the clinician teaches the child to make similar ratings for other emotions. Throughout future sessions, including during exposure exercises, the clinician asks the child to identify various emotions experienced either at the time of the trauma or presently in session, including providing an appropriate label and intensity.

A final skill taught during this component is linking emotion identification to the use of active coping skills. It is helpful for the clinician to connect this final skill to the teaching of rating the intensity of emotions. In essence, the child is asked to consider at what point on the continuum of intensity it is most helpful to begin implementation of coping skills. Ideally, the modulation of emotion will begin at the point where the child begins to experience some discomfort that might affect functioning and prior to the point where the child feels overwhelmed with emotion. After identifying this point, the clinician can then connect this skill to the use of the previously learned relaxation skills as well as introduce and teach additional coping skills that may be indicated.

8   Cognitive Coping and Processing Skills

Trauma can be a very scary and confusing experience for children and they will often try to make sense of what happened. This process can lead to inaccurate thoughts and distorted beliefs because of a child's limited knowledge and previous life experience. Cognitive coping encourages the individual to use various techniques to explore their thoughts and challenge and/or change those unhelpful cognitions. In this component of treatment, the clinician continues to work separately with the child and caregiver to make sure they understand the relationship between thoughts, feelings, and behaviors, and to begin identifying any unhelpful thoughts.

Cognitive coping begins with the clinician providing psychoeducation on thoughts and internal dialogue. This can be difficult as some children may not realize the things they are saying to themselves are actually thoughts. The clinician may start by encouraging the child to share his or her thought processes about a normal, everyday occurrence such as getting ready for school in the morning. Once the child has a good understanding of this concept, the clinician is then able to discuss different kinds of thoughts and problematic ways of thinking. The cognitive triangle is a helpful tool when teaching cognitive coping to children and adolescents (*see* Fig. 2). This is a simple visual diagram that can be used to illustrate the connection between thoughts, feelings and behaviors. The clinician can provide different scenarios and ask the child to identify various thoughts one may encounter, and then explore how the different thoughts may lead to different feelings and, ultimately, different behaviors.

A commonly used example is a child who is pushed from behind while in the hallway at school. The child is asked to offer an explanation for why the other child pushed him or her. The most common answer is that the other child was being aggressive, which prompts feelings of anger and potentially retaliatory behavior. The clinician then prompts the child to offer other explanations for the pushing. Many children have difficulty identifying such alternative explanations, but with guidance from the clinician, they can learn to explore other options, such as the child tripped and accidentally bumped into other people. Such an interpretation does not lead to a feeling of anger, but rather perhaps concern for the child or, at most, irritation. The behavior in response to these new feelings is rarely aggressive and may include helping behavior. At the end of this exercise the clinician has a useful diagram of how changing one's attributions for an event can lead to improved mood and behavior.

An Activating Event

(Example: Child pushed from behind at school)

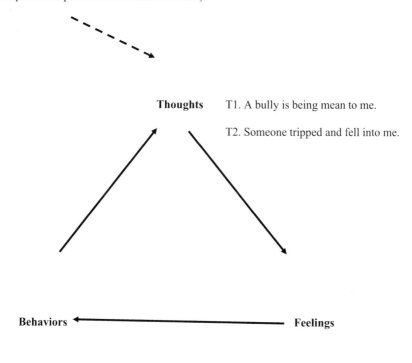

Thoughts   T1. A bully is being mean to me.

T2. Someone tripped and fell into me.

Behaviors ←————————————— Feelings

B1. Punch the bully and fight.   F1. Anger

B2. Help the other child up off the floor.   F2. Concern for other child

Fig. 2 A cognitive triangle example

9   Trauma Narrative and Cognitive Processing

Developing a factual and coherent account of the child's traumatic experience(s) is at the heart of TF-CBT. This exercise provides a focused exposure experience that is particularly helpful in cases of significant posttraumatic avoidance [17]. The specific medium of the narrative may take various forms, including drawing pictures with captions to describe the events in the picture, the youth dictating to the clinician who transcribes the account, and scripting a puppet show, among any number of other options. The clinician should allow the youth to decide the specific medium but also ensure that the child is capable of completing the narrative in the chosen format. Regardless of the manner in which the narrative is constructed, it is critically important that the account is factual to the best of the youth's recollection. Throughout this process the clinician asks probing questions to elicit the child's thoughts, feelings, and physical sensations during their experience, as well as the more objectively factual details of what occurred. Identifying the youth's reactions serves to detect potential maladaptive and/or conflictual thoughts as well as enhance the exposure process.

Constructing the narrative typically begins by having the youth review various aspects of himself or herself, such as favorite foods and memories, and then identifying the traumatic experiences for which the narrative will be developed. This demonstrates that the trauma is merely one small piece of the child's larger life and sets the stage for discussing what occurred. In most cases, the narrative is then described much like any other account: a beginning that describes what occurred before the trauma, a middle that includes the details of the trauma, and an ending that discusses what has occurred in the youth's life since and as a result of the trauma. There are other structures the narrative may take, particularly in cases of complex and sustained trauma histories where it may be difficult for the youth to provide a coherent discrete account. The interested reader is referred to Kliethermes and Wamser [23].

During the construction of the narrative, the clinician remains cognizant of the youth's level of distress and ascertains the child's identification and rating of the intensity of feelings. If the youth begins displaying concerning reactions, then the clinician should prompt the child to implement previously learned coping skills until distress is reduced to a sufficient degree. At that time, narrative construction may continue. However, if the clinician successfully utilized exposure activities throughout the earlier components of TF-CBT, it is typically the case that avoidance is significantly reduced from baseline by the time that narrative construction begins. Consider, for instance, this exposure progression of a 16-year-old female, who we will call Mary, who was raped by an unknown assailant at a party:

Intake: During assessment, Mary only nodded or shook her head when answering questions regarding the trauma. She refused to speak about it or provide any information. Exposure processes at intake involved the clinician repeatedly using the words "rape" and "sexual assault" to ask questions so that Mary heard the words multiple times.

Session 1: Mary continued to refuse to discuss the rape, but the clinician focused on delivering rapport-building activities as well as psychoeducation. The clinician chose materials from the National Center for PTSD (www.ptsd.va.gov) that discussed information about PTSD symptoms and Mary was asked to identify which symptoms she was experiencing. The clinician again repeatedly used the words "rape" and "sexual assault" to discuss the nature of her symptoms, such as nightmares and daytime intrusive thoughts. Mary continued to use head nods to confirm the nature of her symptoms. The clinician provided a treatment rationale for TF-CBT and discussed the reasons for exposure exercises. Mary looked intently at the clinician and asked if this is the reason for repeatedly using the "R-word."

Session 2: The clinician began teaching relaxation skills with a specific emphasis on Mary using the skills at moments when she thinks about the rape as well as when confronting other reminders. Mary was asked what other reminders might be present in her life, to which she described passing "the house where the 'r-word' occurred almost every day." This clinician then restated the phrase using the word "rape," to which Mary smiled and said "correct." The clinician pressed why, if that is the correct word, she would not say it herself. This resulted in a discussion of what the word "rape" meant to Mary and how scary it was for her to use the word in reference to herself. By the end of the session, Mary was successfully saying that she was raped at the party.

Sessions 3–8: Relaxation, affect modulation, and cognitive coping training proceeded as discussed above with the clinician increasing the level of exposure in accordance with Mary's progress. This included connecting the use of the skills to her passing the house where the rape occurred, which subsequently resulted in reduced anxiety, as well as discussing how the skills could be used when she thought about other reminders, such as sexual topics in general, or pointed details of her experience, including the voice of the perpetrator.

Sessions 9–14: By the time that narrative construction began in session 9, Mary was able to provide a coherent account of the day she was raped up to and including events that occurred at the party. As she began to recount the minutes prior to the assault, she displayed significant avoidance and distress. At this point the narrative process proceeded more slowly due to the utilization of coping skills throughout and a particular focus on eliciting her thoughts, feelings, and physical sensations during these important moments. It should be highlighted, however, that the clinician refrained from processing and attempting to restructure maladaptive cognitions until after the narrative construction was complete (see below). In session 10, the process was so difficult that Mary was only able to provide two sentences, both of which she wrote instead of verbalizing. However, the clinician read these sentences aloud so she would hear them. The following session she was able to verbally read the sentences herself, although with some discomfort. By session 12 she was able to say them with greater ease, and displayed relatively little challenge with this aspect of her experience by session 14.

Following completion of the narrative, the clinician utilizes cognitive restructuring techniques to modify any maladaptive cognitions identified throughout the process of treatment. These techniques should only be implemented after the completion of the trauma narrative as there may be multiple justifications underlying the child's belief. For instance, a child may blame him or herself for a tragic car accident because (a) he dreamed about such an accident prior to its occurrence, (b) he was distracting the driver in some way before and/or during the accident, and (c) after the accident, he ran away from the car as opposed to checking on the driver. However, these various contributing pieces to the overall self-blame cognition will likely be revealed during different portions of the trauma narrative. Attempting to address each thought as it occurs in the development of the narrative may result in a significantly longer period of time required for treatment and the clinician may be unaware and miss other rationales for the child's self-blame. In addition, completion of the narrative may offer any number of potential pieces of contradicting information that the clinician may use to challenge maladaptive cognitions.

To effectively accomplish the task of cognitive restructuring, it is important to distinguish between thoughts that are blatantly inaccurate and thoughts that may be accurate, but unhelpful, as such a delineation may help the clinician determine how best to proceed. For instance, in the case discussed above, Mary held the accurate, but unhelpful, thought that she would not have been raped had she not attended the party. It is, indeed, a factual statement that avoiding the party would have prevented the rape, but it is unhelpful to view avoiding social functions as necessary to preserve one's safety. In such instances, the clinician will likely meet with more success if the focus is on restructuring her thoughts around a particularly useful concept, in this case the issue of consent, which transcends location and time. In other cases, inaccurate thoughts may be directly challenged, such as her belief that she allowed the rape to occur, even though she repeatedly asserted that she did not want to have sex, fought the attacker, and screamed for help.

Caregivers are actively involved throughout treatment, and the trauma narrative component holds challenges for the clinician when navigating caregiver involvement. Although conjoint parent–child sessions are frequently used throughout TF-CBT (see below), this is not the case during this portion of treatment. The standard form of delivery is for the clinician to meet with the youth for the first half of session and meet with the caregiver during the second half. Hearing the child's account of their traumatic experience can be emotionally overwhelming for some parents and prompt maladaptive cognitions, such as self-blame or thinking that one has failed to protect one's child. The clinician should share the child's narrative alone with the caregiver each session to allow the caregiver to process their own thoughts and feelings. Doing this in separate portions of the session allows the caregiver to experience and discuss their reactions without worrying about the child's responses to seeing the parent in distress. Such an experience may cause the child to discontinue or resist completing the narrative for fear of upsetting the parent and the parent may be hesitant to discuss his or her own thoughts and feelings in front of the child.

10 In Vivo Exposure

Following the imaginal exposure of the trauma narrative process, the implementation of in vivo exposure exercises allows the child to demonstrate mastery over reminders in the physical world. The clinician works with the child and caregiver to develop a plan to systematically confront trauma reminders that elicit distress in the child. Completion of this plan often requires the child and caregiver to perform prescribed activities outside of session, and the clinician is advised to take appropriate steps to increase the likelihood of success. For instance, scheduling the day and time the activities will be performed may increase the likelihood that they are completed. In addition, the clinician prepares the child and caregiver for these exercises by considering various outcomes, how to handle different reactions using previously learned coping skills, and the importance of not reinforcing avoidance.

The in vivo exposure plan will be tailored and specific to each individual case. A child who was sexually abused by an uncle that drove a red car, for instance, may display significant distress at the sight of red cars. The in vivo plan in this case may involve driving into a parking lot and slowly passing red cars and building up to parking next to a red car for a period of time until the child reports significantly reduced levels of distress. Other common forms of anxiety typically treated with exposure activities are similarly addressed at this point in treatment. For instance, a common concern with younger children is separation anxiety. The clinician can work with the caregiver to develop a plan of systematically increasing the allowable distance between the caregiver and child in graduated increments.

11 Conjoint Parent–Child Sessions

Conjoint parent–child sessions are useful for a number of purposes within TF-CBT. Although placed in sequence near the end of the protocol, the clinician often employs conjoint sessions throughout treatment. Parents are actively involved at each phase of TF-CBT, with the caregiver learning the same skills taught to the child. For early sessions providing psychoeducation and coping skills training, the clinician may conduct conjoint sessions in place of seeing the parent only and allow the child to teach the caregiver what he or she just learned from the clinician. This is particularly useful with younger children who enjoy the opportunity to teach new skills to the parent. Conjoint sessions may also be helpful for demonstrating various parenting skills, improving communication between the parent and child, and during the psychoeducation component.

After narrative construction is completed, the child shows significant improvement, and the parent is fully aware of the details of the narrative and has processed their own reactions, a session-long conjoint session is held. During this session, the child shares the narrative with the caregiver, who remains supportive and asks scripted questions designed to convey a sense of acceptance and pride in the child's work. The clinician should prepare the caregiver and child for this activity during prior sessions and eliminate the potential for surprises to the extent possible. This session is meant to build social support for the child as he or she sees clearly that the traumatic experience can be discussed with the caregiver and that the caregiver will listen and remain supportive. The clinician should take a more facilitative, less-directive role during this discussion.

12   Enhancing Safety and Developmental Trajectory

Following the completion of exposure activities, and prior to the completion of treatment, the clinician reviews with the child and caregiver potential challenges that may arise in the future as well as teaches the child and caregiver safety skills to prevent future trauma exposure. Any number of techniques may be used to enhance safety, depending on the child's and family's needs and trauma experiences. A common technique is to develop a safety plan that the child can learn and use in the future if needed. Such plans may include listing specific people who the child can trust to speak with during upsetting or scary moments, as well as how to contact these individuals. In addition, classes of individuals that may be helpful to the child, such as teachers and police officers, may be included in the safety plan. Safety skills specific to the child's trauma exposure may be taught, such as body safety skills in the case of sexual abuse or an emergency plan in the case of house fires or disasters. Caregivers should be included in the development and teaching of safety plans and skills to the child, and the clinician should consider what caregiver-specific skills in this respect may be helpful to include.

It is important to consider future events that may cause the child to remember and think about the traumatic event. Anniversaries of natural disasters are often accompanied by television reports and discussions by family members and acquaintances. In situations of perpetrated violence, birthdays and holidays may be particularly difficult when the perpetrator is no longer present in the child's life. It is helpful for the clinician to process such future events with the child and caregiver and devise an appropriate plan to cope with these circumstances. Some situations may call for particular sensitivity and preparation. For instance, in the cases of sexual abuse or assault, future consensual sexual activity or intimate discussions with a romantic partner regarding one's history may be problematic. The clinician should discuss such situations with the youth, including when it may be appropriate and inappropriate to share such details with a partner (such a conversation may not be developmentally appropriate with younger children).

13   Supervision and Training

In many ways, TF-CBT resembles any number of other CBT interventions, including the use of psychoeducation, coping skills training, cognitive restructuring, and teaching the caregiver behavior management skills. However, completing a trauma narrative can be a difficult skill to learn for many clinicians. To artfully and effectively implement such exposure techniques, there are a number of nuances and issues to which the clinician must attend. The development of a web-based training platform improved the dissemination of didactic knowledge regarding the components of TF-CBT (www.musc.edu/tfcbt); however, completing such a course does not necessarily result in the development of practical skill in implementation. There is evidence that many clinicians who consider themselves trained in TF-CBT do not implement the TF-CBT protocol with fidelity, with the trauma narrative being the most frequently omitted component [24]. In addition, a recent field trial with community-based clinicians found that the integration of a clinician's preferred treatment techniques (e.g., non-directive play therapy, art therapy) into the TF-CBT protocol was common and actually served to reduce the effectiveness of treatment in some instances [25]. These findings suggest the importance of receiving appropriate training and supervision in the implementation of the TF-CBT protocol. A certification program is now available in the delivery of TF-CBT, and training programs are increasingly available (www.tfcbt.org/).

14   Discussion

Many interventions applied with children exposed to trauma seek the designations of "trauma-informed" and "evidence-based." This effort is important to clinicians and the settings in which they work because many referring agencies, such as pediatrician offices, child protective services, family court, law enforcement, and other community agencies, often require that the children and families they refer receive an intervention that carries these designations. For example, the National Children's Alliance accreditation standards require that children's advocacy centers provide or refer children and families with psychological needs to individuals or agencies that deliver trauma-informed, evidence-based care. However, these terms are often used arbitrarily and can refer to many different interventions that have a very limited evidence base, raising questions about quality of care these children and families may receive. For over three decades, TF-CBT has been evaluated in rigorous clinical trials research to build its evidence base as a trauma-informed treatment for children and families presenting with mental health concerns. The quality of this evidence has led to TF-CBT's designation as a "well-established" intervention for children exposed to trauma [26], a top-tier designation with strict criteria that requires extensive, high-quality research and is reserved for only the most effective interventions available for a specific patient population. Moreover, the research on TF-CBT has led to its designation as a recommended, first-line intervention for children exposed to trauma and diagnosed with PTSD [3, 27], as up to 75% of all children first diagnosed PTSD no longer meet criteria after receiving TF-CBT [16, 28]. Thus, TF-CBT is an intervention that can meet a clinician's or agency's goals to provide the most reliable and effective therapeutic benefits for a range of children and families seeking care following a trauma.

There are now widespread efforts to disseminate TF-CBT into community practice settings, the places where children who have been traumatized most often access mental health care. This access greatly assists a clinician's ability to get trained in TF-CBT. Moreover, training in TF-CBT requires much less time when compared to other approaches. In addition to the online training mentioned above, clinicians complete a live, 2-day training in the treatment protocol followed by a year of telephone consultation to implement the TF-CBT model into the practice or agency in which it is being provided. Ongoing supervision is a critical piece to ensuring that TF-CBT and its therapeutic effects are being delivered with model fidelity in the community. Research is showing that the positive effects of TF-CBT when conducted in research trials can be replicated in community settings when there is successful fidelity to the treatment protocol as it is prescribed [5, 29]. Thus, delivering all of the components of the TF-CBT protocol, often in the sequence in which they are described in the treatment manual, can help achieve positive effects for PTSD and co-occurring conditions for children and families seen in the community.

While the evidence-base is encouraging, there are still many areas for improvement in the delivery of TF-CBT. For instance, are all of TF-CBT's components necessary to achieve clinically significant improvements in PTSD? Attrition rates in community practice settings are often very high [30] and having an evidence base to suggest which of TF-CBT's components are most influential in improving PTSD would allow clinicians to prioritize delivery of these components before patients and families drop out. Also, many children who have been traumatized present with psychological concerns other than PTSD. Is TF-CBT as effective in treating the diverse range of mental health concerns common to children who have been traumatized when they are not experiencing PTSD? Finally, prior TF-CBT research has looked at overall average improvements in mental health outcomes following trauma exposure. Can the delivery of TF-CBT be personalized to fit the preferences and individual needs of each patient that is served? Treatment algorithms have been developed to help clinicians determine which type of intervention and at what dose is likely to provide the maximum benefit from TF-CBT [31]. Thus, there are many exciting areas of future research that aim to enhance the effectiveness and delivery of TF-CBT in real-world settings and according to individual patient characteristics.

Overall, TF-CBT was developed based on sound theory and clinical techniques that have been used and evaluated with many different children and adolescents across diverse family situations, trauma types, age ranges, and racial backgrounds. Time and time again, TF-CBT has helped to provide successful therapeutic gains in a developmentally and culturally competent manner without any additional adverse effects when compared to alternative treatments. This makes TF-CBT an effective, versatile, and safe intervention that is readily available for implementation by clinicians across many different mental health provider settings.

References

1. Cohen JA, Mannarino AP, Deblinger E (2017) Treating trauma and traumatic grief in children and adolescents, 2nd edn. Guilford Press, New York, NY

2. Allen B, Gharagozloo L, Johnson JC (2012) Clinician knowledge and utilization of empirically-supported treatments for maltreated children. Child Maltreat 17:11–21

3. Cohen JA, The Work Group on Quality Issues (2010) Practice parameter for the assessment and treatment of children and adolescents with posttraumatic stress disorder. J Am Acad Child Adolesc Psychiatry 49:414–430

4. Cohen JA, Mannarino AP (1996) A treatment outcomes study for sexually abused preschool children: initial findings. J Am Acad Child Adolesc Psychiatry 35:42–50

5. Cohen JA, Mannarino AP, Iyengar S (2011) Community treatment of posttraumatic stress disorder for children exposed to intimate partner violence. Arch Pediatr Adolesc Med 165:16–21

6. Deblinger E, Lippmann J, Steer RA (1996) Sexually abused children suffering posttraumatic stress symptoms: initial treatment outcome findings. Child Maltreat 1:310–321

7. Scheeringa MS, Weems CF, Cohen JA et al (2011) Trauma-focused cognitive-behavioral therapy for posttraumatic stress disorder in three-through six year-old children: a randomized clinical trial. J Child Psychol Psychiatry 52:853–860

8. Bigfoot DS, Schmidt SR (2012) American Indian and Alaska Native children: honoring children-mending the circle. In: Cohen JA, Mannarino AP, Deblinger E (eds) Trauma-focused CBT for children and adolescents: Treatment applications. Guilford Press, New York, NY

9. De Arellano MA, Danielson CK, Felton JW (2012) Children of Latino descent: culturally modified TF-CBT. In: Cohen JA, Mannarino AP, Deblinger E (eds) Trauma-focused CBT for children and adolescents: Treatment applications. Guilford Press, New York, NY

10. Murray LK, Skavenski S, Kane JC et al (2015) Effectiveness of trauma-focused cognitive-behavioral therapy among trauma-affected children in Lusaka, Zambia: a randomized clinical trial. JAMA Pediatr 169:761–769

11. Mowrer OH (1960) Learning theory and the symbolic processes. Wiley, New York, NY

12. Beck AT, Rush AJ, Shaw BE et al (1979) Cognitive therapy of depression. Guilford Press, New York, NY

13. Ehlers A, Clark DM (2000) A cognitive model of posttraumatic stress disorder. Behav Res Ther 38:319–345

14. Foa EB, Kozak MJ (1986) Emotional processing of fear: exposure to corrective information. Psychol Bull 99:20–35

15. Resick PA, Schnicke MK (1992) Cognitive processing therapy for sexual assault victims. J Consult Clin Psychol 60:748–756

16. Cohen JA, Deblinger E, Mannarino AP et al (2004) A multi-site, randomized controlled trial for children with sexual abuse-related PTSD symptoms. J Am Acad Child Adolesc Psychiatry 43:393–402

17. Deblinger E, Mannarino AP, Cohen JA et al (2011) Trauma-focused cognitive-behavioral therapy for children: Impact of the trauma narrative and treatment length. Depress Anxiety 28:67–75

18. Jensen TK, Holt T, Ormhaug SM et al (2014) A randomized effectiveness study comparing trauma-focused cognitive-behavioral therapy with therapy as usual for youth. J Clin Child Adolesc Psychol 43:356–369

19. Ormhaug SM, Jensen TK, Wentzel-Larsen T et al (2014) The therapeutic alliance in treatment of traumatized youths: relation to outcome in a randomized clinical trial. J Consult Clin Psychol 82:53–64

20. Deblinger E, Steer RA, Lippmann J (1999) Two-year follow-up study of cognitive-behavioral therapy for sexually abused children suffering post-traumatic stress symptoms. Child Abuse Negl 23:1371–1378

21. Holmes MM (2000) A terrible thing happened. Magination Press, Washington, DC

22. Cohen JA, Mannarino AP, Deblinger E (eds) (2012) Trauma-focused CBT for children and adolescents: Treatment applications. Guilford Press, New York, NY

23. Kliethermes M, Wamser R (2012) Adolescents with complex trauma. In: Cohen JA, Mannarino AP, Deblinger E (eds) Trauma-focused CBT for children and adolescents: treatment applications. Guilford Press, New York, NY

24. Allen B, Johnson JC (2012) Utilization and implementation of trauma-focused cognitive-behavioral therapy for the treatment of maltreated children. Child Maltreat 17:80–85

25. Allen B, Hoskowitz NA (2017) Structured trauma-focused CBT and unstructured play/experiential techniques in the treatment of sexually abused children: a field study with practicing clinicians. Child Maltreat 22:112–120

26. Dorsey S, McLaughlin KA, Kerns SEU et al (2017) Evidence base update for psychosocial treatments for children and adolescents exposed to traumatic events. J Clin Child Adolesc Psychol 46:303–330

27. Bisson JI, Ehlers A, Matthews R et al (2007) Psychological treatments for chronic post-traumatic stress disorder. Systematic review and meta-analysis. Br J Psychiatry 190:97–104

28. Cohen JA, Mannarino AP, Perel JM et al (2007) A pilot randomized controlled trial of combined trauma-focused CBT and sertraline for childhood PTSD symptoms. J Am Acad Child Adolesc Psychiatry 46:811–819

29. Webb C, Hayes A, Grasso D et al (2014) Trauma-focused cognitive behavioral therapy for youth: effectiveness in a community setting. Psychol Trauma 6:555–562

30. De Haan AM, Boon AE, de Jong JTVM et al (2013) A meta-analytic review on treatment dropout in child and adolescent outpatient mental health care. Clin Psychol Rev 33:698–711

31. Lang JM, Ford JD, Fitzgerald MM (2010) An algorithm for determining use of trauma-focused cognitive–behavioral therapy. Psychother Theor Res Pract Train 47:554–569

Chapter 6

Cognitive Behavioral Therapy with Angry and Aggressive Youth: The Coping Power Program

Sara L. Stromeyer, John E. Lochman, Francesca Kassing, and Devon E. Romero

Abstract

This chapter will discuss cognitive behavioral therapy for anger and aggression in youth, focusing on the Coping Power Program. The contextual social-cognitive model of anger and aggression, as well as a variety of risk factors, will be discussed, which present a conceptual framework for the Coping Power Program. An overview of other CBT programs for anger and aggression in youth will be presented, followed by a detailed description of the Coping Power intervention, including sample dialogue and handouts. The Coping Power Program includes both a child and a parent component and was originally developed for late elementary to early middle school children. Coping Power addresses topics including children's emotion awareness, ability to cope with anger and frustration, and social problem-solving skills, as well as parenting skills related to building positive family relationships and managing children's behavior. A number of adaptations have examined the Coping Power Program with individuals instead of groups, upwardly and downwardly extended age groups, altered intervention length, and different settings, all collectively indicating that the intervention can be flexibly applied. Research on Coping Power's efficacy and effectiveness will be discussed, followed by implications for dissemination and optimization of the intervention. Information about intervention manuals, workbooks, and training is available at www.copingpower.com.

Key words Aggressive behavior, Anger, Cognitive behavioral therapy, Coping Power program, Contextual social-cognitive model, Adaptations, Dissemination, Optimization

1 Introduction

The Coping Power program is an effective cognitive behavioral intervention for anger and aggression in youth. This chapter will first highlight the contextual social-cognitive model that the program is based on, as well as a variety of risk factors for anger and aggression. Empirical support for the Coping Power program will be presented, including considerations for dissemination and optimization. Finally, key components of cognitive behavioral interventions for anger and aggression will be discussed, followed by a detailed description of the Coping Power program.

Robert D. Friedberg and Brad J. Nakamura (eds.), *Cognitive Behavioral Therapy in Youth: Tradition and Innovation*, Neuromethods, vol. 156, https://doi.org/10.1007/978-1-0716-0700-8_6, © Springer Science+Business Media, LLC, part of Springer Nature 2020

1.1 Anger and Aggression in Youth

Anger and aggression are often considered within the broader category of externalizing problems, which have been consistently linked with negative outcomes, such as delinquency, substance use, conduct problems, academic difficulties, and poor adjustment [1]. As a result, childhood anger and aggression have been a focus of preventive intervention and treatment in order to disrupt trajectories of further development of rule- and norm-violating behaviors later into adolescence and adulthood.

Some level of anger and aggression may reflect appropriate responses to threatening environmental contexts that serve to protect an individual; however, it is necessary to distinguish between these situations and those that are atypical and/or inappropriate. The developmental context is also important to consider, given that younger children are more likely to exhibit anger when a goal is blocked, whereas older children are more likely to exhibit anger, and even aggression, when they perceive threats to their self-esteem [2].

2 Theoretical and Empirical Foundations

2.1 The Contextual Social-Cognitive Model

Childhood aggressive behavior is thought to result from a set of contextual and personal factors and potentially lead to adolescent and adult antisocial behavior [3]. The contextual social-cognitive model of prevention highlights two main mediators of adolescent antisocial behavior that may be targets for intervention: child-level factors (e.g., low social competence and social-cognitive skills) and parent-level factors (e.g., poor caregiver involvement and discipline).

The contextual social-cognitive model of prevention was derived from research supporting a model of social information processing [4]. Children who exhibit higher levels of aggressive behaviors have been shown to demonstrate deficits and cognitive distortions during both phases of the social-cognitive model: *appraisal* and *problem solution*. During *appraisal*, children struggle to accurately encode social information and interpret social events and intentions of others. Research has demonstrated that children with deficiencies during the appraisal stage recall fewer relevant event details, use fewer cues to form interpretations of others, and are more likely to attend to hostile versus neutral cues [5]. These children may also approach social interactions with a hostile attribution bias, or assumptions that others act in hostile or provocative manners [5].

During *problem solution*, children may develop maladaptive solutions to potential problems and overestimate the success of aggressive solutions to their perceived problems. Research suggests that children who exhibit aggressive behavior are more likely to offer aggressive and direct-action solutions than verbal problem solutions (e.g., verbal assertion, compromise; [6]). It is hypothesized that aggressive children are more likely to offer aggressive solutions in part due to their expectations that aggression will lead to desired outcomes [5].

In addition to children's social-cognitive processes, such as those described above, childhood aggressive behavior may also develop from early contextual parental factors. Such parental factors may include harsh discipline, poor parental problem solving, vague commands, and poor supervision or monitoring of children's behaviors [7]. Research has also consistently linked children's externalizing behaviors with low levels of parental warmth and positive involvement and high levels of punitive discipline (e.g., spanking, verbal aggression; [8]). In addition to direct effects of the parenting context on children's aggressive behavior, parent and broader family processes may also affect children's social-cognitive processes. For example, many of the parental risk factors described above (i.e., harsh and inconsistent discipline, low parental involvement, vague commands, and poor monitoring) have been linked with deficiencies in emotion regulation and social-cognitive processing, which may contribute to aggressive behavior [9]. Research has also demonstrated that the relationship between child and parent-related contextual risk factors is bidirectional, with ineffective parenting leading to children's externalizing behaviors, which in turn may also lead to the deterioration of parenting practices in response to difficult and problematic child behaviors [10].

2.2 Other Risk Factors

In addition to children's social-cognitive processes and parental context, there are other child-centered and contextual factors that have been shown to influence the development of anger and aggression. For example, several biological and temperamental factors, including infant fussiness, activity level, and positive affect, predict later displays of conduct problems [11]. Aggressive behavior in children has also been associated with certain patterns of physiological functioning (e.g., lower resting heart rate; [12]) and neurobiology (e.g., dysfunctions of the amygdala and prefrontal cortex; [13]). While previous research has highlighted associations between biological factors and aggressive behavior, these relationships are often moderated by environmental factors. For example, research has demonstrated that children with certain genetic risk factors are more likely to develop conduct problems in adolescence *only* in the presence of environmental risk factors (e.g., parental neglect, interparental violence, and inconsistent discipline; [14]). Thus, it is clear that contextual factors can impact the relationships between biological risk factors and externalizing behavior.

In addition to the parental factors described above, there are other family contextual factors that have been shown to influence aggressive behavior. For example, family structure (e.g., coming from a single-parent household), family poverty, marital conflict, and weak attachments to parents have been associated with conduct problems in children [15]. Other family factors, such as parental criminality, substance abuse, and depression have also been associated with aggressive behavior [16]. Beyond the family context, neighborhood and community characteristics, such as neighborhood disadvantage have been supported as risk factors for the development of aggressive behavior [17]. Furthermore, school factors, such as low-quality classroom context, have been linked with children's disruptive behavior in classroom settings when combined with other risk factors (e.g., attention problems; [18]). Given these multifactorial risk factors and processes for the development of anger and aggression and children, the Coping Power program involves multiple targets and components.

2.3 Efficacy and Effectiveness

To date, research has demonstrated a wide range of support for the Coping Power Program. In an initial study examining efficacy, 183 aggressive boys were randomly assigned to receive the full Coping Power program (child and parent components), the child component only, or no treatment [19]. Following intervention, boys in both Coping Power groups displayed significantly fewer hostile attributions of other's behavior and expectations that aggression would lead to a successful outcome, as well as a higher internal locus of control. At one-year follow-up, participants (particularly in the full Coping Power program) demonstrated lower rates of covert delinquent behavior, lower parent-reported substance abuse, and improved teacher-rated school behavioral functioning than the control group [20]. In a subsequent effectiveness study, 245 aggressive boys and girls participated in the full Coping Power program. Following intervention, participants demonstrated lower ratings in parent- and teacher-rated proactive aggression, as well as improvements in social competence, problem-solving, and anger coping skills [21]. Furthermore, at one-year follow-up, Coping Power resulted in lower rates of substance use and self-reported delinquency compared to the control group [9].

Further effectiveness studies have examined the Coping Power Program's effects in a variety of populations and international samples. For example, the Coping Power intervention has demonstrated positive effects with aggressive deaf children [22] and preventative effects on early adolescent substance use and delinquent behavior in Dutch children with disruptive behavior disorders [23]. In another Dutch sample of children 9- to 16-years-old with mild to borderline intellectual disabilities, Coping Power led to reduced children's externalizing behavior and improved parenting behavior and parent–child relationships [24]. With a sample at higher risk for violence exposure in Pakistan, Coping Power demonstrated reduced aggressive behavior and improvements in behavior, social skills, and social cognitive processes [25]. In Sweden, researchers have examined combining Coping Power with Parent Management Training, noting reductions in behavior problems and improvements in prosocial behavior, as well as parenting skills [26]. Finally, Italian researchers have downwardly extended Coping Power to preschool children, finding lower levels of parent- and teacher-rated behavioral problems post-intervention [27]. Given these findings, Coping Power has demonstrated the ability to be effectively applied with different samples.

2.4 Dissemination and Optimization

2.4.1 Dissemination

When assessing the effectiveness of interventions, it is especially important to consider an intervention's ability to be disseminated to the real-world, or widely circulated to a broader audience. Characteristics of the training process, and of the clinician and organizational characteristics can influence implementation of programs and should be examined within an implementation science perspective. In a prevention-oriented dissemination trial, counselors from 57 elementary schools in Alabama were randomly assigned to one of three conditions: Coping Power-Intensive Training, Coping Power-Basic Training, or Care-as-Usual [28]. Counselors in both training conditions received three days of training prior to the intervention as well as monthly additional trainings throughout the intervention. The Intensive Training group also had access to individualized problem solving via e-mail or a telephone hotline, as well as received feedback based on audio recordings of individual sessions. Training intensity was found to have a significant impact on outcomes, with the Coping Power-Intensive Training group showing significantly greater reductions in teacher-, parent-, and self-reported externalizing behaviors and greater improvements in social and academic behaviors than the other groups [28]. Significant improvements in children's behavior only occurred in the Intensive Training groups, emphasizing the importance of intensive training and availability of feedback to counselors throughout the intervention. A follow-up study of this dissemination trial found that children with counselors who received the Coping Power-Intensive Training had better outcomes in language arts grades than children with counselors in the other groups after 2 years [29]. Clinician and organizational characteristics also affected implementation, as clinicians who had personality traits of being more conscientious and agreeable, and who came from schools with less rigid managerial control and more principal and teacher support, implemented the program with greater quality and sustained their use of the program in later years more completely [30, 31].

2.4.2 Optimization

A critical next step in intervention research, after the evidence-base of an intervention has been established, is to develop methods for optimizing the intervention to enhance its ability to be successful with more subgroups of clients, and then to rigorously test those adaptations. A variety of adaptations are currently being explored for Coping Power. These include integration of Coping Power with other interventions and adaptations to the delivery format and setting.

Program integration. A prevention study is being completed underway to test the effects of integrating contemplative practices (mindfulness, yoga, and compassion-building practices) with the cognitive behavioral practices in traditional Coping Power (CP). This study examines whether the Mindful Coping Power (MCP) program further enhances program effects on child and parent emotional reactivity relative to standard CP. MCP is being directly compared to CP in a randomized feasibility trial with 96 child/parent dyads. MCP has been well-received by child and parent participants. Children have improved their emotional, cognitive and behavioral regulation better in MCP than in CP, and self-reported stronger improvements in their reactive aggression [32].

Clinicians and researchers recognize the difficulty in engaging parents in preventive interventions and maintaining their involvement across time. One proven approach to enhancing parent engagement is the Family Check-Up [33]. In the Family Check-Up, families are assessed and then receive feedback on strengths and areas of concern that can be targeted for change. The Coping Power parent component has recently been adapted for tailored, modular use, to address only the specific areas of need identified for a family, using the Family Check-Up [34]. Seven modules have been created from the Coping Power parent component, which can be used alone or in combination. These include modules for academic support in the home, stress management, positive attention and praise, antecedents, consequences, and family cohesion and problem-solving. This form of tailored, or adaptive, intervention is designed to offer parents only the training relevant to their concerns and may also reduce the overall length of intervention.

Adaptations to the delivery format and setting. A central concern for parents and clinicians in schools and mental health settings has been the length of time required to implement evidence-based prevention programs, reducing schools' likelihood of adopting the programs and reducing parents' attendance and engagement in the program. Interventions often need to be modified to meet the realities of treating children and families in applied settings [35]. One area of difficulty may be delivering lengthy evidenced-based programs efficiently, yet effectively. Two randomized controlled trials have examined outcomes of an abbreviated version of the Coping Power Program that can be implemented in one school year (24 child sessions and 10 parent sessions). Three years after the end of intervention, abbreviated Coping Power in comparison to the control condition, had more generalized long-term effects on reducing children's externalizing behavior problems regardless of the degree of parental attendance and also showed reductions in impulsivity traits and callous-unemotional traits [36].

To make the intervention even briefer, while incorporating technology that can be engaging to children, a recent randomized prevention trial with 96 child/parent dyads tested the feasibility and preliminary effects of a hybrid web and in-person delivery format for Coping Power [37]. This more efficient format allows the full content to be delivered in a smaller number of in-person sessions, while children and parents have access to the content of the full Coping Power program through interactive web and video content, including the *Adventures in Captain Judgment* animated video series. This much briefer, hybrid version of Coping Power was found to have beneficial preventive effects by eliminating the significant increase in children's conduct problems in the school setting displayed by control children. The size of these effects was similar to the effects of the longer version of Coping Power. The website materials appear to successfully engage children. Parents' use of the website predicted children's changes in conduct problems.

Another area that may pose a challenge is the delivery format. There is often a concern of iatrogenic effects when adapting an intervention targeting disruptive behavior for a group delivery format. Due to Coping Power typically being delivered in small groups, a large-scale study was conducted to investigate how children fare if they received Coping Power in a group versus individual format [38]. Results indicated both intervention delivery methods led to similar significant reductions in parent-rated externalizing problems through a one-year follow-up period. However, although teacher-rated externalizing problems also declined significantly for both intervention conditions, the reductions were significantly greater for children receiving Coping Power in an individual format. The main effect was moderated by children's baseline levels of inhibitory control. Children with fewer problems with inhibitory control responded in similar positive ways to either the group or individual format. Children who were more oriented to social rewards (indexed by an oxytocin receptor gene) and who more emotionally dysregulated (indexed by the autonomic nervous system indicators) also have weaker teacher-rated outcomes through a 1-year follow-up [39, 40]. These results are encouraging in terms of effectively adapting an evidence-based program to the specific needs of an applied setting, whether it be working individually or in small groups with children.

3 Intervention

There are a number of cognitive behavioral therapy programs that have been developed for anger and aggression in youth. Common elements include emotion awareness and regulation strategies, skills training and practice, and caregiver involvement in behavior

3.1 Overview of Cognitive Behavioral Therapy for Anger and Aggression

management. Although this chapter focuses on the Coping Power Program, other effective programs deserve mention as well. Several programs initially evolved out of the Hanf [41] approach and include both parents and children. For example, The Incredible Years program is designed for children aged 3–8 years and has both child and parent components with videotape modeling and group discussion [42]. Helping the Noncompliant Child was developed to address noncompliance in children aged 3–8 and is typically employed with individual families and involves in-session modeling [43]. Parent-Child Interaction Therapy, for 2- to 8-year-olds, contains many of the same components as Helping the Noncompliant Child, with a particular focus on the parent-child relationship [44]. Some interventions focus on parents, such as Parent Management Training [45] and the Triple P-Positive Parenting Program [46]. Parent Management Training is employed with parents of children aged 3–12 and involves weekly meetings plus phone contact, while Triple P is a multilevel intervention ranging from universal prevention to individual treatment for serious conduct problems, with extensions from birth to age 16. Other programs focus on children, such as Problem-Solving Skills Training [47], which implements cognitive behavioral skills with youth aged 7–13 years. Additionally, the Anger Coping Program is a group, school-based cognitive behavioral intervention for aggressive children from fourth to sixth grade [48]. The 18-session Anger Coping Program was eventually expanded to become the Coping Power Program, which will be discussed in detail below.

3.2 Coping Power Program

Coping Power is an evidence-based, multi-component, cognitive behavioral preventative intervention program based on the contextual social-cognitive model [19]. It was developed for preadolescent and early adolescent children (late elementary to early middle school) displaying significant anger and aggressive behavior, and thus at-risk for the development of later conduct problems and substance use. Support for the Coping Power program has been demonstrated through both efficacy and effectiveness studies.

Coping Power includes a child component, which consists of a 34-session individual or group intervention, and a complementary 16-session parent group component. Both the child and parent components are designed to be delivered concurrently over a 16- to 18-month period. In its original format, children participate weekly in a group at school and also meet with the therapist monthly on an individual basis. The child and parent program manuals are highly structured, providing detailed instructions and clear guidance for disseminating and completing session objectives and activities.

3.2.1 *Child Component*

Early sessions (Sessions 2–3) of the Coping Power child component [49] focus on long-term and short-term goal setting. Throughout the program, clearly observable and appropriate behavioral goals are identified by the child with the assistance of the therapist and monitored on a daily basis by the teacher or parent with the use of goal sheets. Each session, the child can earn points on the basis of their success. Use of goal-sheet tracking encourages generalization of treatment effects to the classroom or home. Other main areas of focus include organization and study skills (Session 4, 24), emotional awareness (Sessions 5–6), anger management and self-regulation (Sessions 7–11), perspective-taking to reduce hostile attributions (Sessions 12–15), social problem-solving and consequence identification (Sessions 16–22, 25), entering prosocial peer groups and cooperating with peers (Sessions 26–28), and the management of peer pressure (Sessions 29–33). A description of specific Coping Power units and activities that address angry and aggressive children's deficits in emotion awareness, self-regulation, perspective taking, and social problem solving are provided below.

Awareness of feelings and physiological arousal related to anger. The two sessions included in this unit of the intervention program are devoted to teaching children to recognize the physiological cues of varying levels of anger and to utilize a range of coping skills to manage anger more effectively. This unit begins with a focus on identifying the behavioral (e.g., pushing, screaming), cognitive (e.g., "He did that on purpose", "I hate my sister"), and physiological (e.g., heart racing, face gets hot) components of various types of emotions (e.g., sad, happy, afraid), especially their experience of anger. Using a thermometer analogy, children learn to assess various levels of anger intensity they may experience (e.g., annoyance, frustration, rage) and different types of problems that trigger these different levels of anger. Children then practice noticing and describing situational triggers and anger intensities (e.g., "I felt medium anger when my classmate started to tease me about my shoes at lunch", "I felt a little bit annoyed when my baby sister kept crying when I was trying to sleep", "I was fuming hot when my cousin broke my brand new video game"). During this time of the intervention program, children are provided thermometer worksheets for homework (Fig. 1). This assignment is used to track feelings of anger each day, for one week. Children are asked to identify anger intensity they experience each day and to identify the general trigger for this feeling.

Practice anger coping and self-control. The remaining five sessions of the unit described in the previous section focus on teaching and practicing emotion regulation strategies such as distraction, relaxation, and coping self-statements. Distraction techniques are taught as a means to divert attention away from anger-provoking

MONDAY

| INTENSITY
(CIRCLE)

VERY HIGH

HIGH

MEDIUM

LOW

VERY LOW | 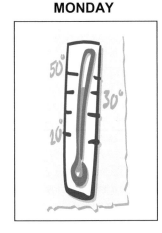 | WHY AM I
ANGRY?

——————————

——————————

——————————

——————————

—————————— |

Fig. 1 Segment of anger thermometer worksheet

situations. Guided imagery, deep breathing, and progressive muscle relaxation techniques are used to promote regulation of physiological responses and prevent anger escalation. Finally, coping self-statements (e.g., "It's not worth getting angry"; "I'm not going to let this get to me"; "I'll grow up, not blow up"; "I can keep my cool and handle this") are taught through a series of in vivo activities designed to be increasingly more realistic. Coping statements are things that we say to ourselves (i.e., thoughts that we think to ourselves) that help us stay calm or calm down.

Children first use anger management strategies during puppet role-plays and finally use the skills during taunting tasks with peers. To further illustrate, during the puppet role-plays, children use puppets to tease another child's puppet. The child holding the puppet being teased is instructed to state aloud the coping statements the puppet is using to avoid anger escalation. After each turn the group leader can then ask one or more of the following questions: (1) "What was the puppet thinking or saying to himself/herself?" (2) "What level of anger did the puppet experience during the taunting?" (3) "What coping statements did the puppet use to cope with his/her anger?" (4) "Did the puppet use different coping strategies or coping statements for different levels of anger?"

Homework for these subsequent sessions uses a variation of the thermometer worksheet to reinforce new skills as they are learned. For example, during week 2 of this unit, in addition to the homework prompts described above, children are also instructed to identify what they did to cope. For week 3 of this unit, children are asked to identify more specifically the coping statements they used to cope.

Perspective Taking. Role-playing and discussion activities are used to demonstrate how difficult it often is to accurately determine and understand others' intentions, over the course of four sessions. As an activity, the clinician can (a) have children role-play different characters in an ambiguous situation or (b) show an image of a scenario involving one or more individuals to facilitate discussion. During these activities, children are encouraged to consider the different viewpoints of each of the characters portrayed. Although the primary focus of these activities is on peer situations, aggressive children also often misperceive teachers' intentions regarding classroom management, teacher expectations, and disciplinary procedures. As a homework assignment, children are asked to schedule and conduct an interview with a teacher of their choosing. The reason for this assignment is to assist children in understanding that teachers also have different perspectives.

Social problem-solving skills. After children learn to manage their anger and take time to view problem situations from others' perspectives, they are taught additional steps for solving problems. Children are introduced to a problem-solving model throughout four sessions in which problem-solving steps are summarized by the acronym PICC. Problem solving steps of the PICC model include: (1) Problem Identification, which entails noticing that there is a problem, picking the problem apart, and describing the problem as specifically as possible, (2) Brainstorming a variety of Choices to solve the problem, and (3) identifying the possible Consequences of each choice and using that information to determine which choice to try first. A series of seven to eight sessions addresses how the PICC model could be used in common social problem situations with teachers, peers, and family. This unit of the intervention involves the steps and the use of the PICC problem-solving model with children's own problems. Brainstorming discussions, as well as hands-on activities and PICC worksheets (Fig. 2), are used to introduce the range of choices, or possible solutions, that children have in most social problem situations, and then to introduce the range of positive and negative consequences that result from these various choices.

3.2.2 Parent Component

The Coping Power parent component [50] focuses on academic support in the home (Sessions 1–2, 12), parent stress management (Sessions 3–4), identifying positive and negative child behaviors and praising appropriate child behaviors (Session 5), ignoring minor disruptive behaviors (Session 6), giving clear and effective instructions (Session 7), establishing age-appropriate rules and expectations (Session 8), giving effective and appropriate discipline and punishment (Sessions 9–10), and family cohesion and communication (Sessions 13–15). In the program, parents learn skills to support and promote the social-cognitive and problem-solving skills that their children learn in the child component. For example, parents learn to apply the PICC model to problems in the home context so that skills learned in the child component will be reinforced at home. Similar to the child component, parents also learn stress and mood management and the influence that thoughts have

The problem is: _____

_____

Possible choices/solutions: **Consequences of the choices/solutions:**

1._____   1._____

2._____   2._____

3._____   3._____

The choice I think will work out best is:

_____

_____

Reporting back: Try out your choice and describe how it actually went

_____

_____

Fig. 2 PICC model problem solving worksheet

on how a person feels, acts, or responds to a situation. Descriptions of specific Coping Power activities that foster positive family relationships and manage children's behavior using basic principles of social learning are provided below.

In the child component, children learn a variety of skills based on social learning theory that they may use to control their behavior and manage their anger in their interactions with others. Relatedly, parent interactions with their children have an influence on their child's behavior. As such, parents are taught the basic principles of social learning. Specifically, the A-B-C (antecedents-behavior-consequences) model is introduced and is used in all subsequent parent sessions. *Antecedents* are the events that happen just before the child's *behavior*, and *consequences* are the events that happen just after the child's *behavior*. As new parenting skills are introduced throughout the program, they are added to the appropriate column on an A-B-C chart. For example, rewarding and praising their child for positive behaviors is added to the C (consequences) section. Other examples of age-appropriate consequences discussed in the program include ignoring minor disruptive behaviors, the use of time-out, work chores, and privilege removal.

Immediately following the introduction of the A-B-C model, parents are asked to consider which problem behaviors they are currently observing in their child and the positive behaviors they

BEHAVIOR	Monday	Tuesday	Wednesday	Thursday	Friday	Saturday	Sunday
Problem Behavior	Happened? YES NO	Happened? YES NO	Happened? YES NO	Happened? YES NO	Happened? YES NO	Happened? YES NO	Happened? YES NO
Positive Behavior	Happened? YES NO Praised? YES NO	Happened? YES NO Praised? YES NO	Happened? YES NO Praised? YES NO	Happened? YES NO Praised? YES NO	Happened? YES NO Praised? YES NO	Happened? YES NO Praised? YES NO	Happened? YES NO Praised? YES NO

Fig. 3 Positive and negative behavior tracking worksheets

would like to see more of. To facilitate discussion, parents complete a checklist of positive (e.g., expresses anger appropriately, follows directions) and negative (e.g., curses, defies authority) child behaviors with reference to their particular child. For homework (Fig. 3), parents refer back to this checklist and select negative behaviors and their positive opposites to track. Parents should indicate whether the behavior was observed and whether praise or some other form of positive consequence was used.

Through discussion and role-plays, parents learn to modify their child's behavior by using positive consequences to reward good behavior. Parents also learn to "catch" their children being good with further emphasis on rewarding and using labeled and unlabeled praise as a positive consequence. Labeled praise is a praise statement that states the good behavior exactly and is useful when teaching a new skill or reinforcing a behavior that the parent wants to increase. Examples are: (a) "You did a good job of taking out the garbage when I asked you to.", (b) "Thank you for playing nicely with your little brother for 30 minutes." and (c) "I like the way you got your homework done on time tonight." Unlabeled praise is a praise statement that tells the adolescent that their behavior was good, but it does not specify the exact behavior that was noticed (e.g., "Good job."; "Thank you.", "I like that."). Both forms of praise are effective, though each may be more or less appropriate to any given situation.

Defiant behavior can lead to negative social interactions and create a negative family environment. Thus, the final component of the A-B-C model covers the use of praise and giving effective instructions as parenting skills that can help increase compliance. Instructions are the antecedents to child compliance or noncompliance. To further illustrate, leaders present examples of "good" instructions (i.e., those that elicit compliance) and "bad" instructions (i.e., those that elicit noncompliance from children). Some characteristics of bad instructions are repeating instructions over and over again and giving instructions in the form of a question rather than in the form of a declaration. A few qualities of good instructions are giving no more than one or two instructions at a time and following instructions with a 10 s period of silence so that the child has an opportunity to comply. Consistent with other sessions, parents discuss, role-play, and complete homework assignments to develop and improve parenting skills.

4   Conclusion

In conclusion, the Coping Power program has been demonstrated as an effective intervention for anger and aggression in youth, with outcomes ranging from reducing existing aggression and disruptive behavior to improving deficits and distortions in social-cognitive processes, to preventative effects on later delinquent behavior and substance abuse. The intervention is highly structured, provides detailed instructions and a selection of activities for completing session objectives, and addresses multiple risk factors associated with aggressive behavior through targeted session content, using a contextual social-cognitive framework. As demonstrated, Coping Power can be flexibly applied with different age groups, populations, and settings, as well as with individuals instead of groups, and with a shorter number of sessions. More recent studies are examining integrating the intervention with a Mindfulness component to boost outcomes, with the Family Check-Up to enhance parent engagement, or with interactive web and video content to increase efficiency. An important consideration for dissemination and optimization is intensive training and regular feedback for clinicians implementing the program, as well as some clinician and organizational characteristics. Information about intervention manuals, workbooks, and training is available at www.copingpower.com.

References

1. Kaplow JB, Curran PJ, Dodge KA et al (2002) Child, parent, and peer predictors of early-onset substance use: a multisite longitudinal study. J Abnorm Child Psychol 30:199–216
2. Lochman JE, Powell N, Clanton N et al (2006) Anger and Aggression. In: Bear G, Minke K (eds) Children's needs III: development, prevention, and intervention. National Association of School Psychologists, Washington, DC
3. Conduct Problems Prevention Research Group (1992) A developmental and clinical model for the prevention of conduct disorder: the FAST track program. Dev Psychopathol. https://doi.org/10.1017/S0954579400004855
4. Crick NR, Dodge KA (1994) A review and reformulation of social information-processing mechanisms in children's social adjustment.

Psychol Bull doi. https://doi.org/10.1037/0033-2909.115.1.74
5. Lochman JE, Dodge KA (1994) Social-cognitive processes of severely violent, moderately aggressive and nonaggressive boys. J Consult Clin Psychol doi. https://doi.org/10.1037/0022-006X.62.2.366
6. Dunn SE, Lochman JE, Colder CR (1997) Social problem-solving skills in boys with conduct and oppositional defiant disorders. Aggress Behav. https://doi.org/10.1002/(SICI)1098-2337(1997)23:6<457::AID-AB5>3.0.CO;2-D
7. Pardini DA, Waller R, Hawes SW (2015) Familial influences on the development of serious conduct problems and delinquency. In: Morizot J, Kazemian L (eds) The development of criminal and antisocial behavior: theory,

research and practical applications. Springer International Publishing, Cham

8. Stormshak EA, Bierman KL, McMahon RJ et al (2000) Parenting practices and child disruptive behavior problems in early elementary school. J Clin Child Adolesc Psychol 29:17–29

9. Lochman JE, Wells KC (2003) Effectiveness of the Coping Power program and of classroom intervention with aggressive children: outcomes at a 1-year follow-up. Behav Ther 34:493–515

10. Pardini DA (2008) Novel insights into long-standing theories of bidirectional parent–child influences: introduction to the special section. J Abnorm Child Psychol 36:627–631

11. Lahey BB, Van Hulle CA, Keenan K et al (2008) Temperament and parenting during the first year of life predict future child conduct problems. J Abnorm Child Psychol 36:1139–1158

12. Lorber MF (2004) Psychophysiology of aggression, psychopathy, and conduct problems: a meta-analysis. Psychol Bull 130:531–552

13. Huebner T, Vloet TD, Marx IVO et al (2008) Morphometric brain abnormalities in boys with conduct disorder. J Am Acad Child Adolesc Psychiatry 47:540–547

14. Foley DL, Eaves LJ, Wormley B et al (2004) Childhood adversity, monoamine oxidase A genotype, and risk for conduct disorder. JAMA Psychiat 61:738–744

15. Moffitt TE, Scott S (2008) Conduct disorders of childhood and adolescence. In: Rutter M, Bishop DVM, Pine DS et al (eds) Rutter's child and adolescent psychiatry, 5th edn. Blackwell, Oxford, pp 543–564

16. Barry TD, Dunlap ST, Cotton SJ et al (2005) The influence of maternal stress and distress on disruptive behavior problems in children. J Am Acad Child Adolesc Psychiatry 44:265–273

17. Fite PJ, Winn P, Lochman JE et al (2009) The effect of neighborhood disadvantage on proactive and reactive aggression. Am J Community Psychol 37:542–546

18. Thomas DE, Bierman KL, Thompson C et al (2009) Double jeopardy: child and school characteristics that predict aggressive-disruptive behavior in the first grade. Sch Psychol Rev 37:516–532

19. Lochman JE, Wells KC (2002a) Contextual social-cognitive mediators and child outcome: a test of the theoretical model in the Coping Power Program. Dev Psychopathol 14(4):945–967

20. Lochman JE, Wells KC (2004) The Coping Power Program for preadolescent boys and their parents: outcome effects at the 1-year follow-up. J Consult Clin Psychol 72:571–578

21. Lochman JE, Wells KC (2002b) The Coping Power Program at the middle school transition: universal and indicated prevention effects. Psychol Addict Behav 16:S40–S54

22. Lochman JE, FitzGerald DP, Gage SM et al (2001) Effects of social-cognitive intervention for aggressive deaf children: the Coping Power Program. J Am Deaf Rehabil Assoc 35:39–61

23. Zonnevylle-Bender MJS, Matthys W, van de Wiel NMH et al (2007) Preventive effects of treatment of DBD in middle childhood on substance use and delinquent behavior. J Am Acad Child Adolesc Psychiatry 46:33–39

24. Schuiringa H, van Nieuwenhuijzen M, Orobio de Castro B et al (2017) Effectiveness of an intervention for children with externalizing behavior and mild to borderline intellectual disabilities: a randomized trial. Cogn Ther Res 41:237–251

25. Mushtaq A, Lochman JE, Tariq PN et al (2017) Preliminary effectiveness study of Coping Power program for aggressive children in Pakistan. Prev Sci 18:762–771

26. Helander M, Lochman J, Högström J et al (2018) The effect of adding Coping Power Program-Sweden to Parent Management Training-effects and moderators in a randomized controlled trial. Behav Res Ther 103:43–52

27. Muratori P, Giuli C, Bertacchi I et al (2017) Coping Power for preschool-aged children: a pilot randomized control study. J Early Interv 11:532–538

28. Lochman JE, Boxmeyer CL, Powell N et al (2009) Dissemination of the Coping Power program: importance of intensity of counselor training. J Consult Clin Psychol 77:397–409

29. Lochman JE, Boxmeyer CL, Powell NP et al (2012) Coping power dissemination study: intervention and special education effects on academic outcomes. Behav Disord 37:192–205

30. Lochman JE, Powell NP, Boxmeyer CL et al (2009) Implementation of a school-based prevention program: effects of counselor and school characteristics. Prof Psychol Res Pract 40:476–482

31. Lochman JE, Powell NP, Boxmeyer CL et al (2015) Counselor-level predictors of sustained use of an indicated preventive intervention for aggressive children. Prev Sci 16:1075–1085

32. Miller S, Boxmeyer CL, Lochman JE et al (2018) Optimizing the Coping Power preventive intervention for children with reactive aggression by integrating mindfulness. Paper

presented in a symposium (S. Miller, Chair), at the Annual Meeting of the Society for Prevention Research, Washington, DC

33. Dishion TJ, Kavanagh K (2005) Intervening in adolescent problem behavior: a family-centered approach. The Guilford Press, New York, NY

34. Herman KC, Reinke WM, Bradshaw CP et al (2012) Integrating the family check-up and the parent Coping Power program. Adv School Ment Health Promot 5:208–219

35. Stirman SW, Crits-Christoph P, DeRubeis RJ (2004) Achieving successful dissemination of empirically supported psychotherapies: a synthesis of dissemination theory. Clin Psychol Sci Pract 11:343–359

36. Lochman JE, Baden RE, Boxmeyer CL et al (2014) Does a booster intervention augment the preventive effects of an abbreviated version of the Coping Power program for aggressive children? J Abnorm Child Psychol 42:367–381

37. Lochman JE, Boxmeyer CL, Jones S et al (2017) Testing the feasibility of a briefer school-based preventive intervention with aggressive children: a hybrid intervention with face-to-face and internet components. J Sch Psychol 62:33–50

38. Lochman JE, Dishion TJ, Powell NP et al (2015) Evidence-based preventive intervention for preadolescent aggressive children: one-year outcomes following randomization to group versus individual delivery. J Consult Clin Psychol 83:728–735

39. Glenn AL, Lochman JE, Dishion T et al (2018) Oxytocin receptor gene variant interacts with intervention delivery format in predicting intervention outcomes for youth with conduct problems. Prev Sci 19:38–48

40. Glenn AL, Lochman JE, Dishion T et al (2019) Toward tailored interventions: sympathetic and parasympathetic functioning predicts responses to an intervention for conduct problems delivered in two formats. Prev Sci 20:30–40

41. Hanf C (1969) A two-stage program for modifying maternal controlling during mother-child (M-C) interaction. Paper presented at the meeting of the Western Psychological Association, Vancouver, BC

42. Webster-Stratton C, Reid M (2003) The incredible years parents, teachers, and children training series: a multifaceted treatment approach for young children with conduct problems. In: Kazdin AE, Weisz JR (eds) Evidenced-based psychotherapies for children and adolescents. Guilford, New York, NY, pp 224–240

43. McMahon RJ, Forehand RL (2003) Helping the noncompliant child: family-based treatment for oppositional behavior, 2nd edn. Guilford, New York, NY

44. Brinkmeyer MY, Eyberg SM (2003) Parent-child interaction therapy for oppositional children. In: Kazdin AE, Weisz JR (eds) Evidenced-based psychotherapies for children and adolescents. Guilford, New York, NY, pp 204–223

45. Patterson GR, Reid JB, Jones RR et al (1975) A social learning approach to family intervention: families with aggressive children, vol 1. Castalia, Eugene, OR

46. Sanders MR (1999) Triple p-positive parenting program: towards an empirically validated multilevel parenting and family support strategy for the prevention of behavior and emotional problems in children. Clin Child Fam Psychol Rev 2:71–90

47. Kazdin AE (2003) Problem-solving skills training and parent management training for conduct disorder. In: Kazdin AE, Weisz JR (eds) Evidence-based psychotherapies for children and adolescents. Guilford, New York, NY, pp 241–262

48. Larson J, Lochman JE (2011) Helping schoolchildren cope with anger: a cognitive-behavioral intervention, 2nd edn. Guilford, New York, NY

49. Lochman JE, Wells KC, Lenhart LA (2008) Coping Power child group program: facilitator guide. Oxford University Press, New York, NY

50. Wells KC, Lochman JE, Lenhart L (2008) Coping Power parent group program: facilitator guide. Oxford University Press, New York, NY

Chapter 7

Adapting Cognitive Behavioral Therapy for Children with Autism Spectrum Disorder and Comorbid Anxiety and Obsessive-Compulsive Disorders

Kelly N. Banneyer, Rachel Fein, and Eric A. Storch

Abstract

Research has demonstrated many children and adolescents with autism spectrum disorder (ASD) display symptoms of anxiety and often meet criteria for anxiety as well as obsessive-compulsive disorders (OCD). However, the presentation of these symptoms varies across individuals. Although cognitive behavioral therapy (CBT) is the recommended evidence-based treatment for anxiety disorders and OCD, many providers trained in CBT feel ill-equipped to work with children and adolescents with comorbid ASD and anxiety as the symptoms of anxiety may present differently relative to neurotypical children and adolescents. This chapter provides an overview of how symptoms of anxiety and obsessive-compulsive and related disorders manifest in children and adolescents with ASD. Further, specific guidelines for adapting CBT to meet the needs of this population are discussed.

Key words Cognitive behavioral therapy (CBT), Autism spectrum disorder (ASD), Anxiety, Obsessive compulsive disorder (OCD), Children

1 Introduction

Autism spectrum disorder (ASD) is a neurodevelopmental disorder characterized by social communication deficits as well as atypical behaviors including restricted and repetitive behaviors, interests, and/or activities [1]. When making a diagnosis of ASD, the individual must display symptoms early in development, and they must currently exhibit impairment across settings and/or areas of functioning. Notably, a diagnosis of ASD may or may not be accompanied by a diagnosis of language disorder and/or intellectual disability. Based on information from the Centers for Disease Control and Prevention (CDCP), ASD occurs in 1 out of 54 children and is more common in males when compared to females (4:1) [2]. Importantly, ASD occurs along a spectrum of severity along multiple domains.

Robert D. Friedberg and Brad J. Nakamura (eds.), *Cognitive Behavioral Therapy in Youth: Tradition and Innovation*, Neuromethods, vol. 156, https://doi.org/10.1007/978-1-0716-0700-8_7, © Springer Science+Business Media, LLC, part of Springer Nature 2020

Although there is no cure for ASD, there are several established treatments that can be implemented with the ASD population. One of the most effective treatments for children with ASD is applied behavior analysis (ABA). Specifically, ABA involves identifying behaviors in need of remediation, defining them and finding objective ways to measure them. Behaviors may include those that are maladaptive (e.g., self-injurious behavior, motor stereotypy, restricted interests) and in need of reduction; whereas, other behaviors may include those that are prosocial (e.g., functional communication, social skills, appropriate play skills) and are in need of augmentation. ABA uses a number of strategies including, but not limited to, prompting, shaping, chaining, task analysis, positive reinforcement, and extinction to increase prosocial behaviors while decreasing maladaptive behaviors. Based on findings from the National Standards Project [1], ABA has years of sound research to support its efficacy in treating children with ASD [3–5].

In addition to the core deficits associated with ASD, many children with ASD display symptoms of anxiety. In fact, some researchers have estimated that ~40% of children with ASD meet full diagnostic criteria for an anxiety disorder [6] while between 11% and 84% of children with ASD display impairing symptoms of anxiety, though not necessarily meeting the diagnostic cutoff for a disorder [7]. With regard to co-occurring anxiety disorders and obsessive-compulsive and related disorders, children with ASD are commonly diagnosed with specific phobia (44–63%) [8, 9], obsessive-compulsive disorder (OCD) (17%), social anxiety disorder (17%), and generalized anxiety disorder (15%) [6].

Although anxiety frequently co-occurs in children with ASD, with equal prevalence among boys and girls [10], the presentation of anxiety symptoms varies across individuals. Pertaining to specific phobia, children with ASD may display unusual phobias compared to neurotypical peers. For instance, phobias related to shots/needles and crowds are most commonly found within the ASD population; however, phobias more commonly found in neurotypical peers (e.g., tunnels, flying, bridges) were less common [8, 11]. Similarly, although the fear of loud noises is less frequent among neurotypical children, approximately 10% of children with ASD report this specific phobia [8, 11]. Additionally, the phobias of children with ASD are often associated with medical visits, animals, or highly specific situations related to the individual [12].

When examining symptoms of OCD seen in children with ASD, Leyfer and colleagues [8] found the most commonly observed compulsions involved repetitive comments and question-asking as well as verbal and behavioral rituals involving another individual. Other researchers have found children with ASD displayed more compulsions related to repeating, touching, tapping, and hoarding while exhibiting less compulsions involving cleaning, checking, and counting [13]. Additionally, some children with ASD exhibit distress or anxiety in response to changes in their routine, engage in repetitive behaviors to alleviate symptoms of anxiety, and/or respond to disruption of repetitive behaviors with distress [14, 15]. Furthermore, distress caused by disruptions in routines, such as transitioning from home to school or from caregiver to caregiver, may often mirror symptoms of separation anxiety disorder [16].

Some findings specific to social anxiety disorder and generalized anxiety disorder have also emerged. Relating to social anxiety disorder, Gillott and colleagues [17] found children with high-functioning ASD (i.e., children with functional communication and broadly average cognitive abilities) were less likely to display fears of negative social evaluation. In contrast, others have observed similar rates of parent-reported self-consciousness and avoidance among neurotypical teenagers compared to those with high-functioning ASD [11, 18]. When reporting on generalized anxiety disorder, authors of a meta-analysis found studies of children with higher mean ages were more likely to report instances of generalized anxiety disorder [6]. Additionally, studies with a higher proportion of patients currently diagnosed with Asperger syndrome had higher prevalence of generalized anxiety disorder in comparison to studies with higher proportions of other autism spectrum disorders. [6].

Interestingly, symptoms of anxiety occur across the full range of intellectual functioning of children with ASD [16]; however, the presentation and rate of anxiety may differ depending on the child's intellectual functioning. Specifically, parents of children with an IQ below 70 typically provide lower ratings on measures of anxiety compared to parents of children with an IQ of 70 or greater. Given how common functional communication deficits are within the ASD population, especially with children with lower IQ, it is not unusual that parents are not endorsing symptoms that require language (e.g., "complains" or "worries") [16, 19]. However, it should be made clear that although children with lower IQ may not verbally express symptoms of anxiety, this does not mean they do not experience symptoms of anxiety. Notably, higher parent ratings of symptoms of anxiety are positively correlated with functional verbal communication, an IQ of 70 or above, as well as higher rates of inappropriate speech (e.g., stereotyped language), irritability, and behavioral difficulties (e.g., hyperactivity) [16]. Additionally, children rated as having the most impairing forms of anxiety were also rated as having higher levels of ASD symptom severity compared to those who were rated as less anxious. These findings suggest that anxiety may potentially exacerbate other behavioral difficulties, or rather, symptoms of anxiety may manifest as core symptoms of ASD or as other behavioral difficulties [16].

2   Theoretical and Empirical Foundations

Exposure-based cognitive behavior therapy (CBT) is the first line treatment for anxiety and obsessive compulsive disorders in children and adolescents (e.g., [20]). In recent years, research has examined adapting CBT for a comorbid presentation of ASD and anxiety [21]. A meta-analysis completed in 2013 found the use of CBT to treat anxiety in youth with ASD has a large effect size in comparison to waitlist control or treatment as usual [22]. Additionally, significantly more treatment gains have been observed in group CBT [23] and in family-based group CBT [24, 25] for children with anxiety and ASD in comparison to waitlist control or treatment as usual.

Research in this area has also focused on a modular approach to CBT treatment. Significantly more children with comorbid ASD and clinical anxiety responded to modular CBT in comparison to waitlist control [26]. One such approach is the BIACA program [27]. BIACA is a family-based modular CBT program with both parent and child modules, and it involves a core focus on the development of coping skills and in vivo exposure practice. The BIACA program also includes optional social skills modules. When modular CBT using the BIACA program was compared to an active treatment-as-usual condition, there were significantly more treatment gains in the CBT condition, with large between-group effect sizes. Gains were generally maintained at a 3-month follow-up [28]. Significant gains for youth in the BIACA program were also found when it was implemented in schools and compared against treatment as usual [29]. The program was also adapted and used with a sample of adolescents (aged 11–15 years). In comparison to waitlist and usual care control groups, adolescents in the BIACA program showed significant decreases in rated anxiety, although not in anxiety remission rates [30].

Similar to conducting interventions with typically developing children, the primary treatment modality for the treatment of anxiety and OCD for children with comorbid ASD is exposure-based CBT. Although studies are still examining the effectiveness of strategies for this population, modular-based CBT currently has the greatest evidence-base. Therefore, this chapter will map out a modular CBT approach to treatment.

3   Intervention

This section will serve as a guide and give information about adaptations to exposure-based CBT programs for use with children with comorbid ASD and anxiety or obsessive-compulsive disorders. Primary areas or modules to be incorporated include assessment, psychoeducation, goal-setting, exposure and response prevention (ERP), and treatment termination. Treatment will typically proceed through these areas in the order listed. Adaptations will have a focus on incorporating a behavioral approach, significant parental involvement, and use of reinforcement systems.

3.1 Assessment

As with intervention in any domain, evidence-based treatment will involve measurement-based care. Assessment typically begins with a diagnostic interview, at which time diagnoses are established. During the interview, it will be important to gather information related to the child's cognitive functioning, pragmatic language, and insight. Treatment modifications will be based on functioning in these areas. An anxiety or obsessive-compulsive disorder diagnosis could be made using a semi-structured diagnostic interview and diagnostic criteria for the disorders. When conducting a diagnostic interview, the clinician should be sure to gather information related to the history of the presenting problem, previous assessment or intervention, family background, family mental health history, prenatal and developmental history, medical history, previous and current medication usage, behavioral symptoms, emotional functioning, social functioning, and academic history. It is also possible to use a more structured interview, such as the Anxiety and Related Disorders Interview Schedule for DSM-5 (ADIS-5L) [31] or Children's Yale-Brown Obsessive Compulsive Scale (CY-BOCS) [32] to diagnose an anxiety or obsessive-compulsive disorder. These interview schedules will also include questions related to the domains listed above, so the clinician can be informed about symptoms, strengths, and challenges to provide an accurate diagnosis and form a comprehensive biopsychosocial case conceptualization. After the initial assessment, the clinician should be able to create a running list of anxiety-provoking/distressing triggers for the child. While the child's anxiety or obsessive-compulsive ritual does not need to be fully mapped out at this point, a preliminary list can then be incorporated into active treatment.

Anxiety in children with ASD may look similar to anxiety presentations in typically developing children or there may be distinct anxiety presentations. For example, children with ASD can have "traditional" fears, such as anxiety about separating from parents or fear of dogs. However, there are also "distinct" anxiety presentations that are more often observed in the ASD population, such as anxiety around change in routines and anxiety related to sensory sensitivities. Additionally, some of the anxiety symptomatology may also present as "distinct." For example, anxiety might present as an increase in disruptive behavior or stereotypies [21]. As discussed above, clinicians should be aware of both "traditional" and "distinct" trends when completing assessment.

During the assessment process with this population, it can sometimes be difficult to differentiate between obsessions and restricted interests and between compulsions and inflexibility around routines. Generally speaking, individuals with OCD engage in compulsions or avoidance to alleviate a distressing/anxious thought. Individuals with ASD may have restricted, special interests, but they focus on the activity or topic because it is enjoyable to them. A compulsion brings about relief from distress, but it is not pleasurable.

In addition to a thorough diagnostic intake, measurement-based care will incorporate regular progress-monitoring. Because children with ASD may have limited insight into their anxious or obsessive-compulsive symptoms, regular progress monitoring should incorporate parent-report questionnaires related to these domains. The DSM-5 [1] offers open-access symptom tracking forms that can be suited to this purpose. Clinicians should use results from behavioral questions to guide treatment, which means this assessment needs to be completed on a regular basis throughout treatment. If possible, clinicians may try to incorporate a short symptom and interference checklist at each session, in addition to regularly giving clinician-based ratings of symptom improvement.

3.2 Psychoeducation   The initial treatment session will focus on psychoeducation, with both the child and caregivers present, so that the entire family has a good understanding of anxiety/OCD and of exposure-based CBT. The initial part of psychoeducation should provide the basis for a CBT framework to address anxiety/OCD by discussing the interaction between thoughts, feelings, and behaviors, and the cycle that negatively reinforces anxiety and avoidance. While traditional CBT protocols make use of metaphors to understand the interaction between thoughts, feelings, and behaviors, more direct terms can be used to help children with ASD have an accurate understanding. Use of visuals and/or the child's interests will certainly aid in explaining the anxiety/OCD as well as the treatment to address it. While explaining this cycle, the feelings that occur may take the form of anger or irritation on the part of the child when accommodation is removed. Disruptive behaviors may present as the primary avoidance behavior (e.g., a child has a tantrum or yells at the parent so they do not have to engage in a feared activity) and may interfere with delivery of the core treatment modules. A parent management training protocol such as The Brief Behavioral Intervention for Preschoolers with Disruptive Behaviors [33] or Parent-Child Interaction Therapy [34] may be considered for supplementing our overall described approach with tools for reducing disruptive behavior. Since children with ASD have difficulties with emotion regulation, disruptive behaviors often occur and should be explained as an anxiety reaction within this cycle. While disruptive behaviors may result due to anxious feelings, appropriate structure and consequences should still be put into place in response to the behaviors. Normalizing this reaction and providing the family with preparation and appropriate expectations related to the potential reactions of their child will be critical to treatment.

Some CBT protocols make use of cognitive techniques to try and change maladaptive/anxious thinking. However, the primary aim of exposure-based CBT, especially when working with children with ASD, is to rely on a more behavioral approach. Trying to directly change feelings or thoughts with cognitive restructuring will likely be ineffective. Instead, the clinician should discuss a change in behavior, both in the part of the parent and the child, as the primary mechanism of change. Once the child is engaging in more adaptive behaviors (e.g., confronting feared stimuli), the child's anxious thoughts and feelings will subsequently decrease. Use the following dialogue as an example:

Clinician: *In any situation, people's thoughts, feelings, and actions all work together to impact their reactions and future thoughts, feelings, and actions in that situation. For example, when seeing a dog on the sidewalk, one child may think, "I really like dogs. I want to pet it." After thinking this, the child may feel happy, and their action may be to ask the owner to pet it. If the child is allowed to pet the dog, the child may think, "What a soft dog. I really like it!" and continue to feel happy and excited. The next time this child sees a dog, the child will be likely to engage in similar thoughts, feelings, and actions. However, another child may see a dog and think, "Dogs are so scary! The last time I saw a dog, it chased me!" and then feel afraid. This child may act by running away or asking a parent to remove the dog. When the dog is removed or the child runs away, there will likely be a feeling of relief. This relief strengthens the previous anxious thoughts, feelings, and actions, and the next time the child sees a dog, the same cycle is likely to occur.*

[The clinician may use this opportunity to apply the cycle of thoughts, feelings, and actions to a stimulus that the child enjoys and to a trigger of the child's anxiety. As noted previously, incorporating visuals will likely be beneficial.]

Clinician: [Speaking to child] *If you are feeling afraid, and your mom tells you to calm down or to not be scared, does that help? Do you not feel afraid anymore?*

Child: *No, I still feel scared.*

Clinician: [Speaking to parent] *I'm sure there are many times you have tried to talk through and change your child's feelings. Is this usually effective at making him feel better?*

Parent: *Usually not. I've told him to calm down a million times and nothing changes.*

Clinician: *Talking about feelings usually does not change feelings. The same is true for thoughts.* [Speaking to child] *Have there been times when your Mom has told you there is no need to worry about something because it is perfectly safe?*

Child: *Yes.*

Clinician: *And are you still worried?*

Child: *Yes.*

Clinician: *Trying to change thoughts and feelings by talking about them is not effective. However, what we can do is learn different actions to take to help change this cycle. Once we start to act bravely and change the cycle of anxiety by no longer avoiding the things that scare us, our thoughts and feelings start to be braver as well.*

After this discussion, the family ought to understand that the purpose of exposure practice is to change the child's behaviors from that of avoidance or other compulsive behaviors to brave or "approach" behaviors.

Either during the initial or subsequent treatment session, the clinician will introduce the family to the concept of differential attention as a behavioral parental tool to help shape behavior. The clinician should explain that attention is one of the parent's most powerful tools. When parents pay attention to anxious or disruptive behaviors, such as consistently answering reassuring-seeking questions, the behavior is reinforced, or maintained. Parental attention to these behaviors is thought of as one way anxiety is accommodated. One aim of treatment will be to help the family pay attention to brave and on-task behaviors and ignore anxious behaviors. However, this should be done in a systematic manner so it is not overwhelming for the family. Fading accommodation using differential attention strategies is one concept that will be discussed during ERP.

3.3 Goal-Setting

When working with families of children with ASD, it will be important for all family members to agree on appropriate goals. However, there are certain areas which ERP will not be able to address in this population. For example, "I want my child to be less awkward when interacting with peers," is not an area to be targeted within ERP. If this is an important treatment target for the family or if social skills deficits interfere in treatment delivery, an adjunctive social skills training program, such as The Program for the Education and Enrichment of Relational Skills (PEERS), may be beneficial [35]. Other inappropriate goals may involve a desire to decrease perceived negative emotions. For example, "To not be anxious anymore" is inappropriate for several reasons. First, anxiety is an innate emotion that all people experience and that is important for human safety. It would be unrealistic for an individual to never have another anxious feeling. Anxiety helps maintain an individual's safety and ability to perform when needed (i.e., someone who did not feel any worries about taking tests would never study). Second, while decreasing interference from anxiety is a primary treatment target, a goal should focus on what the child would like to be able to do or achieve.

Consistent with how goals are developed within ABA programs, primary treatment aims or behaviors identified within the context of CBT should be observable and measurable (i.e., operationally defined). Additionally, they should be stated in a positive, strengths-based manner. For example, for a child that has anxiety about falling asleep and has a two-hour compulsive ritual before bedtime, an appropriate goal statement may be: To go to sleep independently before 9 p.m. Another example for a child with contamination OCD might be: To wash hands less than four times per day for 20 s or less.

3.4 Exposure and Response Prevention

3.4.1 Map Out Symptoms

After delivering psychoeducation and helping the family to establish treatment goals, the clinician will start to form a plan for ERP. This is the point when the child's OCD or anxiety symptoms should be more thoroughly mapped out. If the clinician previously developed a list of anxiety-provoking or distressing triggers during the assessment process, these should be reviewed and discussed in terms of associated anxious thoughts (or obsessions) and anxious behaviors (or compulsions). It is important to understand what is driving avoidance or fear to plan ERP. For example, a child who is scared at night and sleeping with parents may be experiencing anxiety rooted in a variety of areas. This fear could be driven by separation anxiety, a specific phobia (e.g., fear of the dark), or generalized worries (e.g., child says different worries keep him/her from falling asleep). This behavior could also derive from an obsessive ritual (i.e., certain characteristics have to be in place so something "bad" does not happen"). In the case of a child with ASD, anxiety may not be present and the current sleeping arrangement is instead a rigid behavior (i.e., this is the way it has to be because it has always been this way). Treatment will differ based on which of these thoughts is driving the fear. For example, if a child is afraid of the dark, the practice should focus on increasing exposure to the dark.

3.4.2 Form Hierarchy

A practice hierarchy should be created based on the family's goal (s) and list of anxiety-provoking situations or triggers. Some children may present with one difficulty, but others may present with a vast list of anxiety-provoking triggers. Each target for ERP will need to have a hierarchy formed. At the beginning of treatment, collaborative hierarchy formation will need to occur in great detail so the family learns how to create and establish a hierarchy. Depending on the insight and the functioning of the child with ASD, an initial target for ERP should be chosen collaboratively with the child and the family. For some families, relying on the parent's perceptions and observations of the child's behavior may be more informative and appropriate. Ideally, the initial target is a frequently occurring situation everyone is motivated to change that is moderately anxiety-provoking or distressing. Practicing a target that causes minimal distress is not necessary, and choosing a target that will cause overwhelming distress will not be productive. The child will not be able to achieve new learning if overwhelmed, and if the child and parent become overwhelmed, they may decide to leave treatment.

3.4.3 Introduce Subjective Measurement Scale

Once a target has been identified, a hierarchy for ERP should be formed collaboratively with the family. This hierarchy is not simply an ordered list of triggers. It should incorporate different steps of increasing exposure to the feared or distressing stimulus. In order to form a meaningful hierarchy, it can be helpful for the child to identify a way to measure their distress in different situations. This may take the form of a 0–10 subjective units of distress scale (SUDS). Given how concrete or literal some individuals with ASD can be, using a smaller ordinal scale that involves a 0–5 hierarchy may be more appropriate, such as the Incredible 5 Point Scale [36]. If the child has difficulty applying numerical ratings to distress, labels such as "easy/medium/hard," colors such as red/-yellow/green, or other visuals could be used. Additionally, if the child has a restricted interest of some kind, using visuals related to the child's interest could serve as a meaningful depiction of the hierarchy. For instance, if a child is interested in weather, a tornado could serve to represent the most distressing situations and a cloudless sky could represent the least distressing situations. As noted previously, the parent may have more insight into how distressing different situations will make the child as well as which type of scale would be the most appropriate.

3.4.4   Establish Reinforcement System/ Token Economy

Children with ASD generally benefit from external motivation to engage in ERP. When ERP practice is introduced, a token economy system should also be discussed in detail. Specifically, target behaviors should be identified and assigned a point value, and children should be rewarded each time they engage in ERP activities. For instance, the target behavior for children with anxiety linked to loud sounds may be to remain in the kitchen while the blender is on for 5 s in order to receive one point. When children engage in this behavior, they will immediately receive the designated point value. In order to make the reinforcement more salient, children can earn points that are visually displayed on a chart or can be given tokens or other tangible indicators of points, such as tickets, coins, or marbles. This system will be most effective if children are able to exchange tokens for rewards they are motivated to earn. Some children enjoy creating a "rewards menu," to which parents can then assign point values. Rewards options may include screen time (e.g., television, computer, video games, phone), ability to choose a family activity, or even physical rewards of toy or games. Each reward option will be assigned a different point value depending on the size of the reward, the ease of accessing the reward, and how motivated children are to earn it. Most importantly, parents must be in agreement for items placed on the reward menu. Families should be informed that token economies are only effective when target behaviors are clearly identified, tokens are provided immediately, point values correctly match the rewards, there is a clear and regular time to "cash-in" rewards, and families follow through consistently.

3.4.5 In-Session Exposure Practice

Once the hierarchy has been formed and a reinforcement system is in place, practice should commence. The hierarchy should form a guideline for practice. To begin, the clinician should engage the child in a practice toward the bottom of the hierarchy. Many clinicians like to progress linearly from the bottom of the hierarchy to the top, although it is not necessary. What is more important is to have the family's engagement in the practice. Oftentimes, clinicians will offer children the opportunity to choose where they prefer to start on their hierarchy using open-ended questions. However, children with ASD may benefit from being provided with close-ended choices. For instance, in the context of symptoms of separation anxiety disorder, a clinician could ask, "Pick one goal. Would you rather begin with mom staying outside the room for 30 s or you staying outside the room for 30 s?"

For most exposure practices, length of time is a variable to consider and should be increased as practice continues. For example, for children with contamination-based OCD, typically touching a "contaminated" or "dirty" object for 1 s is easier than doing so for 30 s. After and during each practice the child should be instructed to "stick with" or endure the practice rather than engage in a safety behavior or compulsion. Many clinicians like to use the SUDS scale during exposure practice to monitor the distress of the child. This helps the clinician to have an idea about how difficult exposure practice is for the child and helps to keep the child engaged in the practice. Sometimes the clinician will observe SUDS scores to decrease in session as multiple practices are completed. However, this is not necessary for treatment benefit. Furthermore, depending on the level of insight, a child with ASD may not be able to accurately articulate or recognize varying levels of distress using this scale.

Most importantly, the majority of each session should focus on repeated exposure practice. Sometimes the same step may be repeated multiple times, but the clinician should work to make sure the child is challenging his or her anxiety/OCD by continuing to engage in more difficult practice, as clinically appropriate. The clinician should also take opportunities to model exposure practice for parents so parents understand how to complete the practice and will be able to do so at home while modeling brave behavior and providing minimal reassurance. While ERP oftentimes goes "beyond" what the child will need to do in everyday life, ERP should never involve an activity that is dangerous.

To give a brief example for a child with separation anxiety, the initial ERP goal may be for the child's parent to leave the treatment room for 30 s. If the child was successfully able to tolerate this practice, it could be repeated several times. Next, the parent could be instructed to leave the room for 45 s, and this practice could be repeated. Additional trials may then consist of the parent leaving for longer and longer periods of time (i.e., 1, 2, 5, 10 min). During

ERP, the child should be aware of the goal and should not be "tricked" into practice. After each trial when parent returns to the room, the child should be praised and the predetermined reward should be given.

3.4.6 Between-Session Practice

Engaging in exposure practice and reducing accommodation between treatment sessions is crucial for treatment progress to occur. Fading may be part of a child's active exposure hierarchy, but separate hierarchies may have to be developed when safety behaviors primarily involve the parents. In these situations, the bottom level of the hierarchy is the current level of accommodation and the top of the hierarchy may be the family's goal. For example, when a parent is attempting to separate from the child before bedtime, the bottom level of the hierarchy might be: Parent stays in child's bed for entirety of night. The goal might be: Parent completes 5-min bedtime routine and leaves room, checking on child only one time during night. Creating a hierarchy to decrease reassurance may also involve fading of accommodating statements. For example, the child may prefer for the parent to repeat the following statement three times before bed: "Goodnight. I love you. I'll stay in my room all night and listen for any noises. You will sleep well and I will protect you." If it is too overwhelming for the child to completely remove this part of the routine, the parent may need to advance through a hierarchy of fading this accommodating statement. Please *see* Table 1 for an example.

Each treatment session should conclude with an agreement of home-based practice. This practice should be concretely stated, and it is helpful to be written out so the child and family have a physical reminder. Some children with ASD like having a chart where they can indicate when they engage in practice each day. Practice may consist of continuing exposures completed in session or may consist of completing exposure that were unable to be completed in the therapy environment. For example, fading accommodation before nighttime is a topic that would be discussed and planned in the lesson, and practice would occur primarily between sessions. Practice should also be very concrete. Vague instructions, such as "Practice exposures at nighttime," are less likely to be completed than instructions such as, "Practice sitting by yourself in your room with the lights off for five minutes each day at 8 p.m." Establishing a time when families review whether they have completed their homework each day as well as a reminder about the token economy (i.e., reward menu, point values, and "cash-in" procedures.) will also be beneficial to treatment. Each follow-up ERP session should begin with a review of practice completed the previous week.

When forming these steps and assigning practice, it will be important to have parents and family members in agreement with the plan because they will be the primary actors of change. Parents

Table 1
Hierarchy for fading avoidance of reassurance-seeking statement

Level of practice	Parent response
Challenge	Silence
Goal Statement	"See you in the morning" (alternative/different statement each night)
Statements in the Hierarchy	"Goodnight" "Goodnight. I love you" "Goodnight. I love you. I'll stay in my room all night and listen for any noises" "Goodnight. I love you. I'll stay in my room all night and listen for any noises. You will sleep well and I will protect you." Repeated one time "Goodnight. I love you. I'll stay in my room all night and listen for any noises. You will sleep well and I will protect you." Repeated two times
Current statement	"Goodnight. I love you. I'll stay in my room all night and listen for any noises. You will sleep well and I will protect you." Repeated three times

Note: The length of time each level is completed will depend on child's discomfort and parent's ability to tolerate discomfort. For example, a certain level may plan on occurring one night but another level may need to be practiced seven nights before a child is ready to move on. A child may also be willing to skip and move ahead to more difficult levels

will also need to have understanding for how to adapt the hierarchy and plan if necessary. For example, the identified practice for the week may to reduce accommodation by having the parent repeat the statement two times instead of three times. A back-up plan should also be in place. Even if the child initially agreed to the plan, it is possible that distress in the moment may be higher than expected. If a back-up plan is not in place, the family may refrain from any practice until the next session. In this instance, the backup plan may be for the parent to repeat the statement three times but to remove a single phrase or a word.

3.5 Treatment Termination

The clinician should regularly monitor progress (via behavioral measures as discussed above) and communicate with the family about progress toward achieving goals. If progress has plateaued or goals have not been met, there are several factors to consider. First, medication management of anxiety/OCD symptoms may be warranted. Second, the clinician should determine whether there are any other factors interfering with treatment. For example, some children with ASD who have some insight into deficits in social skills may experience anxiety due to actual differences and difficulties in comparison to peers. If social anxiety is based on an identifiable real-life trigger (e.g., teasing or bullying), other intervention is warranted, such as communication with the school about establishing a safe environment and the possibility of social skills training for the child. Other potential barriers to treatment progress may be related to families continuing to provide accommodation and reassurance to the child, inconsistent implementation of home-based practice, and/or lack of consistency in implementing the token economy.

If adequate progress is made toward achieving goals, treatment sessions can be spread out and eventually treatment termination will occur. Before terminating treatment, treatment maintenance strategies should be discussed. The family should be aware of how to respond if a new trigger emerges by reducing avoidance/accommodation and creating a hierachy for exposure practice if needed. Many times, establishing intermittent booster sessions can be helpful for the family stay on track with strategies and for the child to maintain progress. Finally, consultation with other providers (e.g., school psychologists or child psychiatrists) may be necessary to ensure ongoing treatment maintenance.

4 Conclusion

As described above, there are several key aspects of adapting exposure-based CBT intervention for children with comorbid ASD and anxiety or obsessive-compulsive disorders. These factors include an emphasis on behavioral aspects of CBT, high incorporation and reliance on parental involvement, and need to establish a reinforcement system.

Children with ASD may have limited insight, poor understanding of metaphors, and/or below average IQ. For any of these reasons, an emphasis on the behavioral aspects of CBT is necessary for treatment. Incorporation of parents into treatment is important in when completing intervention with children with ASD. When working with children with comorbid ASD and anxiety, parents can provide insight into symptom severity and impairment. Additionally, they are critical to establishing goals of treatment, developing ERP hierarchies, as well as assisting with in-session and between-session ERP practice. Many children, especially those with ASD, may have limited insight into interference from symptoms of anxiety/OCD. Therefore, providing a system of reinforcement to increase motivation to engage in ERP practice in session and outside of session is necessary for treatment engagement and achievement of treatment goals.

References

1. American Psychiatric Association (2013) Diagnostic and statistical manual of mental disorders (DSM-5®). American Psychiatric Association, New York, NY

2. Maenner MJ, Shaw KA, Baio J, et al (2020) Prevalence of autism spectrum disorder among children aged 8 years — autism and developmental disabilities monitoring network, 11 Sites, United States, 2016. MMWR Surveill Summ 69(No. SS-4):1–12

3. Lovaas OI (1987) Behavioral treatment and normal educational and intellectual functioning in young autistic children. J Consult Clin Psychol 55:3–9

4. McEachin JJ, Smith T, Lovaas OI (1993) Long-term outcome for children with autism who received early intensive behavioral treatment. Am J Ment Retard 97:359–372. discussion 373–91

5. Sallows GO, Graupner TD (2005) Intensive behavioral treatment for children with autism: four-year outcome and predictors. Am J Ment Retard 110:417–438

6. van Steensel FJA, Bögels SM, Perrin S (2011) Anxiety disorders in children and adolescents with autistic spectrum disorders: a meta-analysis. Clin Child Fam Psychol Rev 14:302–317

7. White SW, Oswald D, Ollendick T et al (2009) Anxiety in children and adolescents with autism spectrum disorders. Clin Psychol Rev 29:216–229

8. Leyfer OT, Folstein SE, Bacalman S et al (2006) Comorbid psychiatric disorders in children with autism: interview development and rates of disorders. J Autism Dev Disord 36:849–861

9. Muris P, Steerneman P, Merckelbach H et al (1998) Comorbid anxiety symptoms in children with pervasive developmental disorders. J Anxiety Disord 12:387–393

10. Worley JA, Matson JL, Sipes M et al (2011) Prevalence of autism spectrum disorders in toddlers receiving early intervention services. Res Autism Spectr Disord 5:920–925

11. Kerns CM, Kendall PC (2012) The presentation and classification of anxiety in autism spectrum disorder. Clin Psychol Sci Pract 19:323–347

12. Evans DW, Canavera K, Kleinpeter FL et al (2005) The fears, phobias and anxieties of children with autism spectrum disorders and down syndrome: comparisons with developmentally and chronologically age matched children. Child Psychiatry Hum Dev 36:3–26

13. McDougle CJ, Kresch LE, Goodman WK et al (1995) A case-controlled study of repetitive thoughts and behavior in adults with autistic disorder and obsessive-compulsive disorder. Am J Psychiatry 152:772–777

14. Bodfish JW, Symons FJ, Parker DE et al (2000) Varieties of repetitive behavior in autism: comparisons to mental retardation. J Autism Dev Disord 30:237–243

15. Volkmar F, Cook EH, Pomeroy J et al (1999) Practice parameters for the assessment and treatment of children, adolescents, and adults with autism and other pervasive developmental disorders. J Am Acad Child Adolesc Psychiatry 38:32S–54S

16. Hallett V, Lecavalier L, Sukhodolsky DG et al (2013) Exploring the manifestations of anxiety in children with autism spectrum disorders. J Autism Dev Disord 43:2341–2352

17. Gillott A, Furniss F, Walter A (2001) Anxiety in high-functioning children with autism. Autism 5:277–286

18. Russell E, Sofronoff K (2005) Anxiety and social worries in children with Asperger syndrome. Aust N Z J Psychiatry 39:633–638

19. Witwer AN, Lecavalier L (2010) Validity of comorbid psychiatric disorders in youngsters with autism spectrum disorders. J Dev Phys Disabil 22:367–380

20. James AC, James G, Cowdrey FA et al (2013) Cognitive behavioural therapy for anxiety disorders in children and adolescents. Cochrane Database Syst Rev 6:CD004690

21. Kerns CM, Renno P, Storch EA et al (2017) Anxiety in children and adolescents with autism spectrum disorder: evidence-based assessment and treatment. Cambridge, MA, Academic

22. Sukhodolsky DG, Bloch MH, Panza KE et al (2013) Cognitive-behavioral therapy for anxiety in children with high-functioning autism: a meta-analysis. Pediatrics 132:e1341–e1350

23. Sofronoff K, Attwood T, Hinton S (2005) A randomised controlled trial of a CBT intervention for anxiety in children with Asperger syndrome. J Child Psychol Psychiatry 46:1152–1160

24. Chalfant AM, Rapee R, Carroll L (2007) Treating anxiety disorders in children with high functioning autism spectrum disorders: a controlled trial. J Autism Dev Disord 37:1842–1857

25. Reaven J, Blakeley-Smith A, Culhane-Shelburne K et al (2012) Group cognitive behavior therapy for children with high-functioning autism spectrum disorders and anxiety: a randomized trial. J Child Psychol Psychiatry 53:410–419

26. Wood JJ, Drahota A, Sze K et al (2009) Cognitive behavioral therapy for anxiety in children with autism spectrum disorders: a randomized, controlled trial. J Child Psychol Psychiatry 50:224–234

27. Wood JJ, Drahota A (2005) Behavioral interventions for anxiety in children with autism. UCLA Law Rev

28. Storch EA, Arnold EB, Lewin AB et al (2013) The effect of cognitive-behavioral therapy versus treatment as usual for anxiety in children with autism spectrum disorders: a randomized, controlled trial. J Am Acad Child Adolesc Psychiatry 52:132–142.e2

29. Fujii C, Renno P, McLeod BD et al (2013) Intensive cognitive behavioral therapy for anxiety disorders in school-aged children with autism: a preliminary comparison with treatment-as-usual. School Ment Health 5:25–37

30. Storch EA, Lewin AB, Collier AB et al (2015) A randomized controlled trial of cognitive-behavioral therapy versus treatment as usual for adolescents with autism spectrum disorders and comorbid anxiety. Dep Anx 32 (3):174–181

31. Brown TA, Barlow DH (2014) Anxiety and related disorders interview schedule for DSM-5 (ADIS-5l)-lifetime version: client interview schedule 5-copy set. Oxford University Press, Oxford

32. Scahill L, Riddle MA, McSwiggin-Hardin M et al (1997) Children's Yale-Brown obsessive compulsive scale: reliability and validity. J Am Acad Child Adolesc Psychiatry 36:844–852

33. Axelrad M, Chapman S (2016) The brief behavioral intervention for preschoolers with disruptive behaviors: a clinical program guide for clinicians. MedEdPORTAL. https://doi.org/10.15766/mep_2374-8265.10376

34. McNeil CB, Hembree-Kigin TL (2010) Parent-child interaction therapy. Springer Science & Business Media, New York, NY

35. Laugeson EA, Frankel F (2010) Social skills for teenagers with developmental and autism spectrum disorder: the PEERS treatment manual. Routledge, New York, NY

36. Buron KD, Curtis M (2012) The incredible 5-Point Scale: the significantly improved and expanded second edition: assisting students in understanding social interactions and controlling their emotional responses. AAPC Publishing, Shawnee

Chapter 8

Cognitive Behavioral Therapy with Substance Using Youth

Molly Bobek, Brad Donohue, Nicole P. Porter, Alexandra MacLean, and Aaron Hogue

Abstract

This chapter focuses on evidence-based cognitive behavioral therapy (CBT) treatment for adolescent substance use disorders. The chapter provides information on the primary intervention modalities for treating adolescent substance use (ASU): individual/group and family-based. A core elements approach yielding six core elements of CBT is described, and pragmatic CBT interventions are outlined that can be incorporated in multiple modalities. A family-based CBT model for ASU, family behavior therapy, is described and its eight primary intervention components illustrated. Four worksheets for clinical practice are included, and we conclude by discussing the clinical spectrum of family involvement in CBT for youth with ASU.

Key words Substance use, Adolescents, Core elements, Family behavior therapy, Cognitive behavioral therapy, Parents

1 Introduction: Overview of Adolescent Substance Use

Substance use is highly common among adolescents in the USA. Recent data indicate that among twelfth graders surveyed annually, 64% have used alcohol, 45% have used marijuana, 38% have used cigarettes, and 49% have used other illicit drugs [1]. Troublingly, the Substance Abuse and Mental Health Services Administration [2] estimates that nationwide approximately 125,000 adolescents enroll in outpatient substance use treatment each year; however, this is only about 9% of the youth who are deemed in need of such services [3].

While adolescent experimentation with alcohol and drugs is considered to be developmentally normative behavior, research shows that those who begin their use at an early age are at greater risk for developing a substance use disorder (SUD) [4] and are also at risk for other emotional and behavioral problems [5]. In particular, adolescents who use substances demonstrate increased risk for internalizing and externalizing problems, attention-deficit/

Robert D. Friedberg and Brad J. Nakamura (eds.), *Cognitive Behavioral Therapy in Youth: Tradition and Innovation*, Neuromethods, vol. 156, https://doi.org/10.1007/978-1-0716-0700-8_8, © Springer Science+Business Media, LLC, part of Springer Nature 2020

hyperactivity disorder, oppositional defiant disorder, and conduct problems (e.g., [6]). Adolescents with SUDs are more likely to sustain injuries due to accidents, overdose, suicide and violent crime [7, 8]. Additionally, about 60% of youth involved in the juvenile justice system are estimated to be in need of SUD treatment [9]. During adolescence, experimentation with substance use and associated behavioral problems takes place in conjunction with significant biological and psychological changes [10, 11]. Without effective intervention, adolescent substance use (ASU) and SUDs can have far-reaching consequences.

This chapter is intended to provide a basic description of CBT interventions that have garnered strong empirical support across the primary intervention modalities for treating ASU: individual/ group and family-based. Our intention is to establish a foundation for selecting and implementing pragmatic CBT interventions within any combination of individual/group and family treatment sessions. We begin with a rationale and overview of a core elements approach to individual/group CBT for ASU. We then describe six core elements of individual/group CBT for ASU along with clinical applications of each techniques. Next, we describe a family-based CBT model for ASU, Family Behavior Therapy, presenting its primary intervention components. We conclude by discussing the clinical spectrum of family involvement in CBT for youth with ASU, including future possibilities for implementation and dissemination research.

2   Theoretical and Empirical Foundations: Core Elements of Individual/Group CBT

2.1   Rationale for Core Elements of Individual/Group CBT for ASU

The most recent comprehensive literature review of evidence based interventions for adolescent substance use concluded that ecological family therapy and CBT possess the strongest levels of empirical support [12]. Both approaches have excellent efficacy evidence for adolescent SUDs in both research and community settings (*see* also [13–17]), and each is represented by several manualized models. Additionally, providers report that both approaches are highly valued in everyday practice [18, 19].

Implementation challenges such as costs, complexity, and difficulties in sustainment and ongoing quality assurance are common to manualized treatments of many kinds [20, 21]. In response, some experts in mental health (e.g., [22, 23]), substance use (e.g., [24]), and behavioral health policy [25] advocate an alternative strategy to complement manualized treatments: Focus on core elements of evidence-based interventions (EBIs) that represent a reduced set of intervention techniques common to multiple treatments for a given behavioral health problem. Whereas many proprietary manualized treatments are disorder-specific, distilled core elements are intended to be granular, and flexible to address multiple disorders. The techniques that emerge from a core elements approach equip providers with key options that can be helpful to clients presenting with diverse, comorbid, or emerging clinical problems [22, 26], a considerable advantage for treatment planning with youth who evidence complicated challenges.

Germane to the advantages of core elements of EBIs for ASU, core practice elements of three manualized ecological family therapy models that have strong empirical support for adolescent conduct and SUDs have been conceptually distilled. The approach involved a conceptual distillation of treatment manuals and protocol materials [27], followed by an empirical distillation utilizing observational analyses of high-fidelity treatment sessions from each model [28]. Four core elements were delineated, each consisting of several treatment techniques representing multiple models: adolescent engagement, relational reframing, relational focus, and interactional change.

Because of its roots in behavioral psychology, CBT interventions can specifically target affect regulation, motivation and reward, and cognitive control processes that predispose and sustain adolescent antisocial behavior (*see* [29, 30]). The practice elements of CBT are an excellent match for addressing the neurobehavioral and other developmental challenges underlying ASU. The six core elements described below are derived from the CBT treatment strategies and techniques contained in individual-based and group-based treatment protocols or books aimed at treating adolescent conduct or substance use problems, identifying clinical strategies that appeared to be (1) common across existing protocols, (2) theoretically salient to the CBT approach, and (3) designed to target symptoms of ASU. The core element distillation process is described in greater detail in Hogue and colleagues [28]. Primary clinical sources for the distillation process included the CBT-based models supported for adolescent disruptive behavior [31] and substance use [12], respectively [32–36]. All six elements can be used in both individual and group treatment settings, with only slight adaptations required. Moreover, the elements can also be deployed creatively and in various formulations to anchor sessions that include family members, particularly in FBT for ASU, described in greater detail below.

3 Interventions

3.1 Core Elements of Individual/Group CBT with Clinical Examples

3.1.1 Individual/Group Element 1: Functional Analysis of Substance Use Behaviors

Functional analysis interventions are typically based on semi-structured interviews (e.g., [34]) with teens and provide basic information about the antecedents and consequences of their substance use. A primary purpose of functional analysis is to uncover information about reinforcers (consequences that increase the likelihood of a substance use behavior occurring again) of behavior problems that therapists can utilize throughout treatment to assist interventions in being personally relevant and effective for each client. Functional analyses help teens understand their behavior so they can make optimal decisions in the future. Functional analysis usually begins with therapists asking adolescents to describe a typical situation in which a substance use behavior occurs. Therapists then pursue follow-up questions to detail a characteristic chain of antecedents–behavior–consequences (i.e., A-B-Cs) associated with the situation, working to identify stimuli that give rise to the problem (antecedents) and reinforcers that strengthen its continued expression. As seen in Fig. 1, therapists can engage in a semi-structured functional analysis interview with teens by exploring triggers, the substance use behavior itself, and both the positive and negative consequences of the substance use behaviors. For example, functional analysis with a teen who used marijuana five times weekly revealed that free time with peers who enjoy marijuana, conflict with parents, and frustration with school and

FUNCTIONAL ANALYSIS WORKSHEET

Antecedents (Triggers, Thoughts, Feelings)	Behavior (Substance Use)	Positive Consequences	Negative Consequences

Fig. 1 Functional analysis worksheet

boredom were all antecedents to marijuana use. This use was sub-sequently reinforced by increased positive mood, social facilitation (e.g., fun with friends, bonding and building of relationships with admired peers), and reduced anxiety, particularly somatic symptoms.

3.1.2 Individual/Group Element 2: Prosocial Activity Sampling

In prosocial activity sampling the aim is to introduce and determine potential rewards of prosocial behaviors that are capable of com-peting with or replacing behaviors that facilitate substance use. Therapists aim to support teens in restructuring their everyday environments, a form of "stimulus control" in which they (a) avoid high-risk persons and situations with established links to their substance use and (b) seek new, positive outlets for social and recreational activities. Therapists may nominate activities meant to directly counter aspects of the A-B-Cs detailed in functional analysis (e.g., enrolling in a midnight sports league that conflicts with less healthy peer habits) and/or introduce activities with potential new positive rewards.

Specific techniques for increasing prosocial activities include: helping adolescents identify activities to sample and assigning good candidates as homework; reviewing sampled activities to assess their reward values and problem-solve future attempts to participate in the given activity; and using systematic encouragement (brain-storming, role-playing and feedback, enlisting help from family and friends in session or via phone) to help teens maintain their commitment to engaging with promising options. In clinical prac-tice, a teen who indicated during assessment that she had spent increasingly less time making art, which was at one time an absorb-ing activity for her, presented an opportunity for art-focused pro-social activity scheduling. The therapist invited the teen to complete in-session research on opportunities to take art classes to increase her skills and to make art with younger children, given her strong relationships with her much younger siblings. Although the client was dismissive during initial brainstorming, when a spe-cific opportunity to volunteer in an elementary school arts program arose, she expressed an openness to trying it. She pursued the opportunity, and the therapist checked in about her experience and praised her ongoing involvement in subsequent sessions, as well as inviting the teen's caregiver to support her through praise as well.

3.1.3 Individual/Group Element 3: Cognitive Monitoring and Restructuring

Cognitive interventions refer to the process by which adolescents are encouraged to monitor their cognitions and to become aware of how cognitions influence emotions and behaviors. That is, the role that cognitions play within the "cognitive triangle" of thinking, feeling, and doing. Cognitive interventions can be directed at specific thoughts or more global thinking habits. Therapists often proceed by first exploring teens' core beliefs and attitudes about the world: why things work the way they do, why people are the way they are, and how they believe others view them. Therapists then attempt to demonstrate that core beliefs play a major role in determining feelings and actions, including those feelings and actions linked to substance use.

Therapists can help teens learn to identify and interrogate their dysfunctional thoughts rather than implicitly accepting them. To help teens restructure cognitions, therapists attempt to illustrate how dysfunctional thoughts may be associated with unwanted behavioral tendencies, entrenched emotional positions, or problematic biases that may be anxiety-provoking and/or unhelpful. Examples of cognitive restructuring interventions include speaking aloud and role-playing negative self-talk in session (i.e., externalization); teaching clients to question the evidence used to maintain or strengthen problematic beliefs; helping clients re-attribute personal versus external responsibility for negative outcomes in a more balanced and/or realistic manner; decatastrophizing problematic behaviors or situations; and helping adolescents understand the cognitions or feelings of others (i.e., perspective-taking). As seen in Fig. 2, therapists can engage in cognitive restructuring using a thought log with teens in which they can record the date, time, and place of their unhelpful thoughts, the resulting emotions, evidence that supports the thoughts and evidence that challenges their truthfulness, and develop and record more helpful thoughts.

For instance, one teen completed a worksheet designed to highlight links between the thought "I'm a loser" and her depressed mood and alcohol use. The client was also open to externalization interventions such as "when the loser thought arrives," as well as interventions to depersonalize her history of damaged attachments with caregivers that led to her maladaptive cognitions.

3.1.4 Individual/Group Element 4: Emotion Regulation Training

Negative mood states and generalized distress are prime risk factors for adolescent substance use. For these reasons, affect regulation interventions are an essential component of the CBT approach. At a fundamental level, all CBT models for adolescent substance use contain coping-focused interventions designed to help teens recognize and modulate their impulsiveness, depressed and anxious moods, reactivity to stress, and anger.

Anger management falls squarely in line with the overarching A-B-C framework: Triggers that are direct (e.g., observing someone aggress against you) and indirect (e.g., inferring someone is angry or disappointed with you) give rise to defensive or hostile cognitions and physiological arousal, which lead to emotional and behavioral expressions of anger and aggression. To counter these reactions A-B-C chain therapists promote awareness of situations that trigger anger; recognition of the signs of mounting arousal;

Cognitive Restructuring Thought Log

Date/Time /Place	My unhelpful thought	The emotions or feelings	Evidence the thought is true/realistic/ helpful	Evidence the thought isn't true/isn't helpful	A more helpful thought is...

Fig. 2 Cognitive restructuring thought log

strategies for escaping anger-inducing situations before arousal and/or behaviors escalate, such as self-imposed "cooling off" periods; and increased perspective-taking and empathy for the actions of others. Emotional regulation usually includes relaxation training, which can be used to cope with stress, tension, and anxiety as well as anger. Relaxation techniques aim to prevent or reduce arousal levels that render adolescents susceptible to compromised reasoning and decision-making, difficulty with concentration, sleep problems, and reliance on substance use to moderate stress.

In the context of treatment sessions, one teen in treatment for problematic marijuana use was able to improve his ability to notice when his anxiety was becoming difficult to manage, and signal his parents and certain trusted friends to let them know whenever he experienced distress. Relaxation training was conducted with the teen and his parents together in session to support him in coping more effectively with worry and conflict with his parents. Progressive muscle relaxation involved teaching the teen to tense and relax isolated muscles [37] so that the teen could better recognize signs of tension and convert these sensations into relaxation during stressful situations at home that often led to substance use.

3.1.5 Individual/Group Element 5: Problem Solving Training

Training in problem solving is a cornerstone CBT activity for enhancing adolescent self-efficacy as well as life skills. A conventional starting point is introducing the fundamental processes of decision-making, which involves understanding how everyday decisions exert a powerful impact on life quality and direction (a direct and often predictable connection between choices and short- or long-term consequences). The basic formulation for problem solving involves teaching youth to adopt a systematic approach that begins with disaggregating problems into component parts/goals for which manageable solutions are more readily monitored and achieved. It is critical that therapists identify and focus on decisions and problems that have immediate, significant relevance in the life of a given client.

Conventional steps in effective problem solving include: definition of the problem (including an understanding of preferred outcomes of the problem situation), brainstorming solutions for achieving preferred outcomes, evaluating the relative merits of the generated solutions, practicing chosen solution(s) in session utilizing behavioral rehearsal and feedback, and monitoring and revising implementation of the solution(s) at home. This process is outlined in the worksheet in Fig. 3. Therapists can record the problem, attempted solutions, and pros and cons of attempted and other possible solutions, in session, while inviting teens to complete the results of their attempted solutions either as homework or in subsequent sessions.

In therapy with a teen client evidencing problematic alcohol use, therapeutic interventions focused on collaborative inquiry of decisions the client might make differently that would immediately enhance her life. She identified preferring to spend time with her boyfriend alone and avoid his social circle. She explored the pros and cons of various solutions to being encouraged or even coerced by her boyfriend to go out to places where alcohol was served, role-played sharing her preferences with her boyfriend, and asked her grandmother to provide transportation home to facilitate the new arrangement. Through an iterative process across several sessions with role plays in each, she determined that leaving gatherings early was not effective and decided to try spending time with a different friend instead.

3.1.6 Individual/Group Element 6: Communication Training

Communication training is a CBT cornerstone for helping youth avoid negative interactions that create problems and often lead to substance use. Strong communication skills can foster healthier interactions with significant others in support of achieving treatment goals. Although training usually involves generic principles of effective communication (e.g., using "I" statements when engaged in difficult conversations), it remains essential to tailor training to each client's real-world circumstances. This can be accomplished by asking clients to recount typical or recent conversations with

PROBLEM SOLVING LOG

What is the Problem			
What would be a good outcome?			
What are possible solutions and their pros and cons?	Solution #1:	Pros:	
		Cons:	
	Solution #2	Pros:	
		Cons:	
	Solution #3	Pros:	
		Cons:	
What is the action plan for chosen solution?			
What was the outcome?			
What are next steps to revise plan or choose new solution			

Fig. 3 Problem-solving log

specific persons (e.g., parent, teacher, probation officer), allowing for collaborative review of the communication anatomy of both benign and problematic conversations. During this interactive process the client's communication skills and habits can be identified, along with opportunities for improving unfavorable outcomes by using alternative or newly learned skills. Importantly, during this

Drug and Alcohol Refusal Skills Reminders and Practice

When someone offers you alcohol or drugs, keep the following communication skills in mind:

- It is OK to say "no" and often good to say "no" first.
- Invite yourself to speak in a clear and firm way.
- Make direct eye contact.
- Suggest an alternative:
    - Something else to do
    - Something else to eat or drink
- Change the subject
- Avoid vague answers if you can
- Increase your awareness of guilty or worried feelings when refusing marijuana, and use coping skills to manage those feelings.
- If you want to, you can ask the person offering you marijuana to stop doing so and not to do so in the future

Listed below are some examples of people who might offer you drugs or alcohol in the future. Give some thought to how you will respond to them, and write your responses below each item.

Someone close to you who knows you are making changes in your life:
_____
_____

A school friend:
_____
_____
_____

A coworker (if you have a job):
_____
_____

A new acquaintance or a romantic interest:
_____
_____

A person at a party with others present:
_____
_____

A relative at a family gathering:
_____
_____

Fig. 4 Alcohol and drug refusal skills communication worksheet

process therapists also model effective skills (e.g., active listening, collaborative questioning, respectful disagreement). Figure 4 is a worksheet therapists can complete in session with teens that supports them in building communication skills in refusing drugs and/or alcohol. One teen client who had received communication skills training with his parents reported in session that when his parents were able to use "I" statements, he was able to have more empathy for them. At the same time, when the client shared how he felt pressured by his parents, he reported experiencing his parents as more empathic toward him.

4   Family Behavior Therapy for ASU

4.1   Overview of Family Behavior Therapy for ASU

Dating back to the 1980s, family-based treatments have amassed enormous empirical support (*see* [38] *for a review*) in part due to the central role that the family environment has in the development and sustainment of adolescent substance use [39]. Family-based treatments have been shown not only to reduce adolescent substance use but also improve other aspects of adolescent and family functioning, including behavior problems and parenting skills. Among the family-based treatment models that are shown to be effective is family behavior therapy (FBT) [40]. In randomized clinical trials this intervention has demonstrated positive outcomes in adolescent [41], emerging adult [42], and mature adult [43] populations, including populations explicitly assessed to evidence substance misuse and common co-occurring problems, such as conduct disorders and child neglect [44, 45]. As described below, FBT is a comprehensive, family-based treatment that includes widely utilized components of behavioral treatment (e.g., contingency management, stimulus control), emphasis on skill acquisition through reinforcement, and family member involvement to support generalization of skills taught in session to the home and other environments.

Treating Adolescent Substance Use Using Family Behavior Therapy [32], a step-by-step guide to utilizing FBT with adolescents that includes case examples and worksheets, is the foundation for the eight main FBT interventions described in this chapter. Treatment integrity, which refers to ensuring that evidence-based interventions are implemented in accordance with the main principles and procedures of the given model [21], is a critical consideration for delivering standardized treatment models. A key aspect of treatment integrity in the FBT model pertains to its therapeutic style and strategy: an emphasis on positive feedback and encouragement that is specific and descriptive, a facilitative communication style that avoids blame, creative and focused engagement with caregivers and significant others, and an emphasis on learning-by-doing via role plays and use of homework assignments.

Some of the key interventions of FBT need to be implemented at the beginning of treatment, whereas the order of interventions that follow can be determined in collaboration among therapist, adolescent, and caregiver. Prior to implementing any of the eight main interventions, a vital first step is orienting adolescents and caregivers to the nature and expectations of FBT, as well as creating an opportunity for family-unique goals to be articulated. Assessments of the adolescent's substance use behavior, along with satisfaction scales for both the adolescent and the parent (e.g., Life Satisfaction Scale [46] and Parent Satisfaction Scale [47]), can be used to better understand the needs of the youth and measure treatment's impact in domains of substance use and related behavior problems. The orientation process also serves to build an understanding of and commitment to participation in treatment sessions. A key distinction of FBT from other behavioral approaches to adolescent substance use problems is that at least one caregiver is encouraged to be present for every session. Additional significant others may participate in FBT to the extent that they are a resource for the adolescent with regard to accomplishing treatment goals, particularly the goal of reducing substance use.

4.2 Main FBT Intervention Components and Selected Protocol Materials

4.2.1 FBT Intervention 1: Consequence Review

An essential intervention of FBT is consequence review, which serves as both an assessment tool and a means to build motivation for substance-using youth (a difficult population to engage in treatment) to work toward therapeutic goals. Youth are invited to examine their perceived unpleasant consequences of substance use as well as the positive or reinforcing aspects of avoiding substance use. Consequence review operates on the premise that the negative consequences of substance use are typically ignored or avoided within the family due to their aversive associations. Indeed, conversations about substance use with significant others are often confrontational and accusatory, which can lead youth to actively justify positive aspects of use. Therefore, this intervention begins with the therapist unconditionally soliciting and listening to the youth's concerns about the negative consequences of substance use without judgment or empathy. After an appreciation of negative consequences is sufficiently gained, the therapist clarifies youth responses, empathizes with reviewed concerns, and solicits positive consequences associated with substance avoidance. A therapeutic style of nonjudgmental curiosity is vital to soliciting meaningful information and understanding regarding how substance use has led to negative consequences, changes in relationships, and so on. A facilitative and authentically nonconfrontational manner also invites adolescents to elaborate on their experiences. Anytime a youth shares openly about their substance use, it is important for the therapist to validate the feelings that emerge and reflect appreciation for the fact that the youth is sharing. The therapist also makes connections between what the adolescent is sharing to what other youth commonly report as negative consequences of use. The therapist should also maintain a stance of curiosity about the perceived value of the substance use for the adolescent, and an empathic position in exploring possible alternatives to substance use, for example, more goal-directed behavior. Throughout treatment, if an adolescent or caregiver is struggling to maintain motivation, the consequence review can serve as a touchstone of reasons to continue with the goals of therapy.

4.2.2 FBT Intervention 2: Level System	A key feature of FBT that distinguishes it from individual/group CBT for ASU is the participation of caregivers. The FBT level system intervention encourages youth to determine and attain goals with the support of their families. The goal is to build a contingency management system in which adolescents earn rewards from their caregivers consequent to performing behaviors that are incompatible with substance use and other undesired behaviors. The therapist can assess current systems for rewarding behavior within the family, including how consistently rewards are provided or how meaningful current rewards are to the adolescent. Implicit in this intervention are principles of behaviorism: If caregivers positively reinforce behavior that is incompatible with substance use, this substance use can be extinguished. In this system, daily points may be earned for the completion of desired behaviors. Reinforcers are generated from both youth and caregiver (s) separately using a menu of options. Reinforcers must be desired by the youth and within the caregiver's control (e.g., money, clothes, time together, having a friend spend the night). There are three levels of desired behavioral goals, with each level progressively being more difficult to achieve. Behavioral goals must be achieved in order to obtain points for the day (i.e., no signs of substance use, school and/or work attendance, making curfew). Points can be exchanged for rewards at the end of each day. When all goals are achieved for the week, youth are permitted to advance to the next level. If any targeted substance use is indicated the youth is demoted to the first level. The level system is structured to be progressive, with increasing improvement matched to increased rewards.
4.2.3 FBT Intervention 3: Treatment Plan	The FBT preparatory interventions of orientation, consequence review, and level system prepare the therapist and family for developing a treatment plan. Ideally, preparatory interventions increase interest and motivation for both youth and caregiver to achieve therapeutic goals and commit to the FBT process. Youth and caregiver are provided with a summary of the five skill-based FBT interventions (described below). After they have a basic understanding of those five interventions, youth and caregiver can separately rank the priority of the skills areas, after which the therapist can average the rankings to determine the order in which the skills are taught and practiced. Higher priority skills are implemented prior to lower priority skills (successive and cumulative implementation) and are thus reviewed more frequently than lower rated

interventions as treatment progresses. This treatment planning process emphasizes flexibility and client autonomy and supports retention in the often challenging course of family treatment.

4.2.4   FBT Intervention 4:
Reciprocity Awareness

Reciprocity awareness can be understood as a way of repairing the often difficult relationships in families in which a teen is using drugs or alcohol. Problems in these immediate, intimate relationships invariably contribute to adolescents' emotional and behavioral problems. The way that FBT creates healing in relationships is by creating an environment in which everyone in the family can express appreciation for family members' past actions, and family members can respond to this expression of appreciation positively and provide reassurance that the acknowledged behaviors will occur in the future. The premise of reciprocity awareness is that gratitude and appreciation are powerful reinforcers for the performance of non-problem behavior and incompatible with problem behavior. Functional relationships are conceptualized to occur when there is an equitable exchange of "perceived reinforcement" among family members. Therefore, therapists assign positive statement exchanges for homework throughout treatment, so that positive behaviors are not forgotten or taken for granted. This intervention provides substantial comfort that may feel unexpected for families under stress, and it can allow for relationships to be powerful in creating and sustaining meaningful change.

4.2.5   FBT Intervention 5:
Positive Request

Just as communication training is a core element of individual/group CBT for ASU, the skill of positive request is a key feature of communication training in FBT. Many families lack skillful ways to ask for positive reinforcers and to communicate effectively to manage potential problems. The positive request intervention teaches family members how to ask effectively for what they want, and how to respond more effectively to others asking for what they want. As described in other intervention components, FBT holds as a premise that family members can positively reinforce one another toward more skillful behavior, including decreased substance use, and positive request creates opportunities for this to be explicit between members. The therapist teaches family members how to make specific requests, for which help is offered and alternatives can be suggested. The use of positive requests can be supported and encouraged by therapists when communication problems arise in later sessions, or when family members report conflict outside of session. Youth and significant others are taught that some components of the request are more important than others, and when something is especially desired it may be important to attempt as many of the components as can be remembered. Therapists usually invite family members to practice positive requests in session and at home; homework assignments are reviewed in subsequent sessions.

4.2.6   FBT Intervention 6: Environmental Control

Environmental control, also understood as stimulus control, involves the family assisting youth in changing their environment to support reduced drug use and troublesome behavior. Just as the individual/group core elements of functional analysis of behavior and prosocial activity scheduling aim to increase awareness and make changes to environmental factors that support the youth's substance use, environmental control involves the youth and caregiver collectively in avoiding triggers and participating in safe environments that are incompatible with problem behaviors. The therapist initiates a rationale for this exploration and then asks both youth and caregiver which stimuli in the youth's life are "safe." Usually, youth and caregiver separately develop a list of people, places, emotions, and activities that put the youth "at risk." Youth are queried to review things that are liked and disliked about the generated stimuli so that the therapist can determine factors that may be used later in therapy to motivate the youth toward prosocial behavior. The goal is for the entire family to decrease exposure to at-risk stimuli and increase exposure to safe stimuli. The last part of the environmental control intervention involves scheduling a family activity that has appeal and novelty for everyone in the family, in which family members can use earlier acquired communication skills as applicable to plan, participate in, and discuss in a subsequent session when reviewed as a homework assignment.

In subsequent sessions, therapists review safe and at-risk stimuli with the youth, as well as with significant others when reviewing at-risk stimuli. Youth are queried to review actions and thoughts that were performed to avoid substance use and troublesome behavior throughout the prior week. Skills and positive affect are descriptively praised, and when difficulties are experienced, therapists use problem-solving and role-playing to determine future actions likely to assist goal achievement. That is, when substance use or other problem behavior occurs, youth and caregivers are taught to non-emotionally generate methods of decreasing future problem behavior if the situation were to recur, and whenever possible, to role-play skills learned in therapy that are similar to the cognitive and emotion regulation skills described above for individual/group CBT. For example: thinking of negative consequences if substance use were to occur, muscle relaxation, choosing a previously generated alternative behavior, and thinking of positive consequences likely to result from avoidance of drug use.

4.2.7   FBT Intervention 7: Self-Control

Self-control is an intervention meant to increase youth's capacity to recognize thoughts and feelings that lead to problem behavior, including drug use, and then to interrupt the intensity of these triggers with thoughts and behaviors that support healthier behaviors. Self-control can be taught to caregivers as well as youth, and caregivers have an opportunity to model the use of this skill with their youth, which both creates more opportunities for learning and improves the supportive nature of the relationship. The intervention involves first learning to recognize an earlier time period in which something could have been done to prevent the undesired behavior from happening if the youth "wanted" to avoid the problem. They are then taught to perform a series of theoretically grounded skills, including thought stopping or refocusing attention, reviewing one negative consequence to self and other, diaphragmatic breathing, problem solving, and imagery specific to successfully managing the undesired behavior and being rewarded for doing so. Youth are encouraged to practice self-control skills at home in subsequent weeks. Self-control may be utilized during environmental control sessions, in sessions focused on other FBT skills, or anytime during which learned skills can support healthy behaviors in the context of difficult emotions and experiences.

4.2.8   FBT Intervention 8: Job-Getting Skills

A signature feature of FBT is the acquisition of skills to support youth becoming employed. Whereas other FBT interventions target substance use that itself interferes with the youth employment, job-getting skills operates from a premise that employment increases youth self-esteem, autonomy, and capacity to be independent. The goals of the intervention are to motivate youth to pursue work and to teach skills necessary to request a job interview and ultimately obtain employment. Skills that feel fundamental and essential for adult life are often valued by youth and can increase interest in the therapeutic process. Therapists rely heavily on the use of role play and modeling with caregivers, with liberal praise for all efforts made by the youth.

5   Discussion

Spectrum of Family Involvement in Youth CBT for ASU: Possibilities and Next Steps for Intersections of Individual/Group Treatment, Family Skills Training, and Systemic Family Therapy

The level of family involvement in adolescent substance use treatment can be thought to lie on a continuum. Family-involved treatment, at the most basic level, seeks to inform family members about the nature of substance misuse in terms of biological, psychological, and behavioral consequences through psychoeducation. This type of intervention differs from family therapy in that the point of intervention is not the family system but rather educating individual family members [48], which ultimately may have somewhat limited clinical effectiveness. At the other end of the spectrum lies systemic family therapy, which approaches adolescent substance use in the context of the family system and intervenes at the level of family relationships, also taking into account other systems with which the family may be in contact such as juvenile justice and child welfare [49]. This type of treatment was developed to address immediate treatment needs and resolve conflict through understanding and repairing family relationships. For a number of reasons, some outlined earlier in this chapter, systemic family therapy is not always feasible and/or available (*see* [12]). As described below, other family-based therapy models exist that vary along this continuum of family involvement and that offer techniques derived from manualized models. Family skills training, which can be understood as occupying the middle of the continuum, is a multi-component behavioral approach that combines behavioral parent training (focusing on cognitive, affective, and behavioral changes in the parent) with youth social and life skills training, along with practice sessions [50]. Further along the continuum lies FBT, which involves family members in a comprehensive manner that is anchored by multiple members participating in (nearly) every session.

Like FBT, Adolescent-Community Reinforcement Approach (A-CRA) is a behavioral approach that lies on the spectrum between family psychoeducation and systemic family therapy. ACRA is a 12-week behavioral intervention that seeks to increase positive prosocial activities for adolescents who use substances. The intervention begins with the therapist building rapport and, like FBT, conducting a functional analysis of substance use behaviors and social behaviors. Client self-assessments, including satisfaction scales, are used to develop and monitor treatment goals. Techniques used include both overarching (i.e., prosocial priming and reinforcing) and specific methods (i.e., learning to make positive requests). The main premise informing A-CRA is that the adolescent's community can be used to reinforce non-using behaviors and encourage prosocial behaviors. Therapists teach skills such as relapse prevention, problem solving, and communication. Initially, caregivers and adolescents are seen separately and then join together [51–53].

Research suggests that ethnic and culturally diverse populations tend to have less access to, and engage less in, behavioral health services and receive lower quality healthcare treatment compared to their White counterparts (e.g., [54, 55]). It is vital that culturally comprehensive and meaningful treatment be provided to adolescents with ASU. Brief semi-structured interviews for ethnic consideration in therapy have been integrated within family-based treatment involving emerging adults (*see* Donohue et al. [42]), but not ASU. Some approaches include adapting treatments to cultural contexts by providing patients with therapists of similar ethnicity [56] Culturally Informed and Flexible Family-based Treatment for Adolescents (CIFTA) is a model intended to be offered over 14 weeks that operates from systemic family therapy principles and integrates themes relevant to Hispanic families. Therapists deliver CIFTA using a modular and flexible approach that typically includes about half of the sessions alone with the adolescent, and the other half with the parent alone or the family together. The therapeutic work with family members integrates interventions such as motivational interviewing and family skills training along with psychoeducation modules that include parenting, drug education, risky sexual behavior, and stress due to acculturation [57, 58].

Multicomponent treatments refer to intervention models or packages that contain multiple intervention components, some of which include family therapy. Multicomponent treatments seek to leverage multiple treatment mechanisms to address the expansive set of developmental risk and protective factors that influence ASU, maximizing both the intensity and diversity of interventions delivered. Hogue and colleagues [12] designated two multicomponent treatments as having the highest level of empirical support, Well-Established: motivational enhancement therapy plus cognitive behavioral therapy (MET/CBT) and MET/CBT plus family therapy. A third Well-Established treatment, A-CRA, was categorized as a singular CBT model but can legitimately be classified as a multicomponent treatment that includes family involvement to some extent, as discussed above. Another three multicomponent treatments were deemed Probably Efficacious: family therapy plus contingency management (CM), MET/CBT + CM, and MET/CBT + family therapy + CM. A major takeaway from Hogue and colleagues, and similar empirical reviews (e.g., [59]), is that multicomponent treatments for ASU continue to accumulate an impressive empirical record for reducing ASU and co-occurring problems among youth.

An important distinction should be made between multicomponent treatments versus multidomain treatments, which contain at least one feature or component that specifically targets a co-occurring disorder or behavioral problem other than ASU, such as later adaptations of Family Behavior Therapy [44]. Treatment packages that target multiple domains of adolescent behaviors and emotions are critical for evidence-based ASU practice, given that the vast majority of youth receiving substance use services also present with comorbid mental health disorders [60]. A particularly promising kind of multidomain treatment package is Combined treatment, in which ASU behavioral treatment is coordinated with pharmacological treatment for a given mental health disorder (e.g., [61, 62]) for attention disorders; [63] for mood disorders). A second kind is Integrated treatment, in which at least one behavioral component targeting ASU is coordinated with at least one behavioral component targeting a co-occurring problem. Integrated models can proceed with concurrent delivery, implementing the two components simultaneously throughout treatment (e.g., [64, 65]), or with sequential delivery in which treatment focuses first on one disorder and then the other (e.g., [66, 67]).

In adolescents, several specific aspects of family relationships interact with individual youth characteristics to produce substance use risk, including lack of family structure, behavioral monitoring, or communication, low cohesion, negative attitudes, and poor parental supervision and influences [68]. Therefore, caregiver involvement in ASU treatment can promote: (1) parenting skills to support development of family cohesion, appropriate supervision, and conflict resolution; (2) substance use psychoeducation to ensure understanding of the genetic cycle of addiction and the importance of modeling appropriate behavior and attitudes; and (3) quality communication skills [68]. Caregivers also play a crucial role in making logistical arrangements that support treatment, including scheduling and attending appointments and providing transportation.

For all these reasons, a next step in family-based treatment planning for clients with substance use problems includes gaining an understanding of how barriers to treatment may hinder family involvement. Common barriers include denial, poor family motivation, prior negative treatment experiences, parental substance use and/or mental health disorders, and cultural differences [69]. Caregivers and other concerned significant others (CSO) may also experience difficulty in balancing employment, childcare, and transportation. Strategies for addressing potential barriers start with a careful, family-specific assessment to identify factors that may impede treatment engagement. Open and frequent communication on the therapist's part, as well as flexibility in scheduling, are well documented strategies for fostering family engagement in treatment [70]. Treatment planning can then progress to include identification of family-focused and family-relevant treatment goals, provision of family-wide substance use psychoeducation, validation of the thoughts and feelings of all family members, and acknowledgement of parental and CSO stress and burden. Family-informed treatment engagement and intervention strategies such as those described in this chapter can help reduce the enormous treatment gap and elevate the success of ASU interventions across the developmental span.

References

1. Johnston L et al (2016) Monitoring the future national survey results on drug use, 1975–2015: overview, key findings on adolescent drug use. http://www.monitoringthefuture.org/pubs/monographs/mtf-overview2015.pdf.

2. SAMHSA Office of Applied Studies (2007) The OAS report: a day in the life of american adolescents: substance use facts. Government Printing Service, Rockville, MD

3. Haughwout SP et al (2016) Treatment utilization among adolescent substance users: Findings from the 2002 to 2013 national survey on

drug use and health. Alcohol Clin Exp Res 40 (8):1717–1727

4. Center for Behavioral Health Statistics and Quality (2015) Behavioral health trends in the United States: results from the 2014 national survey on drug use and health (hhs publication no. Sma 15-4927, nsduh series h-50). Substance Abuse and Mental Health Services Administration, Rockville, MD

5. The National Center on Addiction and Substance Abuse (2011) Adolescent substance use: America's #1 public health problem. The National Center on Addiction and Substance Abuse, New York, NY

6. Chan YF et al (2008) Prevalence and comorbidity of major internalizing and externalizing problems among adolescents and adults presenting to substance abuse treatment. J Subst Abuse Treat 34(1):14–24

7. Keyes KM et al (2015) Effects of minimum legal drinking age on alcohol and marijuana use: evidence from toxicological testing data for fatally injured drivers aged 16 to 25 years. Inj Epidemiol 2(1):1–10

8. Wong SS et al (2013) The risk of adolescent suicide across patterns of drug use: a nationally representative study of high school students in the United States from 1999 to 2009. Soc Psychiatry Psychiatr Epidemiol 48 (10):1611–1620

9. Young DW et al (2007) A national survey of substance abuse treatment for juvenile offenders. J Subst Abuse Treat 32(3):255–266

10. Hernandez L et al (2015) Moderators and mediators of treatments for youth with substance use disorders. Oxford University Press, New York, NY

11. Steinberg L (2007) Risk taking in adolescence: new perspectives from brain and behavioral science. Curr Dir Psychol Sci 16(2):55–59

12. Hogue A et al (2018) Evidence base on outpatient behavioral treatments for adolescent substance use, 2014–2017: outcomes, treatment delivery, and promising horizons. J Clin Child Adolesc Psychol 47(4):499–526

13. Baldwin SA et al (2012) The effects of family therapies for adolescent delinquency and substance abuse: a meta-analysis. J Marital Fam Ther 38(1):281–304

14. Chorpita BF et al (2011) Evidence-based treatments for children and adolescents: an updated review of indicators of efficacy and effectiveness. Clin Psychol Sci Pract 18(2):154–172

15. Henggeler SW, Sheidow A (2012) Empirically supported family-based treatments for conduct disorder and delinquency in adolescents. J Marital Fam Ther 38:30–58

16. Riedinger V et al (2017) Effects of systemic therapy on mental health of children and adolescents: a meta-analysis. J Clin Child Adolesc Psychol 46:880–894

17. Tanner-Smith EE et al (2013) The comparative effectiveness of outpatient treatment for adolescent substance abuse: a meta-analysis. J Subst Abuse Treat 44(2):145–158

18. Cook JM et al (2010) What do psychotherapists really do in practice? An internet study of over 2,000 practitioners. J Psychother Pract Res 47:260–267

19. Gifford EV et al (2012) How do components of evidence-based psychological treatment cluster in practice? A survey and cluster analysis. J Subst Abuse Treat 42(1):45–55

20. Gallo K, Barlow D (2012) Factors involved in clinician adoption and nonadoption of evidence-based interventions in mental health. Clin Psychol Sci Pract 19(1):93–106

21. Hogue A et al (2013) Making fidelity an intramural game: localizing quality assurance procedures to promote sustainability of evidence-based practices in usual care. Clin Psychol Sci Pract 20(1):60–77

22. Chorpita BF et al (2007) Understanding the common elements of evidence-based practice: misconceptions and clinical examples. J Am Acad Child Adolesc Psychiatry 46:647–652

23. Weisz JR et al (2011) Kernels vs. ears, and other questions for a science of treatment dissemination. Clin Psychol Sci Pract 18 (1):41–46

24. Magill M et al (2016) The alcohol intervention mechanisms scale (aims): preliminary reliability and validity of a common factor observational rating measure. J Subst Abuse Treat 70:28–34

25. Institute of Medicine (2015) Psychosocial interventions for mental and substance use disorders: a framework for establishing evidence-based standards. The National Academies Press, Washington, DC

26. Barth R et al (2014) Translating the common elements approach: social work's experiences in education, practice, and research. J Clin Child Adolesc Psychol 43(2):301–311

27. Hogue A et al (2017) Distilling the core elements of family therapy for adolescent substance use: conceptual and empirical solutions. J Child Adolesc Subst Abuse 26 (6):437–453

28. Hogue A et al (2019) Core elements of family therapy for adolescent behavior problems: empirical distillation of three manualized treatments. J Clin Child Adolesc Psychol 48 (1):29–41

29. Conrod PJ et al (2010) Brief, personality-targeted coping skills interventions and survival as a non-drug user over a 2-year period during adolescence. Arch Gen Psychiatry 67(1):85–93

30. Conrod JP, Nikolaou K (2016) Annual research review: on the developmental neuropsychology of substance use disorders. J Child Psychol Psychiatry 57(3):371–394

31. McCart MR, Sheidow AJ (2016) Evidence-based psychosocial treatments for adolescents with disruptive behavior. J Clin Child Adolesc Psychol 45(5):529–563

32. Donohue B, Azrin N (2012) Treating adolescent substance abuse using family behavior therapy: a step-by-step approach. John Wiley & Sons, Hoboken, NJ

33. Henggeler SW et al (1998) Multisystemic treatment of antisocial behavior in children and adolescents. Guilford Press, New York, NY

34. Godley SH et al (2016) The adolescent community reinforcement approach: a clinical guide for treating sustance use disorders. Chestnut Health Systems, Bloomington, IL

35. Lochman JE et al (2008) Coping power: child group facilitator's guide. Oxford University Press, Oxford

36. Webb C et al (2002) The motivational enhancement therapy and cognitive behavioral therapy supplement: 7 sessions of cognitive behavioral therapy for adolescent cannabis users. Center for Substance Abuse Treatment, Substance Abuse and Mental Health Services Administration, Rockville, MD

37. Schoepp G., Eustace J. (2008) Essentials of relaxation training for children & adolescents. Publisher CiteSeer (PDF online)

38. Deas D, Thomas SE (2001) An overview of controlled studies of adolescent substance abuse treatment. Am J Addict 10:178–189

39. Repetti RL et al (2002) Risky families: family social environments and the mental and physical health of offspring. Psychol Bull 128 (2):330–366

40. National Institute on Drug Abuse (2014) Principles of adolescent substance use disorder treatment: a research-based guide. National Institute of Drug Abuse, North Bethesda, MD

41. Azrin N et al (1994) Youth drug abuse treatment: a controlled outcome study. J Child Adolesc Subst Abuse 3(3):1–16

42. Donohue B et al (2018) Controlled evaluation of an optimization approach to mental health and sport performance. J Clin Sport Psychol 12 (2):234–267

43. Azrin N et al (1996) Follow-up results of supportive versus behavioral therapy for illicit drug use. Behav Res Ther 34(1):41–46

44. Azrin NH et al (2001) A controlled evaluation and description of individual-cognitive problem solving and family-behavior therapies in dually-diagnosed conduct-disordered and substance-dependent youth. J Child Adolesc Subst Abuse 11(1):1–43

45. Donohue B et al (2014) A controlled evaluation of family behavior therapy in concurrent child neglect and drug abuse. J Consult Clin Psychol 82(4):706

46. Donohue B et al (2003) Initial reliability and validity of the life satisfaction scale for problem youth in a sample of drug abusing and conduct disordered youth. J Child Fam Stud 12(4):453–464

47. Donohue B et al (2001) Satisfaction of parents with their conduct-disordered and substance-abusing youth. Behav Modif 25(1):21–43

48. Treatment CFSA (2004) Substance abuse treatment and family therapy. CFSA, Washington, DC

49. Boustani M et al (2016) Family-based treatments for adolescent substance abuse: how scientific advances yield new developmental challenges. The Oxford handbook of adolescent substance abuse. Oxford University Press, Oxford

50. Kumpfer KL, Alvarado R (2003) Family-strengthening approaches for the prevention of youth problem behaviors. Am Psychol 58 (6-7):457

51. Godley SH et al (2001) The assertive aftercare protocol for adolescent substance. In: Innovations in adolescent substance abuse interventions. https://doi.org/10.1016/B978-008043577-0/50034-1

52. Godley MD et al (2002) Preliminary outcomes from the assertive continuing care experiment for adolescents discharged from residential treatment. J Subst Abuse Treat 23(1):21–32

53. Godley MD et al (2007) The effect of assertive continuing care on continuing care linkage, adherence and abstinence following residential treatment for adolescents with substance use disorders. Addiction 102(1):81–93

54. Abdullah T, Brown TL (2011) Mental illness stigma and ethnocultural beliefs, values, and norms: an integrative review. Clin Psychol Rev 31(6):934–948

55. Cummings JR et al (2011) Racial/ethnic differences in treatment for substance use disorders among us adolescents. J Am Acad Child Adolesc Psychiatry 50(12):1265–1274

56. Coleman HL et al (1995) Ethnic minorities' ratings of ethnically similar and european american counselors: a meta-analysis. J Couns Psychol 42(1):55

57. Santisteban DA et al (2011) Preliminary results for an adaptive family treatment for drug abuse in hispanic youth. J Fam Psychol 25(4):610

58. Santisteban DA, Mena MP (2009) Culturally informed and flexible family-based treatment for adolescents: a tailored and integrative treatment for hispanic youth. Fam Process 48 (2):253–268

59. Brewer S et al (2017) Treating mental health and substance use disorders in adolescents: what is on the menu? Curr Psychiatry Rep 19 (1):5

60. Dennis M et al (2004) The cannibis youth treatment (cyt) study: main findings from two randomized trials. J Subst Abuse Treat 27:197–213

61. Hogue A et al (2017) A clinician's guide to co-occurring adhd among adolescent substance users: comorbidity, neurodevelopmental risk, and evidence-based treatment options. J Child Adolesc Subst Abuse 26(4):277–292

62. Riggs PD (2011) Randomized controlled trial of osmotic-release methylphenidate with cognitive-behavioral therapy in adolescents with attention-deficit/hyperactivity disorder and substance use disorders. J Am Acad Child Adolesc Psychiatry 50(9):903–914

63. Riggs PD et al (2007) A randomized controlled trial of fluoxetine and cognitive behavioral therapy in adolescents with major depression, behavior problems, and substance use disorders. Arch Pediatr Adolesc Med 161 (11):1026–1034

64. Letourneau EJ et al (2017) First evaluation of a contingency management intervention addressing adolescent substance use and sexual risk behaviors: risk reduction therapy for adolescents. J Subst Abuse Treat 72:56–65

65. Suarez L et al (2012) Supporting the need for an integrated system of care for youth with co-occurring traumatic stress and substance abuse problems. Am J Community Psychol 49 (3-4):430–440

66. Adams ZW et al (2016) Clinician perspectives on treating adolescents with co-occurring post-traumatic stress disorder, substance use, and other problems. J Child Adolesc Subst Abuse 25(6):575–583

67. Rohde P et al (2014) Sequenced versus coordinated treatment for adolescents with comorbid symptoms and substance use disorders. J Consult Clin Psychol 82:342–348

68. Velleman RD et al (2005) The role of the family in preventing and intervening with substance use and misuse: a comprehensive review of family interventions, with a focus on young people. Drug Alcohol Rev 24 (2):93–109

69. Suveg C et al (2006) Parental involvement when conducting cognitive-behavioral therapy for children with anxiety disorders. J Cogn Psychother 20(3):287

70. Gilbert JM et al (2004) Substance abuse treatment and family therapy. A treatment improvement protocol (TIP) series 39. Substance Abuse and Mental Health Services Administration, Rockville, MD

Chapter 9

Cognitive Behavioral Therapy with Eating Disordered Youth

Jamal H. Essayli and Kelly M. Vitousek

Abstract

Cognitive behavioral therapy (CBT) is currently recognized as the most effective treatment for adults with eating disorders (EDs); however, few studies have examined CBT's efficacy for younger patients. In this chapter, we describe how to adapt CBT for children and adolescents with EDs. Similar to the approach used with adults, CBT for younger patients involves an array of interventions designed to modify beliefs and behaviors related to eating, weight, and shape. These include developing a clinical formulation, psychoeducation, enhancing motivation, open weighing, meal planning, self-monitoring, cognitive restructuring, exposure therapy, and relapse prevention. Individual CBT for children and adolescents with EDs should be augmented with family involvement, with the frequency and format of family sessions varying as a function of the patient's age, symptom pattern and severity, and other considerations. Particularly during the early phase of treatment for underweight patients, family sessions focus on coaching the patient's caregivers to support their child with normalized eating and weight restoration. As patients become more active participants in treatment and behavior change, family involvement is reduced gradually.

Key words Cognitive behavioral therapy (CBT), Eating disorders, Children, Adolescents, Anorexia nervosa, Bulimia nervosa, Family therapy

1 Introduction

Eating disorders (EDs), including anorexia nervosa (AN), bulimia nervosa (BN), and binge-eating disorder (BED), are complex psychological conditions associated with a host of serious medical complications [1] and high relapse rates [2]. This is particularly true of AN, which ranks among the most dangerous psychiatric disorders [3]. Patients with EDs also demonstrate distinctive features that further complicate the clinical picture and can thwart successful treatment, including positive valuation of core symptoms and ambivalence about the prospect of change, as well as the psychological consequences of semi-starvation [4, 5].

Individuals with AN and BN, and a substantial proportion of those with BED, hold core beliefs about the importance of weight and shape as measures of self-worth, which contribute to the development of stereotyped behaviors designed to control body weight

Robert D. Friedberg and Brad J. Nakamura (eds.), *Cognitive Behavioral Therapy in Youth: Tradition and Innovation*, Neuromethods, vol. 156, https://doi.org/10.1007/978-1-0716-0700-8_9, © Springer Science+Business Media, LLC, part of Springer Nature 2020

[6, 7]. These behaviors typically include dietary restraint, binge eating, and/or compensatory behaviors such as self-induced vomiting, laxative abuse, and excessive exercise. These symptoms can emerge across ED diagnoses: while AN is the only ED characterized by subnormal body weight, purging may occur in individuals with AN or BN, and binge eating in those with AN, BN, or BED [8]. The diagnostic categories of "other specified feeding or eating disorder" (OSFED) and "unspecified feeding or eating disorder" describe individuals who experience disordered eating but do not meet full criteria for AN, BN, or BED. In contrast to BED, AN and BN typically emerge during adolescence or young adulthood [9, 10]. Although EDs can occur in prepubescent children, most pediatric patients with EDs are adolescents: the great majority of patients with AN are at least 13 years old, while those with BN tend to be 15 years or older [11–13]. Due to space limitations and the later characteristic age of onset for BED, the current chapter will focus on adapting CBT for children and adolescents with AN, BN, and their subthreshold presentations included in OSFED.

2   Theoretical and Empirical Foundations

Cognitive behavioral theorists have proposed that the overvaluation of weight and shape and their control represent the "core psychopathology" of EDs [6, 14]. Individuals with EDs learn to evaluate themselves largely or exclusively on the basis of dietary restraint, weight, and/or body shape. These core beliefs result in disordered eating- and weight-related "safety" behaviors, including severe calorie restriction, food avoidance, rigid dietary rules and rituals, excessive body checking, and body avoidance [15]. Safety behaviors reduce anxiety in the short term but maintain anxiety over the long-term, as the non-occurrence of the feared outcome (e.g., weight gain) is attributed to safety behaviors such as avoidance of specific foods or conformity to dietary rituals [16]. Dietary restraint and rigid rules about eating in turn predispose individuals with EDs to subjective overeating and objective binge-eating episodes, which are commonly associated with intense distress and the subsequent employment of harmful compensatory measures such as further dietary restraint, fasting, self-induced vomiting, laxative abuse, diuretic misuse, and excessive exercise [6]. Compensatory behaviors are maintained through negative reinforcement, as they serve to reduce eating- and weight-related anxiety [15, 17]. Additionally, successful dietary restraint and weight loss can be positively reinforced by feelings of euphoria, accomplishment, pride, power, identity, and self-control [4, 5, 18–20].

Cognitive behavioral therapy (CBT) is recognized as the most effective treatment for adults with BN, producing symptom remission more reliably and durably than the alternative psychological and pharmacological modalities tested, although a substantial minority of patients experience persistent symptoms [2, 21–23]. While few studies have evaluated CBT for adolescents with BN [24], many patients in randomized controlled trials (RCTs) are young adults, and it seems reasonable to hypothesize that the approach may be similarly effective for individuals who are just a few years younger [25, 26]. CBT for AN, like all examined interventions for the disorder, yields lower rates of recovery relative to treatments for BN [27].

Family-based treatment (FBT), initially referred to as the Maudsley Model or Maudsley Family Therapy [28–30], is the most well-studied psychological treatment for adolescents with AN [24, 27]. Although FBT proponents recommend conjoint family therapy in which the patient, parents, and siblings attend all sessions together [30], the evidence suggests that slightly better outcomes are attained when the patient and parents are seen separately [31–33]. Both FBT and CBT emphasize that the family is not to blame for the child's illness, and agree that parents are responsible for ensuring their child's safety. In contrast to CBT, however, FBT construes the ED as an invading force that has overtaken the child's life, and initially directs parents to take full control over their child's eating behaviors. Decision-making about eating and exercise are progressively transitioned back to the patient as she or he improves [30]. Clinical trials evaluating FBT for AN have produced results comparable [33] or superior [34] to those investigating CBT for adolescents with AN [35]. RCTs directly comparing various versions of family therapy and CBT for AN and BN have yielded mixed results [36–38].

Cognitive behavioral experts have consistently recommended involvement of family members in treatment [17, 35, 39], and existing CBT manuals for EDs provide some guidance on how to adapt treatment for younger patients [14, 40, 41]. Despite consensus that families should be included in the treatment of children and adolescents with EDs, the optimal degree and format of family participation remains unclear. In this chapter, we provide recommendations for augmenting CBT with variable levels of family involvement. CBT with family involvement may be able to capitalize on the strengths of both individual CBT and FBT, and support outcomes that match or exceed those associated with either.

3   Intervention

There is a strong basis for postulating that CBT is well matched to the prominent cognitive and behavioral features of EDs. Distorted beliefs about eating, weight, and shape play a central role, avoidance of anxiety-provoking situations and behaviors maintains ED symptoms, and convictions about the value of dietary restraint and thinness contribute to ambivalence about change. Motivational enhancement strategies and cognitive interventions that emphasize collaboration and individual decision-making are often necessary to persuade patients to initiate distressing behavior change and improve nutritional status, which in turn are essential for the higher-order cognitive work that follows. Throughout the course of treatment, cognitive, motivational, and exposure-based interventions are implemented to facilitate corrective learning (e.g., more accurate information about the relationship between eating and weight) and to modify patients' core beliefs, values, and assumptions about eating, weight, and shape [14, 17, 40, 41].

CBT's focus on collaborative empiricism, individual decision-making, and corrective learning differ from the external control model that characterizes FBT, which is currently the best established approach for younger patients with AN. Children and adolescents may be less responsive to CBT interventions designed to enlist patients in initial behavior change, as they have lower levels of cognitive maturity, are less likely to have experienced the adverse long-term effects of EDs, and may be less influenced by abstract cognitive and psychoeducational material. Because of the serious health risks associated with low weight, irregular eating, and dangerous compensatory behaviors, it is crucial that individuals with severe EDs normalize weight and dietary intake. These considerations raise questions for CBT practitioners treating children and adolescents with EDs. How do we reconcile the use of external control strategies to support weight restoration with CBT interventions that encourage collaboration, respect individual decision-making, and focus on the patient's personal beliefs and values?

In fact, CBT for younger patients overlaps with the approach often used with adult cases receiving intensive outpatient or hospital care. During the early phase of treatment, which is focused on weight restoration in AN and normalizing erratic eating behaviors across all EDs, CBT-informed outpatient and inpatient programs share an emphasis on *constrained* decision-making. Consistent with CBT principles, therapists and staff work to maximize patient autonomy, responsibility, and choice within the confines of "treatment non-negotiables" [42], which include regular eating, weight restoration and maintenance, and reduction or elimination of purging and laxative abuse. When discussing treatment non-negotiables, providers maintain an empathetic, validating, and inquisitive therapeutic style, and encourage patients to share their viewpoints and ask questions.

In an outpatient setting, both CBT and FBT rely on the patient's primary caregivers[1] to exert some external control over the child's eating behaviors. Unlike FBT, however, CBT maintains

---

[1] Hereafter, we refer to the primary caregivers as "parents" for simplicity.

a persistent interest in understanding the young person's perspective on her or his symptoms and involves the patient in the decision-making process as much as possible without compromising her or his physical or mental health. While FBT assumes that all patients should receive the same degree of family participation in treatment, CBT recommends that the format and level of family involvement be individualized according to each patient's need for external support. As patients with EDs begin to normalize eating behaviors and those with AN restore weight, family involvement decreases gradually.

The first part of this section describes individual CBT interventions for EDs, most of which are discussed in more detail elsewhere [14, 17, 40, 41]. Because of space limitations, we place greater emphasis on material that is specific to the distinctive features of EDs and on treatment elements that have received less attention in traditional CBT protocols for EDs. The second part of this section provides practical recommendations and considerations for augmenting individual CBT for EDs with a flexible degree of family involvement.

3.1 Individual CBT Interventions for Children and Adolescents with EDs

3.1.1 Assessment, Medical Evaluation, and Determining Appropriate Level of Care

Prior to initiating treatment, individuals with EDs should receive a comprehensive psychological evaluation that includes assessment of ED features [17]. Reliable and valid assessment measures include the Eating Disorder Examination structured interview (EDE) [43], the self-report questionnaire version of the EDE [44], and the third edition of the Eating Disorder Inventory [45]. Before or during the psychological assessment process, the therapist should obtain parental consent to contact the child's pediatrician or family physician. A medical evaluation should be conducted by a physician familiar with the medical findings and complications of eating disorders, particularly if the child is underweight, rapidly losing weight, vomiting, and/or engaging in laxative abuse. Some of the physical and physiological variables that might be assessed include weight and height; heart rate and blood pressure; levels of phosphorus, magnesium, potassium, and glucose; thyroid function; estradiol or testosterone levels; bone mineral density; hydration; and gastrointestinal function [46, 47]. Alarming results may indicate hospitalization to ensure medical stability.

3.1.2 Discussion of Confidentiality

Although all therapists who treat minors should be familiar with ethical dilemmas related to confidentiality, the medical complications associated with EDs warrant special consideration. Specifically, dangerous compensatory behaviors such as laxative abuse and self-induced vomiting, as well as severe dietary restraint and weight loss, constitute risk of harm to self and, depending on the severity, may need to be disclosed to the patient's family.

3.1.3 Setting an Agenda

Consistent with CBT for other conditions, therapists treating EDs start sessions by setting an agenda [14, 41, 48]. Agendas help keep sessions focused and goal-oriented, prioritize topics by importance, and emphasize the collaborative nature of CBT as patients are encouraged to add and modify agenda items.

3.1.4 Clinical Formulation

Several prominent cognitive behavioral therapists recommend prioritizing the development of a visual clinical formulation with each patient at the start of treatment [14, 41]. Developing a clinical formulation helps patients understand the interconnectedness of their eating-related beliefs and behaviors, underscores the importance of modifying eating behaviors, and provides a rationale for CBT. The therapist typically starts by describing the CBT model that conceptualizes the over-evaluation of weight and shape as the "core psychopathology" of EDs. Next, using a blend of Socratic questioning and psychoeducation, the therapist and patient collaboratively identify previous experiences that may have contributed to the patient's current beliefs about eating, weight, and shape. The therapist writes down each of these hypothesized factors inside a box or oval and draws arrows to denote the ways in which these variables contributed to the over-evaluation of weight and shape. The therapist and patient then explore behavioral, cognitive, physiological, and/or emotional consequences (including perceived benefits) related to the over-evaluation of weight and shape, which are subsequently added to the visual clinical formulation. All components of the formulation are then connected with arrows that identify hypothesized antecedents and consequences (*see* [14, 40, 41] for samples). New information should be added to the visual formulation as the patient develops additional insight into her or his ED over the course of treatment.

While there are benefits to developing a joint formulation early in treatment, we recommend that therapists use their clinical judgment to consider whether the formulation should be proposed before or after other treatment components are initiated. For some patients, key interventions such as providing psychoeducation, meal planning, and introducing open weighing may take up all of the first few therapy sessions. In these instances, we often postpone the development of a clinical formulation to a later session.

3.1.5 Psychoeducation

ED-related psychoeducation provides the rationale for prioritizing interventions that facilitate regular eating, corrects misinformation about eating and weight, and sets the foundation for the motivational and cognitive work to follow. Psychoeducational topics are heavily emphasized during the first few CBT sessions, and revisited frequently throughout treatment.

The effects of strict dieting and semi-starvation. Patients with AN should be provided with detailed information about the physiological, behavioral, and psychological effects of semi-starvation

[14, 41, 49], most extensively documented in the Minnesota Starvation Study [50]. Patients (and uninformed providers) often formulate pathological explanations for what are in fact typical human responses to starvation. It is important that a patient with AN recognize that starvation explains why she is cold all the time, why she tears her bread into small pieces, why she devotes an hour to eating an apple, why she spends much of her day reading recipes online, why she is constantly irritable, and why she has lost interest in spending time with her friends and family [5]. The hardwired effects of starvation virtually guarantee that patients will continue to experience unpleasant symptoms (e.g., depression, low energy, food preoccupation, irritability, binge eating) as long as they maintain restrictive eating patterns. Although this information is especially relevant to highly restrictive, underweight patients, all individuals with EDs are likely to manifest attenuated starvation effects. Those who experience subjective loss of control over eating and episodes of binge eating may benefit from learning that these are expected responses to dietary restraint [51, 52] that are unlikely to resolve until they moderate dietary rules and consume regular meals and snacks throughout the day [14, 17, 41].

The ineffectiveness of self-induced vomiting, laxative abuse, and diuretic abuse. Patients who engage in self-induced vomiting, laxative abuse, and/or diuretic abuse should be provided with accurate information about the limited effectiveness of these behaviors for weight control. Many individuals with EDs hold the erroneous belief that these purging behaviors "cancel out" the calories eaten. In reality, research suggests that individuals with BN who induce vomiting eliminate less than 50% of the calories consumed [53]. Laxatives act on the large intestine after most calories have already been absorbed, and are thus an ineffective weight loss strategy [54]. The sensation of weight loss and any decreases in body weight after laxative use are merely the temporary effects of dehydration. Similarly, diuretics facilitate the loss of fluids from the body, with no effect on calorie absorption.

The medical complications associated with EDs. Although non-medical therapists cannot provide medical management to patients, it is crucial that they be familiar with the medical complications associated with EDs, discuss these risks with patients and their families, and recognize when to consult with a physician. Accurate, detailed information about the health consequences of EDs can be accessed readily online [55], in treatment manuals for EDs [41], and in scholarly articles [1, 56].

Myths about food, dieting, and weight. Individuals with EDs are avid consumers of messages related to nutrition and weight loss but are uncritical and biased in interpreting this information. Moreover, many in the general population hold inaccurate beliefs about food and experience eating-related guilt [57, 58]. A quick online search for healthy eating advice yields warnings about the negative effects of gluten on the brain, the addictive properties of carbohydrates, the link between processed foods and poor sleep, and the impact of refined sugars on early aging. Unsurprisingly, patients are often puzzled when ED therapists recommend that they reduce their vegetable intake, avoid low-fat foods, and eat desserts more often. Clinicians must be prepared to correct dietary misinformation and provide both psychological and nutritional rationales for eating a varied diet of carbohydrates, fats, and protein [59].

Working with patients to discern nutritional truth from fiction is complicated by the fact that *certain* weight-related beliefs are valid [60]. For instance, an adolescent with AN may divulge that she dropped 20 lbs. by skipping breakfast, cutting out carbohydrates, and avoiding restaurants, and report fear that changing these dietary rules and rituals will lead to uncontrollable weight gain. It is important that therapists recognize and effectively communicate that it is not these specific dietary rules and rituals that caused the patient's weight loss but rather her consistent decrease in energy intake relative to energy expenditure over time [61]. Similarly, the patient should understand that weight gain after deliberate weight suppression is the result of eating an appropriate amount of food rather than a direct effect of consuming her first meal earlier in the day, reintroducing pizza and pasta into her diet, or eating in restaurants more frequently.

3.1.6 Enhancing Motivation

Many children and adolescents with EDs, particularly AN, enter treatment against their wishes and report low intrinsic motivation to change [5]. Although eating behaviors may improve solely in response to external pressure from parents or hospital staff, individuals who do not increase their internal motivation to recover are prone to poor long-term outcomes and relapse [62–64]. While research has not consistently favored *separate* motivational interviewing sessions for EDs [65], CBT *integrates* motivational enhancement strategies with other cognitive behavioral interventions [5, 41]. Therapists should assess motivation to change on an ongoing basis and implement motivational enhancement strategies as warranted throughout the course of treatment.

Pros and cons of maintaining versus changing eating behaviors. Several CBT manuals for EDs recommend exploring the pros and cons of recovery as a strategy to increase motivation for change [14, 41, 66]. Pros and cons include the perceived benefits and consequences of both maintaining and changing eating behaviors. After exploring pros and cons and providing the patient with relevant psychoeducation about EDs, we often discuss the "package deals" associated with various degrees of recovery [5]. An adolescent with AN who maintains extremely restrictive eating patterns and low body weight may continue to derive feelings of achievement and increased body satisfaction from her ED, but must also endure depression, poor concentration, low energy, and food preoccupation. Partial recovery, which might be achieved by loosening dietary restraint and gaining some weight, characteristically yields improvement in some physical and psychological effects; however, weight preoccupation, avoidance of preferred foods, and anxiety in social eating situations are likely to persist, as well as heightened risk of relapse. While full recovery from an ED involves foregoing the positive feelings derived from controlled eating and weight loss, it also frees the patient from the multitude of negative consequences inextricably tied to an ED.

Projecting into the future. Encouraging patients to envision what their future might look like if they were to maintain versus recover from an ED is another recommended strategy that may enhance intrinsic motivation to change [5, 41]. Therapists should use a collaborative and curious Socratic style to help patients reflect upon the degree to which an ED is compatible with their future goals. The sample dialogue below includes some of the questions we might use when asking patients to project into the near future.

THERAPIST: I want us to spend more time discussing what *you* think about changing your eating and getting back to a healthy weight. I'm curious— what do you think would happen if your parents said that you no longer had to come to therapy and that you could choose to eat in any way you wanted?

PATIENT: I'm pretty sure I would drop out of therapy and go back to the way I was eating before. I hate having to eat so much food and I feel better about myself when I'm losing weight.

THERAPIST: OK, so you anticipate that you would immediately cut down on your eating. In the four months before you and your parents came to treatment, you had lost a lot of weight. What do you think would happen to your weight if you returned to your old eating habits?

PATIENT: I think I would quickly lose all the weight that I've gained in treatment so far.

THERAPIST: That does seem likely. After getting back to what you weighed when treatment began, do you think you would keep trying to lose more weight?

PATIENT: I'm pretty sure I would. My parents stopped me before I could get to my goal weight.

THERAPIST: If you were able to reach your goal weight, do you think you would stop trying to lose weight at that point, or keep trying to lose more?

PATIENT: I'm not sure. I'd be really happy that I got to my goal weight. But I guess I might be tempted to lose a little more weight even after that. It's just that losing weight makes me feel better about myself than almost anything else.

THERAPIST: When do you think your temptation to lose weight might stop?

PATIENT: I don't know. It's hard for me to see that temptation ever going away.

THERAPIST: Even before reaching your goal weight, you started experiencing some medical issues, like feeling dizzy and weak, and blacking out that one time. You also noted that you often felt depressed and had trouble concentrating on things. Based on what we know about how the body is affected by starvation, it's likely that you would have even worse problems of that kind at lower weights. What do you think might happen if no one stopped you from continuing to lose weight?

PATIENT: I guess I could see it leading to more medical problems. It could get pretty bad. I might even have to be taken to a hospital.

THERAPIST: I think you're right. Hospitalization is usually what happens when people lose that much weight. To make sure you were safe, the hospital would refeed you until you got up to a somewhat healthier weight before letting you go. If that were to happen to you, what might you do after getting out of the hospital?

PATIENT: I honestly don't think that even going to the hospital could get me to stop losing weight. I think I would go right back to eating as little as I could and start losing weight again.

THERAPIST: And what might happen if you were to lose more weight again?

PATIENT: I might have to be put back in the hospital.

THERAPIST: So it sounds like not eating enough and losing weight would lead to a cycle of being in and out of the hospital.

PATIENT: Yeah, I guess so. Well, maybe I could get myself to lose *some* weight, but not so much that I would have to go to the hospital.

THERAPIST: How do you think you would feel about still not being at your preferred weight? You said that it's hard to imagine the temptation to lose going away.

PATIENT: I don't really know. I'd probably still be unhappy that I'm not at my goal weight.

THERAPIST: That sounds like a real dilemma. Right now, the pull toward eating less and losing weight is so strong that you can't imagine making the decision *yourself* to eat enough food to stay within a healthy weight range and avoid all those medical consequences, as well as the problems with depression and concentration—but you also hate it when other people take over your life in order to prevent those things from happening. Maybe if you could kind of split the difference, you think, you could keep your eating and weight *partly* restricted and have *fewer* of those negative effects—but you're not sure how you could give up your longing to weigh less.

PATIENT: Honestly, all those options sound bad. It seems like I lose out no matter what I do.

Another helpful exercise encourages the patient to imagine that she or he and the therapist have been transported to some time (e.g., 5 years) in the future and are meeting again to discuss life. Patients are asked to role-play different future versions of themselves, including someone who has fully recovered from an ED and someone who continues to have an ED [41]. We start these role-plays with a lighter tone and questions unrelated to eating and weight to increase interest in the exercise, reduce defensiveness, and assist patients with vividly imaging their future. For example, we often ask patients to start by describing their age, place of residence, college major, job, roommates, friends, relationship with family, dating life, and hobbies from this future perspective. Questions then gradually shift to eating, weight, body image, and mental health, such as "What do you eat on an average day?" "How often do you think about your weight and body shape?" "How do you feel when you go out to eat with friends or family?" "What were some of the most important changes that helped you achieve recovery?" "Is there any part of you that regrets recovering from an

ED?" "How do you handle stressful aspects of life without ED symptoms?" "How has your mood been?" "Is anxiety any better or worse than it was before?" "If you could go back in time and give your old self some advice, what might you say?" The therapist uses similar questions when patients are asked to envision themselves as young adults with EDs. After being "transported" back to the present, the therapist asks the patient to reflect on the exercise, compare and contrast the different future versions of themselves, and discuss potential implications for the patient's decision to change or maintain their eating behaviors [41].

Exploring the patient's values. Control of eating and weight is often *consonant* with the values reported by patients with AN, such as being disciplined, self-restrained, rule-oriented, hard-working, and favoring higher-order goals over primal human urges [5]. Therapists should explore the full range of patients' personal beliefs and values, gently highlighting both consistencies and inconsistencies between various values and behaviors, and encouraging the patient to consider ways to resolve incompatibilities. For example, a therapist might discuss whether values related to achievement are more compatible with maintaining ED symptoms or embarking on the difficult path to recovery. Collaborative and strategic Socratic questioning usually leads patients to reveal some ambivalence about selecting thinness as the ultimate measure of achievement, individuality, and self-worth. When asked, almost all patients state that they would not recommend that their own (hypothetical) children, close friends, or family members develop an ED, suggesting some incongruence between their values and the decision to maintain an ED. Encouraging patients to reflect on societal beauty standards and pressures to be thin may also increase cognitive dissonance between beliefs and behaviors [67, 68] and increase motivation to change.

| 3.1.7 *Open Weighing* | Weighing patients with EDs on a regular basis is necessary to monitor an important indicator of their health and treatment progress. While many therapists who treat EDs choose to keep patients blind to their weight [69, 70], CBT involves *open* weighing: patients see their weight on the scale during therapy sessions once per week and are involved in direct discussion about their weight, including weight goals for patients who are underweight [14, 17, 41]. Individuals with EDs often overestimate the impact of eating on their weight, believe that consuming feared foods and breaking dietary rituals will lead to rapid and substantial weight gain, and interpret any increase in weight as catastrophic. Exposing patients' to their own weight and comparing their predictions about weight gain to the often modest actual changes in weight can help modify distorted beliefs, reduce anxiety, and increase self-efficacy related to eating and weight [71]. Given the high likelihood that patients will |

misinterpret weight information, open weighing should occur during therapy sessions, and patients are discouraged from weighing themselves at home [14].

We typically provide the rationale for open weighing and psychoeducation about weight before the first in-session weigh-in, including information about variables that impact short-term weight fluctuations (e.g., water retention, hormonal changes, and the weight of urine, stools, liquid, and food in the body). Additionally, we convey that weight must be measured over several weeks before drawing conclusions about weight trajectory [41]. After discussing any concerns the patient may have about seeing her or his weight, the therapist constructs a weight graph with the number of weeks on the *x*-axis and weight (e.g., in pounds) on the *y*-axis. Consistent with the open weighing procedures described by Waller and colleagues [41, 71], we ask the patient to predict her or his weight to the nearest whole pound. We recommend rounding weights to the nearest whole number to prevent patients—and *therapists*—from overinterpreting minor weight fluctuations (e.g., 0.3 lbs) as indicative of meaningful weight change. The therapist marks the patient's weight prediction on the graph, weighs the patient on a standing scale with shoes off and clothing on, reads the weight (rounded to the nearest pound) out loud, documents the patient's actual weight on the graph, and discusses the patient's response to seeing her or his weight, including possible reasons for the discrepancy between predicted and actual weight. During each weekly weigh-in, the therapist follows a similar procedure, and graphs the patient's actual weight and cumulative predicted weight over time. For example, if a patient predicts that she has gained 3 lbs. since last week's weigh-in, the therapist would add a data point that is 3 lbs. greater than her *predicted* weight from the week prior. After multiple weeks of treatment, this cumulative predicted weight graph (representing the patient's beliefs about weight) can be contrasted with the patient's actual weight over time.

More often than not, patients overestimate their weight when making a prediction. Exploring the discrepancy between patients' actual versus predicted weight can help modify patients' often erroneous beliefs about food and weight. It may be particularly useful to discuss whether the patient is misinterpreting certain sensations, behaviors, or thoughts (e.g., feeling full or bloated, eating feared foods, poor body image, anxiety or depression) as evidence of weight gain. When a patient accurately predicts or underestimates her or his weight, the therapist should similarly use a Socratic style to explore the patient's beliefs about the meaning of this weight information, question whether the patient is misinterpreting short-term weight fluctuations as uncontrollable long-term weight gain, and discuss the degree to which *certain* beliefs about eating and weight may be accurate. For instance, an adolescent with AN who accurately predicts that she has gained

2 lbs. over the previous week may hold both the *valid belief* that eating regularly will lead to weight restoration and the *mistaken belief* that eating regularly will lead to 2 lbs. of weight gain every week over the next several years.

CBT for AN includes direct discussion of the patient's target weight range [17]. While resumption of menses (in postmenarchal girls who are not on birth control) serves as a useful marker of the patient reaching her *minimum* weight threshold, an adolescent's premorbid body mass index-for-age percentile provides a better estimate of ideal target weight, assuming that the patient demonstrated appropriate eating behaviors and growth prior to the onset of her or his ED [72] (growth charts from the Centers for Disease Control and Prevention can be accessed online at www.cdc.gov/growthcharts). During the weight restoration phase of treatment for AN, patients should gain an average of at least 1 lb. per week. Because of the serious medical complications associated with malnutrition in children and adolescents, we inform patients with AN and their families from the start that any further weight loss or failure to gain weight adequately in outpatient treatment will signal a need for higher levels of care, such as partial hospitalization or inpatient treatment.

3.1.8 Meal Planning

One of the most important CBT interventions for EDs is meal planning, during which the therapist and patient collaboratively decide what, when, and where the patient will eat over the next several days [14, 17, 41]. To help the patient maintain consistent blood sugar, reduce the likelihood of significant hunger and binge eating, increase control over eating, and/or restore weight to a healthy range, the meal plan should include three meals (i.e., breakfast, lunch, and dinner) and one to three snacks per day. Meal plans should specify times and settings for all meals and snacks and note the foods and drinks to be consumed. This information is written down during therapy sessions on a self-monitoring form, also referred to as a food record [73]. Clinicians can use the self-monitoring form in Table 1, an alternative electronic variation [74], or a revised food record that better fits the individual patient. Table 1 illustrates a daily food record completed by a 16-year-old female patient with BN. In this example, the adolescent wrote down the planned foods and beverages for each meal and snack in the "Planned Food/Bev" column, the time of day when she intended to eat in the "Time" column, and the anticipated location for all meals and snacks in the "Place" column.

Meal planning assists patients with developing regular eating behaviors, increasing food variety, modifying rigid dietary rules and rituals, and decreasing binge eating and purging. Moreover, CBT for patients with AN (and those with other EDs who are significantly weight suppressed) involves the additional treatment goal of

Table 1
A sample daily food record completed by an adolescent patient with bulimia nervosa

Time	Place	Planned food/ bev	Actual food/ bev	B/V/L	Thoughts/emotions/other
7:00 a.m.	Home	1 bowl of cereal 1 banana	1 bowl of cereal 1 banana		I feel bad for choosing the sugary cereal. Why did I do that!?
10:00 a.m.	School	1 yogurt	1 yogurt		I'm still full from breakfast. I didn't really want to eat the yogurt but I did anyway
12:00 p. m.	School	1 peanut butter and jelly sandwich 1 small bag of chips 1 juice box	Nothing (threw lunch in trash)		I'm STILL full from breakfast! It's my fault for eating that sugary cereal
3:00 p.m.	Home	8 baby carrots with ranch dressing	Nothing		I feel hungry now but can wait a few more hours until dinner
6:30 p.m.	Home	1 bowl of Alfredo pasta 2 pieces of garlic bread	1 bowl of Alfredo pasta 4 pieces of garlic bread 5 bags of chips 10 cookies	B	I ate more bread than I was supposed to! I left the table and went to my room to eat the chips and cookies I keep there. Felt a little better after throwing the food up
8:00 p.m.	Home	2 cookies	Nothing	V	

weight restoration. Accordingly, providers treating children and adolescents with AN must be familiar with the general calorie content of foods that are selected for the patient's meal plan, as well as the recommended daily calorie intake for the patient based on age, gender, height, and activity level. If a 15-year-old male with AN is adherent to his meal plan but failing to gain weight, the therapist needs to assist him with modifying his meal plan to augment calorie intake. For instance, the patient might decide to increase the size of his breakfast, add a large afternoon snack, and eat more calorie-dense foods for dinner.

The first several weeks of therapy tend to focus intensively on meal planning, which is faded as patients' eating behaviors improve. Many young patients with EDs can be persuaded to begin improving their eating behaviors with little to no external control by family members. Other patients are initially unwilling to change their eating behaviors and/or gain weight on their own, and may require frequent family participation in therapy sessions focused on meal planning during the early stages of treatment. Factors guiding the degree to which parents should assist with meal planning are discussed in more detail in a subsequent section.

3.1.9 Self-Monitoring

After meal planning with the patient, the therapist describes how to complete the rest of the self-monitoring form, which keeps track of what occurs between therapy sessions. Patients are asked to record the foods and beverages they consume throughout the week in the "Actual Food/Bev" column, and the times and locations where they eat in the "Time" and "Place" columns, respectively (*see* Table 1). Binge eating, self-induced vomiting, and laxative use are recorded in the "B/V/L" column: "B" represents a perceived binge-eating episode, "V" represents self-induced vomiting, and "L" represents laxative use. Patients are encouraged to record any notable thoughts or emotions they experienced throughout the week in the "Thoughts/Emotions/Other" column. Additionally, patients should document exercise and other behaviors related to eating and weight (e.g., diet pill and diuretic use) on the self-monitoring form.

To target binge eating and inappropriate compensatory behaviors noted on the self-monitoring form, the therapist works with the patient to uncover behavioral, situational, cognitive, emotional, and physiological antecedents and consequences of these symptoms. This discussion often identifies useful treatment components, such as psychoeducation about vulnerability to binge eating in response to dietary restraint, cognitive restructuring aimed at modifying "all or none" thinking around food, or behavioral strategies that attempt to interrupt the binge–purge cycle [14, 17, 41]. For example, after talking to her therapist about the binge/purge episode recorded in Table 1, the patient identified several contributing factors, including skipping lunch and afternoon snack, keeping food in her room, and evening time. Through this discussion, the patient recognized that skipping meals and snacks increases her vulnerability to binge eat, while she is unlikely to binge or purge during daytime, when she is around others, and when she is outside of her bedroom. To decrease binge eating and vomiting over the next week, the patient agreed to have lunch and an afternoon snack every day, move food out of her bedroom, and engage in a pleasant activity after dinner such as playing video games with her brother.

In general, delaying binge eating or purging for a predetermined amount of time (e.g., 30 min) can reduce anxiety and urges to binge or purge. During this "wait period," patients often find it helpful to participate in an enjoyable activity, read through or listen to reasoned conclusions that have been recorded in therapy [48], and/or engage in explicit decision-making about the choice to binge eat or purge, such as by writing down a pros and cons list.

Keeping track of binge-eating and/or purging frequency, transforming the data into a line graph, and reviewing the graph each week can also heighten motivation to reduce these behaviors. Any interventions planned for the week are written down on the patient's self-monitoring form or other document, and outcomes of these efforts are reviewed with the therapist during the next therapy session. While the amount of time in therapy spent reviewing the self-monitoring form tends to decrease as the patient's eating behaviors improve, patients are usually encouraged to continue with self-monitoring until the last phase of treatment.

3.1.10 Cognitive Restructuring

Cognitive restructuring refers to the process of identifying, evaluating, and modifying patients' thoughts, assumptions, and beliefs. Therapists should adopt a curious, collaborative, and empirical stance toward the patient's perspective and use Socratic questioning to encourage the patient to view interpretations from different angles [5, 75]. Paralleling the formal cognitive restructuring techniques described for other disorders [48, 76], CBT for EDs assists patients with: understanding the relationship between thoughts, behaviors, physiology, and emotions; identifying "automatic" thoughts; examining the accuracy of thoughts; generating more reasoned conclusions; uncovering core beliefs; making behavioral changes that are consistent with reasoned conclusions; and evaluating the outcomes of behaving in different ways [17, 41]. Thoughts related to eating, weight, and shape can be pulled from the patient's food record (*see* Table 1) and conversations in therapy sessions. Common cognitive restructuring questions can be helpful when discussing thoughts, including "What evidence supports/goes against this thought?" "What are some other possible explanations?" "What might you tell a friend who had this thought or was in the same situation?" "What would be the worst/best/most realistic possible outcome?" and "In what ways is this thought helpful/unhelpful?" Cognitive restructuring is integrated throughout the course of CBT for EDs. Whether reviewing psychoeducational material, discussing the patient's reaction to her or his weight, planning meals for the upcoming week, or exploring why the patient chose to skip breakfast over the past few days, the therapist should attend to the patient's views and assist her or him with considering more accurate and adaptive interpretations and conclusions.

3.1.11 Exposure Therapy

Inherent in many CBT interventions for EDs are elements of exposure: therapists openly weigh patients despite their anxiety about this process; discourage safety behaviors such as purging, excessive exercise, frequent weighing, body checking, and dietary rituals; and encourage patients to eat regularly and incorporate feared foods into their diets [14, 41]. To target anxiety associated with eating certain foods, for example, a therapist might assist a patient with developing a hierarchy of feared foods with subjective units of distress ratings, and encourage the patient to choose progressively more anxiety-provoking foods to add to her or his meal plan each week [77, 78]. The way in which exposure is usually conducted for EDs, however, has been narrow in scope and conservative in approach [79]. Unlike CBT protocols for EDs, exposure therapy for anxiety disorders involves *overcorrection*: exposure moves beyond simply "being around" the feared stimulus to more challenging sessions in which the patient's specific fears are amplified or exaggerated [80, 81]. By adopting the principle of overcorrection, therapists have the potential to use exposure in imaginative ways to address a relatively wide range of issues, including ritualized eating, weight-related self-evaluation, body dissatisfaction, and eating-related social situations [82].

In recent years, scientist-practitioners have begun to explore creative adaptations of exposure for EDs. Relatively novel exposure interventions include therapist-guided mirror exposure, which targets body image by instructing individuals to look at themselves in a full body mirror and describe what they see using objective terminology [83]; interoceptive exposure targeting distressing eating-related somatic sensations such as fullness by asking patients to drink a high volume of water [84]; and in vivo exposure targeting anxiety-provoking eating situations such as portioning food out of a large container [85] and eating a high energy food in front of a romantic interest [86].

Extending the principle of overcorrection used in exposure treatment for anxiety disorders suggests even more challenging exposure scenarios designed to amplify patients' fears about a wide range of scenarios [82], which may be particularly useful after patients have normalized eating and increased food variety. For example, to address a patient's fear of negative judgment from others about her food choices, we might develop an exposure session in which the patient eats a calorie-rich pasta with a therapist who orders a salad and makes critical comments about the patient's meal (e.g., "*That's* what you decided to eat?!" "Wow, that looks like it has a lot of fat in it" "I guess you aren't into 'clean' eating like I am," "Aren't you worried about gaining weight?"). The details of exposure sessions should be established collaboratively with the patient in advance. While our patients often report that therapist-assisted in vivo exposure is a particularly beneficial component of their treatment, research evaluating this approach to exposure is warranted before drawing conclusion about its incremental contribution to CBT for EDs.

Exposure therapy is one of the most effective psychological treatments for anxiety disorders [87], and appears to be a logical choice to target fear responses in patients with EDs. In adapting the intervention to EDs, however, clinicians should consider several distinctive features that affect the timing, targets, and implementation of exposure. For example, some ritualistic patterns such as prolonged eating and excessive use of salt and spices are more accurately construed as starvation behaviors rather than ED symptoms, and should be treated through weight restoration rather than response prevention. In addition, many ED patients have intentionally *cultivated* their fear of "fattening" or "unhealthy" foods, and are invested in *retaining* their anxiety as a weight control tactic that facilitates dietary restraint [88]. Psychoeducation about the effectiveness of exposure is unlikely to be persuasive for individuals who are terrified of becoming less afraid. These considerations may need to be addressed through other therapeutic strategies before patients can be engaged as active collaborators in designing and carrying out exposure exercises.

3.1.12 Relapse Prevention

The final therapy sessions tend to focus on relapse prevention. Common relapse prevention interventions include summarizing the progress that the patient has made over the course of treatment, reviewing skills that were instrumental to the patient's positive outcome, identifying areas of continued vulnerability and warning signs of relapse, anticipating potential future difficulties and plans to address these challenges, assisting the patient with differentiating small setbacks from a full relapse, discussing any concerns about ending treatment, and developing a plan for appropriate weight monitoring [14, 17, 41].

3.2 Augmenting CBT for EDs with Family Involvement

3.2.1 Determining the Degree and Format of Family Involvement

CBT for both pediatric anxiety disorders and EDs typically incorporates family into treatment, as therapists encourage parents to help their child participate in between-session exposure to feared stimuli and reduce avoidance behaviors [89]. By definition, exposure-based treatment is anxiety-provoking, and many children require family support to initiate and maintain distressing behavior change. In addition, EDs are characterized by particularly dangerous avoidance and safety behaviors such as extreme dietary restraint and self-induced vomiting for which rapid intervention is indicated.

At the same time, long-term recovery requires patients to make the decision *themselves* to sustain repeated exposure to feared stimuli. Getting over AN depends on a daily commitment to eat normal meals and cope with body image concerns, just as surmounting social anxiety involves ongoing participation in social events. While CBT for children and adolescents often enlists the family to help initiate behavior change, an important treatment goal is to increase patients' own investment in working toward recovery without external pressure. Accordingly, therapists should depend on parental control to the minimum degree needed to facilitate behavior change and ensure the patient's safety, and gradually decrease external direction as the patient makes progress.

Consistent with these principles, CBT for children and adolescents with EDs typically includes a blend of individual and family therapy sessions. The degree of family involvement varies according to the amount and forms of external support the patient needs to participate in treatment and change her or his eating behaviors. The patient's age, symptom presentation and severity, motivation to recover, and progress over time should be used to determine the frequency and content of family sessions. In general, patients who are younger, underweight, and/or reluctant to change require substantially higher levels of family involvement. Weekly family sessions are indicated for most children and adolescents with AN in the initial stages of treatment; in contrast, CBT for adolescents with BN may involve only a few family sessions over the course of treatment.

Therapists must rely on clinical judgment to guide the optimal format of family involvement. Research on FBT suggests that there are advantages and disadvantages to the alternative formats examined, with individual patients responding better to conjoint, separate, and parent-only approaches [33, 90]. For example, families with higher levels of expressed emotion and criticism fare worse in conjoint family therapy [31, 32], perhaps because airing parental frustrations in the child's presence increases conflict during and between therapy sessions.

For the subset of patients who require substantial family involvement to participate in treatment, we typically begin with one individual and one family therapy session per week. In most instances, family sessions are split into two segments, meeting initially with the parents and then with parents and patient conjointly. Factors that influence the proportion of time dedicated to parent-only versus conjoint sessions include the preferences of the parents and patient as well as the clinician's observations of family interactions during sessions. We may move quickly to conjoint sessions when working with a 15-year-old with AN who has a strong relationship with her parents and prefers to participate in family discussions of the recovery plan; however, when treating a 13-year-old who becomes agitated during family therapy, we might decide to keep the patient and parents separated for most of the session.

4   The Family's Role in Supporting the Normalization of Eating Behaviors and Weight Restoration

Psychoeducation and communicating empathy. We provide parents with detailed material about the physical and psychological consequences of EDs, the hardwired effects of semi-starvation, and the reinforcement principles that maintain ED behaviors.

Psychoeducation helps parents develop a more informed and compassionate understanding of their child's predicament and perspective, and conveys the importance of immediate weight restoration and normalized eating. Weight restoration for children with AN, like chemotherapy for children with leukemia, is frightening, aversive, uncomfortable, and difficult—*and* an essential and nonnegotiable part of recovery. From the start of treatment, we assist families with communicating and implementing this sympathetic yet firm message at home. A father might repeatedly tell his 14-year-old daughter with AN that he loves her and is committed to her health, safety, and future—which is precisely why he cannot allow her to eat less than her body requires.

Therapists should also be prepared to provide families with accurate nutritional information. Myths about healthy eating, dieting, and weight are ubiquitous, and many parents inadvertently convey false or oversimplified messages about food to their children. Conscientious parents who know that relatively low-fat, low-sugar diets are recommended for the general public and who work to provide healthy, home-cooked meals to their families are often puzzled by the apparently contrary advice given in treatment. Family members are entitled to clear explanations about why their child with an ED should be encouraged to include fried foods and desserts in her or his daily diet and to eat occasional meals in fast-food restaurants.

Setting clear expectations and increasing supervision. Parents should communicate to their child that eating three meals per day is nonnegotiable, and demonstrate an unwavering commitment to this expectation. Therapists should also help parents understand and convey that their child must eat enough food to support participation in daily life, including school, work, and extracurricular activities. A father may need to sit at the kitchen table with his 17-year-old daughter with AN during mealtimes, and calmly remind her that she cannot leave to attend a social event until she has finished eating. Gradually regaining independence and reducing parental monitoring act as logical rewards for treatment progress, while further decreases in autonomy and reduced access to unsupervised or physically demanding activities are natural consequences of persistent ED behavior.

Collaborative meal planning. In contrast to FBT, which empowers families to make all eating-related decisions for their child [30], CBT actively involves the patient in meal planning from the outset. Most young patients with BN and some with AN can participate effectively in meal planning during individual sessions; for others, decisions are made by the parents during family sessions. In all cases, however, the therapist solicits input from the child as well as the parents. Some components of the meal plan are nonnegotiable, such as eating three meals and at least one snack per day and consuming a sufficient number of calories to support weight gain or maintenance. Within these boundaries, however, the patient is encouraged to make a number of decisions related to what, when, and where to eat. For example, a 14-year-old girl with AN may decide to eat one large snack rather than two smaller snacks each day. Although it is preferable for patients to eat the same foods as other family members, some flexibility may be appropriate in the early stages of recovery. For example, we might encourage parents to allow a 13-year-old son with OSFED to eat pasta on a night when the rest of the family has pizza, if pizza is ranked near the top of his hierarchy of feared foods.

Collaborative meal planning should accompany ongoing discussion about the kinds of comments the patient experiences as helpful and unhelpful, which can help parents use verbal reinforcement more effectively. While some children appreciate praise for improved eating behavior at meal time, others prefer to have parents recognize their efforts to change at other times or in different ways.

5   Conclusion

Although CBT experts consistently recommend involving family in the treatment of children and adolescents with EDs, there is surprisingly little guidance on how to do so. Because ambivalence about change is a prominent feature of EDs, especially AN, external parental pressure is often necessary to facilitate patients' participation in treatment and protect them from ED-related harm, with the degree of parental control varying substantially from case to case. Nevertheless, the CBT model of EDs posits that corrective learning is key to long-term recovery. Accordingly, therapists maximize the patient's autonomy as much as possible, attend closely to her or his perspectives and beliefs, and work to develop internal motivation to change. As the patient's willingness to participate in CBT increases, external control is faded and treatment shifts to self-regulation of eating behaviors, self-selected exposure, cognitive restructuring, and related CBT interventions. While our clinical experience suggests that augmenting CBT for EDs with family involvement in this way increases the treatment's effectiveness, this specific approach has yet to be examined empirically. We hope that this chapter serves as a valuable supplement to more comprehensive CBT protocols and supports further study of the efficacy of CBT for children and adolescents with EDs.

References

1. Mehler PS, Brown C (2015) Anorexia nervosa – medical complications. J Eat Disord 3: e1–e8

2. Steinhausen HC (2009) Outcomes of eating disorders. Child Adolesc Psychiatr Clin N Am 18:225–242

3. Chesney E, Goodwin GM, Fazel S (2014) Risks of all-cause and suicide mortality in mental disorders: a meta-review. World Psychiatry 13:153–160

4. Vitousek KB, Hollon SD (1990) The investigation of schematic content and processing in eating disorders. Cogn Ther Res 14:191–214

5. Vitousek K, Watson S, Wilson GT (1998) Enhancing motivation for change in treatment-resistant eating disorders. Clin Psychol Rev 18:391–420

6. Fairburn CG, Cooper Z, Shafran R (2003) Cognitive behaviour therapy for eating disorders: a "transdiagnostic" theory and treatment. Behav Res Ther 41:509–528

7. Grilo CM, Masheb RM, White MA (2010) Significance of overvaluation of shape/weight in binge-eating disorder: comparative study with overweight and bulimia nervosa. Obesity 18:499–504

8. American Psychiatric Association (2013) Diagnostic and statistical manual of mental disorders, 5th edn. American Psychiatric Association, Arlington, VA

9. Hoek HW, Van Hoeken D (2003) Review of the prevalence and incidence of eating disorders. Int J Eat Disord 34:383–396

10. Hudson JI, Hiripi E, Pope HG Jr et al (2007) The prevalence and correlates of eating disorders in the National Comorbidity Survey Replication. Biol Psychiatry 61:348–358

11. Jaite C, Hoffmann F, Glaeske G et al (2013) Prevalence, comorbidities and outpatient treatment of anorexia and bulimia nervosa in German children and adolescents. Eat Weight Disord-St 18:157–165

12. Smink FR, Van Hoeken D, Hoek HW (2012) Epidemiology of eating disorders: incidence, prevalence and mortality rates. Curr Psychiatry Rep 14:406–414

13. Stice E, Marti CN, Shaw H, Jaconis M (2009) An 8-year longitudinal study of the natural history of threshold, subthreshold, and partial eating disorders from a community sample of adolescents. J Abnorm Psychol 118:587–597

14. Fairburn CG (2008) Cognitive behavior therapy and eating disorders. NY, New York, NY

15. Pallister E, Waller G (2008) Anxiety in the eating disorders: understanding the overlap. Clin Psychol Rev 28:366–386

16. Salkovskis PM (1991) The importance of behaviour in the maintenance of anxiety and panic: a cognitive account. Behav Psychother 19:6–19

17. Garner DM, Vitousek KM, Pike KM (1997) Cognitive-behavioral therapy for anorexia nervosa. In: Garner DM, Garfinkel PE (eds) Handbook of treatment for eating disorders, 2nd edn. Guilford Press, New York, NY, pp 94–144

18. Gale C, Holliday J, Troop NA et al (2006) The pros and cons of change in individuals with eating disorders: a broader perspective. Int J Eat Disord 39:394–403

19. Serpell L, Treasure J, Teasdale J et al (1999) Anorexia nervosa: friend or foe? Int J Eat Disord 25:177–186

20. Skarderud F (2007) Shame and pride in anorexia nervosa: a qualitative descriptive study. Eur Eat Disord Rev 15:81–97

21. Linardon J, Wade TD (2018) How many individuals achieve symptom abstinence following psychological treatments for bulimia nervosa? A meta-analytic review. Int J Eat Disord 51:287–294

22. Waller G (2016) Treatment protocols for eating disorders: clinicians' attitudes, concerns, adherence and difficulties delivering evidence-based psychological interventions. Curr Psychiatry Rep 18:36–43

23. Wilson GT, Grilo CM, Vitousek KM (2007) Psychological treatment of eating disorders. Am Psychol 62:199–216

24. Lock J (2015) An update on evidence-based psychosocial treatments for eating disorders in children and adolescents. J Clin Child Adolesc Psychol 44:707–721

25. Lock J (2005) Adjusting cognitive behavior therapy for adolescents with bulimia nervosa: results of case series. Am J Psychother 59:267–281

26. Wilson GT, Sysko R (2006) Cognitive-behavioural therapy for adolescents with bulimia nervosa. Eur Eat Disord Rev 14:8–16

27. Kass AE, Kolko RP, Wilfley DE (2013) Psychological treatments for eating disorders. Curr Opin Psychiatry 26:549–555

28. Dare C, Eisler I (1997) Family therapy for anorexia nervosa. In: Garner D, Garfinkel PE (eds) Handbook of treatment for eating disorders, 2nd edn. Guilford Press, Chichester, pp 333–349

29. Eisler I (2005) The empirical and theoretical base of family therapy and multiple family day therapy for adolescent anorexia nervosa. J Fam Ther 27:104–131

30. Lock J, Le Grange D (2015) Treatment manual for anorexia nervosa: a family-based approach. American Psychiatric Association, New York, NY

31. Eisler I, Dare C, Hodes M et al (2000) Family therapy for adolescent anorexia nervosa: the results of a controlled comparison of two family interventions. J Child Psychol Psychiatry 41:727–736

32. Le Grange D, Eisler I, Dare C et al (1992) Evaluation of family treatments in adolescent anorexia nervosa: a pilot study. Int J Eat Disord 12:347–357

33. Le Grange D, Hughes EK, Court A et al (2016) Randomized clinical trial of parent-focused treatment and family-based treatment for adolescent anorexia nervosa. J Am Acad Child Psychiatry 55:683–692

34. Lock J, Le Grange D, Agras WS et al (2010) Randomized clinical trial comparing family-based treatment with adolescent-focused individual therapy for adolescents with anorexia nervosa. Arch Gen Psychiatry 67:1025–1032

35. Dalle GR, Calugi S, Conti M et al (2013) Inpatient cognitive behaviour therapy for anorexia nervosa: a randomized controlled trial. Psychother Psychosom 82:390–398

36. Ball J, Mitchell P (2004) A randomized controlled study of cognitive behavior therapy and behavioral family therapy for anorexia nervosa patients. Eat Disord 12:303–314

37. Le Grange D, Lock J, Agras WS et al (2015) Randomized clinical trial of family-based treatment and cognitive-behavioral therapy for adolescent bulimia nervosa. J Am Acad Child Adolesc Psychiatry 54:886–894

38. Schmidt U, Lee S, Beecham J et al (2007) A randomized controlled trial of family therapy and cognitive behavior therapy guided self-care for adolescents with bulimia nervosa and related disorders. Am J Psychiatry 164:591–598

39. Wilfley DE, Kolko RP, Kass AE (2011) Cognitive-behavioral therapy for weight management and eating disorders in children and adolescents. Child Adoles Psychiatr Clin 20:271–285

40. Gowers SG, Green L (2009) Eating disorders: cognitive behaviour therapy with children and young people. Wiley, New York, NY

41. Waller G, Cordery H, Corstorphine E et al (2007) Cognitive behavioral therapy for eating disorders: a comprehensive treatment guide. Cambridge University Press, New York, NY

42. Geller J, Srikameswaran S (2006) Treatment non-negotiables: why we need them and how to make them work. Eur Eat Disord Rev 14:212–217

43. Fairburn CG, Cooper Z, O'Connor M (1993) The eating disorder examination. Int J Eat Disord 6:1–8

44. Fairburn CG, Beglin SJ (1994) Assessment of eating disorders: interview or self-report questionnaire? Int J Eat Disord 16:363–370

45. Garner DM (2004) Eating disorder inventory-3 (EDI-3). Professional Manual, Odessa, FL

46. Anderson LK, Reilly EE, Berner L et al (2017) Treating eating disorders at higher levels of care: overview and challenges. Curr Psychiatry Rep 19:48–56

47. Rock CL (1999) Nutritional and medical assessment and management of eating disorders. Nutr Clin Care 2:332–343

48. Beck JS (1995) Cognitive therapy: basics and beyond. Guilford Press, New York, NY

49. Garner DM, Garfinkel PE, Bemis KM (1982) A multidimensional psychotherapy for anorexia nervosa. Int J Eat Disord 1:3–46

50. Keys A, Brožek J, Henschel A et al (1950) The biology of human starvation I-II. University of Minnesota Press, Minneapolis, MN

51. Polivy J, Zeitlin SB, Herman CP et al (1994) Food restriction and binge eating: a study of former prisoners of war. J Abnorm Psychol 103:409–411

52. Tuschl RJ (1990) From dietary restraint to binge eating: some theoretical considerations. Appetite 14:105–109

53. Kaye WH, Weltzin TE, Hsu LG et al (1993) Amount of calories retained after binge eating and vomiting. Am J Psychiatry 150:969–969

54. Bo-Lynn G, Santa-Ana CA, Morawski SG et al (1983) Purging and calorie absorption in bulimic patients and normal women. Ann Intern Med 99:14–17

55. National Eating Disorders Association (2018) Health consequences. https://www.nationaleatingdisorders.org/health-consequences

56. Westmoreland P, Krantz MJ, Mehler PS (2016) Medical complications of anorexia nervosa and bulimia. Am J Med 129:30–37

57. Gonzalez VMM, Vitousek KM (2004) Feared food in dieting and non-dieting young women: a preliminary validation of the Food Phobia Survey. Appetite 43:155–173

58. King GA, Herman CP, Polivy J (1987) Food perception in dieters and non-dieters. Appetite 8:147–158

59. Dietary Guidelines Advisory Committee (2015) Scientific report of the 2015 dietary guidelines advisory committee: advisory report to the secretary of health and human services and the secretary of agriculture. Dietary Guidelines Advisory Committee, Washington, DC

60. Murray SB, Loeb KL, Le Grange D (2016) Dissecting the core fear in anorexia nervosa: can we optimize treatment mechanisms? JAMA Psychiat 73:891–892

61. Casazza K, Fontaine KR, Astrup A et al (2013) Myths, presumptions, and facts about obesity. N Engl J Med 368:446–454

62. Carter JC, Kelly AC (2015) Autonomous and controlled motivation for eating disorders treatment: baseline predictors and relationship to treatment outcome. Br J Clin Psychol 54:76–90

63. Carter JC, Mercer-Lynn KB, Norwood SJ et al (2012) A prospective study of predictors of relapse in anorexia nervosa: implications for relapse prevention. Psychiatry Res 200:518–523

64. Gregertsen EC, Mandy W, Kanakam N et al (2019) Pre-treatment patient characteristics as predictors of drop-out and treatment outcome in individual and family therapy for adolescents and adults with anorexia nervosa: a systematic review and meta-analysis. Psychiatry Res 271:484–501

65. Dray J, Wade TD (2012) Is the transtheoretical model and motivational interviewing approach applicable to the treatment of eating disorders? A review. Clin Psychol Rev 32:558–565

66. Vitousek KB, Orimoto L (1993) Cognitive-behavioral models of anorexia nervosa, bulimia nervosa, and obesity. In: Kendall PC, Dobson K (eds) Psychopathology and cognition. APA, New York, NY, pp 191–243

67. Stice E, Marti CN, Spoor S et al (2008) Dissonance and healthy weight eating disorder prevention programs: long-term effects from a randomized efficacy trial. J Consult Clin Psychol 76:329–340

68. Stice E, Rohde P, Shaw H (2012) The body project: a dissonance-based eating disorder prevention intervention. Oxford University Press, Oxford

69. Forbush KT, Richardson JH, Bohrer BK (2015) Clinicians' practices regarding blind versus open weighing among patients with eating disorders. Int J Eat Disord 48:905–911

70. Kelly-Weeder S, Kells M, Jennings K et al (2017) Procedures and protocols for weight assessment during acute illness in individuals with anorexia nervosa: a national survey. J Am Psychiatr Nurses Assoc 24:241–246

71. Waller G, Mountford VA (2015) Weighing patients within cognitive-behavioural therapy for eating disorders: how, when and why. Behav Res Ther 70:1–10

72. Golden NH, Jacobson MS, Sterling WM et al (2008) Treatment goal weight in adolescents with anorexia nervosa: use of BMI percentiles. Int J Eat Disord 41:301–306

73. Wilson GT, Vitousek KM (1999) Self-monitoring in the assessment of eating disorders. Psychol Assess 11:480–489

74. Juarascio AS, Manasse SM, Goldstein SP et al (2015) Review of smartphone applications for the treatment of eating disorders. Eur Eat Disord Rev 23:1–11

75. Padesky C (1993) Socratic questioning: changing minds or guiding discovery? Keynote address delivered at the European congress of behavioral and cognitive therapies, London

76. Greenberger D, Padesky CA (1995) Mind over mood: a cognitive therapy treatment manual for clients. APA, New York, NY

77. Hildebrandt T, Bacow T, Greif R, Flores A (2013) Exposure-Based Family Therapy (FBT-E): an open case series of a new treatment for anorexia nervosa. Cogn Behav Pract 21:470–484

78. Steinglass JE, Albano AM, Simpson HB et al (2014) Confronting fear using exposure and response prevention for anorexia nervosa: a randomized controlled pilot study. Int J Eat Disord 47:174–180

79. Vitousek KM (2012) Away from the clinic and off of the script: making better use of behavioral experiments in outpatient therapy. Keynote address given at the 22nd annual renfrew conference, Philadelphia, PA

80. Craske MG, Treanor M, Conway CC et al (2014) Maximizing exposure therapy: an inhibitory learning approach. Behav Res Ther 58:10–23

81. Huppert JD, Siev J, Kushner ES (2007) When religion and obsessive–compulsive disorder collide: treating scrupulosity in ultra-orthodox Jews. J Community Psychol 63:925–941

82. Vitousek KM, Gray JA (2005) Eating disorders. In: Gabbard GO, Beck JS, Holmes J (eds) Oxford textbook of psychotherapy. Oxford University Press, New York, NY, pp 177–202

83. Hildebrandt T, Loeb K, Troupe S et al (2012) Adjunctive mirror exposure for eating disorders: a randomized controlled pilot study. Behav Res Ther 50:797–804

84. Reilly EE, Anderson LM, Gorrell S et al (2017) Expanding exposure-based interventions for eating disorders. Int J Eat Disord 50:1137–1141

85. Glasofer DR, Albano AM, Simpson HB et al (2016) Overcoming fear of eating: a case study of a novel use of exposure and response prevention. Psychotherapy 53:223–231

86. Trottier K, Carter JC, MacDonald DE et al (2015) Adjunctive graded body image exposure for eating disorders: a randomized controlled initial trial in clinical practice. Int J Eat Disord 48:494–504

87. Abramowitz JS (2013) The practice of exposure therapy: relevance of cognitive-behavioral theory and extinction theory. Behav Ther 44:548–558

88. Garner DM, Bemis KM (1982) A cognitive-behavioral approach to anorexia nervosa. Cogn Ther Res 6:123–150

89. Taboas WR, McKay D, Whiteside SP et al (2015) Parental involvement in youth anxiety treatment: conceptual bases, controversies, and recommendations for intervention. J Anxiety Disord 30:16–18

90. Hughes EK, Sawyer SM, Loeb KL et al (2015) Who's in the room? A parent-focused family therapy for adolescent anorexia nervosa. Eat Disord 23:291–301

Chapter 10

Mindfulness-Based Cognitive Therapy for Children

Laila A. Madni, Chelsie N. Giambrone, and Randye J. Semple

Abstract

Mindfulness-based cognitive therapy for children (MBCT-C) is a downward adaptation of the adult mindfulness-based cognitive therapy (MBCT). The child version is a structured group psychotherapy developed to help children aged 8–13 years manage anxiety. MBCT-C engages children in short, experiential group activities; offering interactive, age-appropriate mindfulness practices in child-friendly language. Hands-on activities and creative repetition promote learning. Activities include breath meditations, still and movement-based body awareness practices, and sensory-based activities (i.e., mindful seeing, touching, tasting, smelling, and hearing). One to two therapists lead groups of six to eight children in 12 weekly, 90-min sessions. Like the adult MBCT program, children are expected to practice at home between sessions. Parents or caregivers are invited to participate in two separate sessions and in the child's home practices. The main goals of MBCT-C are as follows:

- To enhance social-emotional resiliency in a safe group environment.
- To learn to distinguish thoughts that are judgmental from those that simply describe or note one's experiences.
- To recognize that judgments often escalate mood disturbances, which can then trigger maladaptive behaviors.
- To promote positive changes in how the child relates to his or her own thoughts, and emotions, and body sensations.
- To cultivate self-acceptance and acceptance of those things that cannot be changed.
- To expand awareness of personal emotional and behavioral choices.

Key words Mindfulness, Children, MBCT-C, MBCT, Anxiety, Meditation, Depression, Acceptance, Awareness, Group therapy

1 Introduction

Anxiety disorders are the most common psychiatric diagnoses for children and adolescents, with lifetime prevalence rates as high as 32% for youth aged 13–17 years [1]. Up to 50% of adolescents' anxiety symptoms begin by age six [2]. Further exacerbating the problem, co-occurring anxiety or depressive disorders are present in

Robert D. Friedberg and Brad J. Nakamura (eds.), *Cognitive Behavioral Therapy in Youth: Tradition and Innovation*, Neuromethods, vol. 156, https://doi.org/10.1007/978-1-0716-0700-8_10, © Springer Science+Business Media, LLC, part of Springer Nature 2020

many of these youth [3]. Anxious children are also more likely to develop additional problems in the future. Data from the National Comorbidity Survey suggest that anxiety disorders are the most common precursors to major depression [4]. Often left untreated, children with anxiety disorders are at higher risk for academic difficulties [5] and substance abuse [6] in adolescence.

Cognitive behavioral therapy (CBT) and psychotropic medications (mainly antidepressants and anxiolytics) are the current "best practice" treatments for childhood anxiety [7]. While psychotherapy and medications independently can be effective, the combination of CBT and medication has been shown to have an additive effect [8]. However, many parents prefer not to medicate their children and, although helpful for some children, CBT is not accessible or effective for many. A randomized clinical trial of CBT for children (aged 8–14 years) with anxiety disorders found that up to one half of children in a CBT group still met criteria for an anxiety disorder post-treatment [9]. A recent meta-analysis [10] found that CBT for children was effective only 59% of the time. Clearly, additional interventions are still needed.

Williams, Teasdale, Segal, and Kabat-Zinn [11] have defined mindfulness in the mindfulness-based cognitive therapy (MBCT) program as "the awareness that emerges though paying attention, on purpose, in the present moment, and nonjudgmentally to things as they are" (p. 47). Mindfulness-based cognitive therapy for children [12] has adopted this definition. MBCT and MBCT-C include some components of CBT however, the theories and interventions are significantly different.

2   Theoretical and Empirical Foundations

CBT is a structured, present-focused, time-limited treatment aimed at reducing symptoms primarily through cognitive restructuring and development of emotion and behavior regulation strategies. The cognitive model suggests that holding inaccurate or distorted thoughts about a situation triggers emotional distress. Also structured, present-focused, and time-limited, MBCT and MBCT-C are based on a different premise—that thoughts, although related to emotions, are not the primary trigger for emotional distress. Rather, this model assumes that the source of emotional distress is the way we relate to our own thoughts. Instead of changing thoughts, when practicing mindfulness, clients learn to recognize that thoughts are simply events in the mind, and cultivate acceptance of all their experiences—including thoughts and feelings—just as they are. *Decentering* is the metacognitive awareness of thoughts and emotions as transient "mind stuff," and is believed to be what allows us to relate to our thoughts in more wholesome ways [13]. Although acceptance and change are seemingly paradoxical goals, acceptance may serve as a catalyst for change. Additionally, Kuyken et al. [14] suggested that MBCT could be helpful in reducing symptoms of depression and depressive relapse by increasing self-compassion, and decreasing reactivity. Schoenberg and Speckens [15] identified neurological changes potentially related to participation in MBCT, including recalibration of cortical circuitry, which allows individuals to disengage from rumination and associated negative cognitive and emotional processes.

3   Mindfulness-Based Cognitive Therapy

Mindfulness-based cognitive therapy (MBCT), the adult version of MBCT-C, was developed by Segal, Williams, and Teasdale [16] in the late 1990s to prevent depressive relapse in remitted patients with previous episodes of major depression. MBCT now has significant research support for a variety of other health conditions as well, including anxiety spectrum disorders, attention deficits, and as an adjunctive treatment for cancer, HIV, and other serious medical conditions [17–19].

MBCT is an 8-week group program that consists of 150-min, weekly classes plus one all-day retreat. The classes teach mindful breath and body practices. Insights are facilitated by guided inquiries that follow each exercise. Mindfulness training is integrated with cognitive behavioral methods, which includes psychoeducation about depressive symptoms, awareness of changing mood states, and planning pleasant activities. Daily home practice is an important part of the program. Participants cultivate mindful awareness of thoughts, emotions, and body sensations with an open, curious, and a nonjudgmental stance. A primary aim of MBCT is to change how the clients relate to their own internal experiences. The 8-week program focuses on moving off automatic pilot mode, which is the experience of mindlessness, or absence of present moment awareness within everyday experiences. We do this by increasing awareness of thoughts, emotions, and body sensations; cultivating present-focused awareness, acceptance, and compassion; and decentering from thoughts and feelings.

4   Adapting Mindfulness Training for Children

Children differ from adults in three ways that are relevant to teaching mindfulness. First, children's cognitive and affective development is at an earlier stage than adults, which limits their abstract reasoning skills. To accommodate this difference, MBCT-C has adopted child-friendly language that is more concrete and somewhat more directive than MBCT. Initial sessions focus on teaching children to observe and differentiate their thoughts, feelings (emotions), and body sensations, while engaged in short mindful breath and body practices. Activities are simple, not competitive, fun, and interesting.

Children also have less capacity for sustained attention than most adults, and can get restless if engaged in one activity for a long time. You need only imagine trying to get a 9-year-old to practice a 45-min seated breath meditation. For this reason, the program was extended from 8 to 12 weeks and each session shortened to 90 min. MBCT-C includes breath meditation, body scan, and mindful movement activities that are similar to those used in the adult program, but are considerably shortened. Each weekly session includes opportunities for children to practice mindful movements or stretching, alternating with breathing and body-awareness practices, sensory-based activities, visualizations, reading, writing, or drawing activities, and interactive group discussions.

Children are dependent on their families in ways that most adults are not. Not only do they depend on caregivers to provide the basic needs required for survival, they depend on adults for emotional, social, and spiritual support. Consequently, parents and caregivers are integrated into the program in four ways. Ideally, they will (a) attend an initial orientation session, (b) maintain contact with the therapist throughout the program, (c) attend a review session at the end of the program, and (d) participate with their child in the home practice activities.

Given that MBCT-C groups are intended for children with social, emotional, or behavioral difficulties, the groups tend to be smaller than adult groups and most classrooms. Depending on the age and dynamics of the group, and the specific needs of the children, we suggest that a single therapist limit the group size to 8 or 10; two co-therapists may feel comfortable working with up to 12 or 15 children. Having a second therapist present allows one to attend to a child who might need individual attention, while the other facilitates the group.

In addition to child-friendly activities, the language used in MBCT-C should be accessible to children and appropriate to their developmental level. Language, concepts, and activities can be further simplified to respond to the emotional and cognitive needs of the group as a whole. Practice activity examples should be relevant to the social or emotional concerns of each group. A group of young teens may want to focus on exploring peer relationships and interpersonal dynamics, for example, whereas a younger group may be more interested in examining hobbies, sports, or family issues.

5   Research Support for MBCT-C

MBCT-C is an evidence-based intervention with growing research support [20]. An early study provided evidence for the overall acceptability and feasibility of mindfulness training adapted for children [21]. Children understood mindfulness concepts, practiced mindful awareness activities, and used them appropriately in their daily lives. The first randomized controlled trial of MBCT-C reported reductions in internalizing and externalizing symptoms, and fewer attentional problems as compared to a waitlisted control group [22, 23]. Other studies of MBCT-C have reported reductions in anxiety [24]; depressive symptoms [25, 26]; externalizing behaviors, oppositional defiant problems, and conduct problems [27]; anger, physical aggression, and hostility [28]; and inattention, conduct problems, and peer relations problems [29]. Strawn et al. [30] conducted an fMRI study of MBCT-C for anxious children at risk for bipolar disorder and found increased activation of brain structures (insula and anterior cingulate) that are associated with fear processing and emotional regulation. A randomized clinical trial of MBCT-C was recently conducted with children hospitalized with cancer [31].

6   Intervention

6.1   Program Overview

MBCT-C [12] consists of twelve 90-min group sessions that use breath, body, and sensory-based activities to teach mindful awareness skills. Rather than sitting still for prolonged breath and body practices, children are encouraged to develop mindfulness using their five senses—engaging with the world around them with "eyes wide open."

Parents or caregivers are encouraged to cultivate personal mindfulness practices and participate in home practice assignments with their child. Written session summaries familiarize parents with the concepts, language, and practices of MBCT-C. They are invited to attend an *Introduction to Mindfulness* session prior to the beginning of the groups. During this 2-h session, they are given a definition of mindfulness and its benefits, and discuss expectations for participating in the program. They engage in guided meditation practices to introduce them to some of the activities and skills their child will be learning over the next few months. At the conclusion of the program, parents meet again with the therapist in a group to discuss their observations of the children's experiences, progress over the previous 12 weeks, or any concerns. Parents and therapists discuss strategies to help each child continue his or her mindfulness practice after the program ends and explore child-specific modifications that might be needed.

MBCT-C is conducted in three phases. Sessions 1–3 are devoted to orienting the children to the concept of mindfulness, identifying program expectations and parameters, and participating in short, experiential mindfulness activities. Early activities focus on establishing awareness of the breath and body. Sessions 4–10 expand the practice of mindfulness using the five senses. Children learn to distinguish thoughts, feelings, and body sensations, understand how these internal experiences relate to each other, and recognize how thoughts and anxiety are connected. These middle sessions also teach ways to differentiate judging from observing (or noting), identify "choice points," and discover new ways to respond to situations—with mindful awareness rather than impulsive, emotional reactivity. Sessions 11 and 12 focus on generalization and continuation of mindfulness practices after the program ends. Children often develop strong connections with other group members, so we offer opportunities in these closing sessions for each child to explore and share his or her thoughts and feelings about being a part of the group and the ending of the group. To conclude the program, we celebrate our shared journey with a small "graduation" ceremony and party.

6.2 Materials and Supplies

In work with children, some supplies and materials are generally needed. These are inexpensive items, many of which may be found in your own home. Yoga or exercise mats are useful for mindful movement and mindfulness of body activities but are not required. A clean, carpeted floor or bath towels work fine. Child-sized meditation cushions can be used for seated meditations, but you can choose to use chairs or small firm pillows. We use meditation bells (bells of mindfulness) to signal the beginning and end of each mindfulness activity. A variety of chimes, bells, gongs, or musical instruments such as triangles may be used for this purpose. Meditation bell apps are available for downloading to smartphones. Since children often benefit from organizational assistance, we give them binders or folders to store their session summaries, handouts, and home practice assignments. You will need an assortment of art supplies (markers, colored pencils, construction paper, tape, etc.) and optionally, cheerful stickers to encourage attendance and home practice assignments.

Most of the sensory-based mindfulness activities require some materials. Many of these items can be found at home or purchased inexpensively. Each session includes written session summaries, poems, stories, or information handouts, and home practice assignments. These printable session materials and a session-by-session list of suggested materials can be found in the MBCT-C treatment manual [12]. Examples of the supplies that can be used are described in Table 1.

Table 1
Partial list of supplies needed to conduct MBCT-C

Session	Supplies needed
1.	A nonbreakable cup and water
2.	One small box of raisins
4.	One clementine orange per child and hand wipes or paper towels
5.	Audio player with five or six short segments of music
6.	Small musical instruments (e.g., finger symbols, castanets, tambourine, maracas, or shaker eggs). Many household items can be used as musical instruments (e.g., wooden chopsticks, water bottle with sand or pebbles, or pot lids) and a "conductor's baton" (e.g., a pencil or short stick)
8.	Several optical illusions and an assortment of blocks or other items to build a three-dimensional abstract structure
9.	Small items with unusual textures (e.g., lava rock, Silly Putty, velvet, or an apple)
10.	Eight to ten scents in small containers (e.g., vinegar, vanilla extract, tuna fish juice, garlic, or aromatic lotions)

6.3   Guiding the Learning Process

An essential component of teaching mindfulness skills is the post-activity inquiry, which begins as an exploration of thoughts, emotions, and body sensations experienced during the mindfulness activity. These therapist-led discussions encourage participants to explore and deepen their understanding of mindfulness. The inquiries typically aim to help children: (a) learn to describe present-moment experiences—in their thoughts, feelings, and body sensations, (b) understand how living with mindful awareness differs from living on "autopilot," and (c) explore how this knowledge might be integrated into everyday life.

Helping children notice when they are judging or describing must be done with care. Children often have no qualms offering blunt opinions, but may take feedback about those opinions as criticism. The aim is not to stop the judgments but rather to see judgments for what they are—just thoughts, beliefs, desires, or expectations—that are not always consistent with reality. We all have a tendency to cling to things we like and push away things we do not like. MBCT-C is based on the idea that unhappiness arises from holding onto beliefs and expectations about an experience that are incongruent with the experience itself.

It is also helpful to validate and normalize the challenges that often arise during the mindfulness activities. Examples include finding it hard to focus on the object of attention, feeling unable to remain still during the quiet activities, or forgetting to do the home practice assignments. Validating the reality of these challenges encourages open, authentic dialogue and helps each child feel comfortable participating (at whatever level they can) in group and home-based activities. The key points and mindfulness practices for each of the 12 sessions are shown in Table 2.

Table 2
Overview of the theme, key points, and mindfulness activities for the 12-session program[a]

Session	Theme	Key points	In-session practices
1	Being on Automatic Pilot	• We live much of our lives on automatic pilot • Mindfulness exists, and it is a different, more helpful way of being in the world	• Getting to Know You • Discovering Awareness in a Cup • What Mindfulness Means to Me • Taking Three Mindful Breaths
2	Being Mindful Is Simple, But It Is Not Easy!	• Living with awareness is not easy, so why are we doing this anyway? • We give attention to the barriers to practice • Understanding the importance of practice • Bringing awareness to the breath and body	• Taking Three Mindful Breaths • Raisin Mindfulness • Mindfully Moooving Slooowly • Taking Three Mindful Breaths
3	Who Am I?	• Thoughts arise in the present but are often about the past or future • Thoughts may not be accurate to the present reality • Thoughts are not facts	• Taking Three Mindful Breaths • Mindfulness of the Body • Hey, I Have Thoughts, Feelings, and Body Sensations! • Listening to the Sounds of Silence • Taking Three Mindful Breaths
4	A Taste of Mindfulness	• We have thoughts, feelings, and body sensations, but these are not who we are • Thoughts, feelings, and body sensations are not exactly the same as the events they describe	• Introduction to Three-Minute Breathing Space • Opening to One Orange • Mindful Yoga Movements • Three-Minute Breathing Space
5	Music to Our Ears	• Thoughts, feelings, and body sensations often color how we experience the world • With our thoughts, we create individual and unique relationships and experiences • Awareness holds it all	• Three-Minute Breathing Space • Do You Hear What I Hear? • Mindfulness of the Body • Three-Minute Breathing Space
6	Sound Expressions	• Practicing mindful awareness helps us recognize that thoughts, feelings, and body sensations influence how we express ourselves • We can choose to express ourselves with mindful awareness	• Three-Minute Breathing Space • Sounding Out Emotions—Mindfully • Mindful Yoga Movements • Three-Minute Breathing Space

(continued)

Table 2
(continued)

Session	Theme	Key points	In-session practices
7	Practice Looking	• Judging is not the same as noting • If we simply observe experiences rather than judge them, the experience may change • We can choose to observe or note our experiences instead of judge them	• Three-Minute Breathing Space • Visualizing with Clarity • Mindful Yoga Movements • Seeing What Is in the Mind's Eye • Three-Minute Breathing Space
8	Strengthening the Muscle of Attention	• Judging often changes how we experience the world • Becoming more aware of judgments may change how we relate to thoughts and feelings • Discovering "choice points"	• Three-Minute Breathing Space • Seeing Through Illusions • Moving Mindfully • Seeing What Is Not There • Three-Minute Breathing Space
9	Touching the World with Mindfulness	• We have little control over most events that occur • We do have choices in how we respond to events • Choice points exist only in the present moment • Bringing greater awareness to this moment, we may see more choice points	• Three-Minute Breathing Space • Being in Touch • Mindfulness of the Body • Three-Minute Breathing Space
10	What the Nose Knows	• We often react to events by moving toward things we like or judge as "good" and moving away from things we do not like or judge as "bad" • Judging an experience may interfere with seeing clearly what is present in each moment • We have choices in how we respond to events	• Three-Minute Breathing Space • Judging Stinks! • Mindful Yoga Movements • Three-Minute Breathing Space
11	Life Is Not a Rehearsal	• Mindfulness is available in everyday life • We can practice mindful awareness using all our senses	• Three-Minute Breathing Space • Thoughts Are Not Facts • Feelings Are Not Facts Either • Raisin Mindfulness • Mindfulness Is… • Three-Minute Breathing Space

(continued)

Table 2
(continued)

Session	Theme	Key points	In-session practices
12	Living With Presence, Compassion, and Awareness	• Mindful awareness can be helpful in our daily lives • Bringing greater awareness to our lives is a personal choice • Living with awareness requires commitment, compassion, and continued daily practice	• Three-Minute Breathing Space • Exploring Everyday Mindfulness • Program Evaluation (optional) • Three-Minute Breathing Space • Graduation Ceremony • Graduation Party! • Three-Minute Breathing Space
	Three-Month Follow-Up	• Support for maintaining a daily practice of mindful awareness	• No session • Therapist mails Letter to My Self and Daily Practice Calendar to each child

[a]Table excerpt from Semple and Lee (2011). Mindfulness-Based Cognitive Therapy for Anxious Children. New Harbinger Publications, Oakland, CA (Reprinted by permission)

6.4 The 12-Session Program

Session 1. Children participate in an "ice-breaker" activity to get to know each other and are introduced to the structure of each session. One optional component is the *Feely Faces Scale*, which is a 7-point self-rating scale with visual representations of feelings that range from very unhappy to very happy. Using the *Feely Faces Scale* reminds children to "check in" with themselves at the beginning and the end of each session, becoming mindfully aware of their emotional state in that moment. Therapists then describe the concept of automatic pilot (autopilot) and engage children in a discussion to elicit examples from their own lives. To experience the differences between functioning on autopilot vs. functioning with mindful awareness, children pass a cup of water hand-to-hand around the group circle, first half-full, then completely full, and then again after darkening the room or asking them to close their eyes. Each time around the circle, the task demands increase and require giving greater attention to smaller details. During the post-activity inquiry, children explore how the experience differed with each new challenge. For example, one child might mention that passing a cup filled to the brim needed more focused attention than passing the half-full cup. Another might have noticed needing to move more slowly. When the room was darkened, some might notice that they were relying more on the other senses, perhaps being attuned to sounds, or even noticing the movement of air as the person seated next to them moved the cup closer. Many comment on the increasing difficulty. The therapist can use this activity as a metaphor for the development of greater mindfulness. Emphasizing the importance of practice and patience in cultivating mindfulness is useful as well. After this introduction to mindfulness, children are invited to draw a picture that represents their understanding of mindfulness. This drawing becomes the front cover of their binder, which will consist of *Feely Faces Scales*, session summaries, home practice assignments, poems, information handouts, and a personal journal. In concluding this first session, the therapist guides children in taking three mindful breaths and invites them to participate in a group poetry reading. At the end, each child can again check in with themselves and record their mood state on the *Feely Faces Scale*. Children are given a written summary of the session and home practices to complete before the next session. Every session ends with distributing a session summary and home practice handouts. These are the general components of each session. For the sake of brevity, the weekly mood check-in, home practice review, opening and closing breathing practices, poetry reading, session summary, and new home practice assignment will not be noted in the individual session descriptions below.

Session 2. Now that children have dipped their toes into the sea of mindfulness, they learn that although practicing mindfulness may be simple, it is often not easy. Learning mindfulness takes practice, so each session consists mainly of experiential activities. Therapists invite children to participate in a few core activities to start this process. Children practice taking three mindful breaths, while focusing on the inflow and the outflow of breath throughout their bodies. During the post-activity inquiry, the therapist elicits descriptions of thoughts, feelings, and bodily sensations observed during the breathing practice. Children then learn to incorporate mindfulness into an ordinary daily activity—mindful eating. Each child practices eating one raisin with full, mindful awareness of the experience—observing the qualities of the raisin—its texture, weight, appearance, taste, and smell. They may notice how the raisin transforms in the mouth and how the body responds to the raisin. Normally active children are not comfortable sitting still for 90 minutes, so each session includes one movement activity. *Mindfully Moooving Slooowly* is a slow movement activity aimed at bringing awareness to the many body sensations of each small movement. Children are invited to walk as slowly and mindfully as they can. The goal is not to get from one place to another but rather to experience each inbreath and each outbreath; to feel the movement of the muscles and the changing sensations in each foot as it contacts the ground; and bring awareness to the moments of stillness between each step. During the inquiry, therapists help children describe discrete sensations and begin to identify when the mind may be judging the activity, lost in another moment, or comparing it to other experiences. The session closes with another practice of taking three mindful breaths.

Session 3. Being mindful of restlessness or disengagement, therapists may wish to extend the three mindful breaths practice for a few additional breaths. Stop the activity if the group becomes restless. Mindful breathing is practiced in every session. Three mindful breaths may be added between any activity to provide a transitional space or a moment of self-calming. Session 3 introduces children to mindfulness of the body through a brief (approximately 10 min), guided body scan, which may be done sitting or lying down. During this activity, children are prompted to direct attention to one part of their motionless body at a time, starting with the toes of one foot, moving up the leg, shifting to the other foot and leg, exploring the feeling of the breath moving through the body, while bringing awareness to sensations in each part of the body. Children then have an opportunity to describe their experiences and explore what made a body scan different than just sitting or lying down with eyes closed.

This also lays the groundwork for the next practice, which is connecting thoughts, feelings, and body sensations. The CBT model suggests that changing thoughts can change feelings, and behaviors. For example, a CBT therapist might encourage a child to identify anxiety-related distorted thoughts, challenge their accuracy and/or helpfulness, and then change the maladaptive thoughts to be more realistic or helpful in managing the anxiety. Alternatively, the MBCT-C model suggests that increasing awareness of thoughts, feelings, and body sensations changes how we respond to them; that decentering from the unhelpful thoughts reduces their emotional impact, and thereby reduces anxiety. For example, a MBCT-C therapist might encourage the same child to sit and breathe with the anxious feelings, bring attention to the associated thoughts and body sensations, recognize them as internal experiences, cultivate nonjudgmental acceptance of these experiences as "just thoughts," "just feelings," or "just body sensations," and then help the child make a conscious choice about how to respond to them.

A brief mindful listening activity is an opportunity to adopt a different perspective by focusing attention on the silence between the sounds of a bell being rung several times. Then a visualization activity encourages children to closely observe their thoughts, feelings, and body sensations. A short inquiry after each of these activities deepens the children's understandings of their internal experiences.

Session 4. In this session, children are introduced to the Three-Minute Breathing Space, which is a core activity borrowed unchanged from MBCT. The Three-Minute Breathing Space will be practiced at the beginning and end of each remaining session. Building on the raisin activity that was done in Session 2, we now focus on mindfulness in the sense of taste. Mindful eating is practiced with an experience of touching, holding, smelling, and

then peeling and eating an orange—while bringing mindful awareness to thoughts, feelings, and body sensations. The therapist
invites the children to compare the experience of mindfully eating
this orange to how they might typically eat (i.e., on autopilot). A
post-activity inquiry follows. After being seated for 15–20 min, the
next activity is a mindful movement practice with simple yoga
postures. The aim is to cultivate mindful awareness of the body in
movement rather than achieving perfect form. Competing with
other group members is also discouraged. An inquiry and discussion of the mindful movement activity follows.

Session 5. In the previous session, children practiced mindfulness of taste. The next two sessions focus on mindfully engaging
with sounds. Session 5 focuses on receptive listening while Session
6 focuses on expressively creating sounds. During Session 5, children practice listening with mindful awareness to five or six segments of music. Select a wide variety of instrumental music
(without lyrics). For example, you might choose segments from a
Jazz rift, African drumming, a Bach sonata, Bossa Nova, Calypso,
and Mariachi.

After checking in with themselves to identify their thoughts,
feelings, and body sensations, they create a title for each piece of
music that reflects their current emotional state. Titles for a single
piece can vary widely—from "wedding march" to "angry monster"
to "funeral service." The post-activity inquiry explores how each
child experienced something different while listening to the same
music. The therapist helps them explore the concept of subjectivity
and how their own thoughts and feelings might influence how they
experience their worlds. They learn that they can transform an
experience simply by attending to it with nonjudgmental awareness. They also begin to learn that awareness is the container within
which all experiences are created. This concept is emphasized as one
key point of this session: "awareness holds it all." The second
mindfulness activity for this session is another brief body scan,
which is followed by the second Three-Minute Breathing Space.

Session 6. This session is also mindfulness of sounds, but is now
focused on the mindful expression of sounds. We offer the caveat
that this session can get somewhat loud. Based on children's previous experiences identifying thoughts, feelings, and body sensations
arising in response to listening to sounds, they will now use their
thoughts, feelings, and body sensations to guide the creation of
sound by conducting their own symphony. Each child is asked to
create a short piece of music that expresses his or her present-
moment experience—in thoughts, feelings, and body sensations.
Emphasize that each child will have a turn being the "conductor."
As conductor, each child assigns each of other children a musical
instrument (or re-purposed household item). Then, without
words, the conductor guides his or her "orchestra" in creating a
piece of music; instructing the musicians to play their instruments

fast or slow, loud or soft—without words—simply by pointing or using a baton. The conductor invites the other children to create sounds that express the conductor's internal experiences. The inquiry also explores the subjectivity of how we interpret our experiences, but mainly focuses on examining the ways we communicate our internal experiences to others. The second mindfulness activity for Session 6 is another mindful movement practice.

Session 7. Sessions 7 and 8 involve mindfulness of seeing. This session focuses on seeing clearly—practicing observing the world with mindful awareness of small details. Two visualization activities offer practice in using the *mind's eye* to see clearly. The first visualization helps them observe the different experiences of judging vs. observing or noting. The therapist guides the children in visualizing a pleasant scene. Afterward, the group inquiry explores the thoughts, feelings, and body sensations of the experience, and identifies what specific elements defined the experience as being pleasant. The aim is to elicit understanding that thoughts *about* the experience (positive judgments) define the experience. Following a mindful yoga activity, the therapist guides the second visualization, which is combined with a drawing task. The task now is to first use the "mind's eye" to see clearly a familiar object in their home. This might be the kitchen stove, a piece of furniture, or a favorite toy. The therapist prompts them to see clearly all the details of the object and then draw it. The children will then compare their drawing with the actual object. The aim is to notice ways the image in the mind's eye might distort the reality of a familiar object. They will explore elements they accurately recalled through the drawing, as well as components they may have changed or omitted entirely. During the home practice review at the beginning of Session 8, we review parts of the drawing that are not true to the reality of the object—drawn from vivid, yet often inaccurate, memories.

Session 8. Mindfully looking at optical illusions helps children learn how their experiences might change when they intentionally shift their attention. Children discover that what they "see" is the actual object in combination with a cognitive interpretation of that object. Practice in redirecting attention is the aim of the *Seeing Through Illusions* activity. One thing they may discover with the rabbit–duck optical illusion, for example, is that both images cannot be seen simultaneously. Most children will initially see only one image, until the other is pointed out to them. Then, it becomes easy to shift attention back and forth to see first the rabbit, then the duck. This is followed by a mindful movement activity. The last mindfulness activity, *Seeing What is Not There*, is a negative-space drawing activity. The therapist creates a three-dimensional abstract structure from blocks or other objects in the room. The structure should have a number of interior angles and open spaces. The children are invited to sit in a circle around the structure and give mindful attention to drawing the spaces between (e.g., spaces where air, light, or shadows pass through). Each child sees the structure from a different angle, and consequently, each drawing will be different. This and the optical illusion activity demonstrate clearly that our felt experiences are influenced by what we choose to attend to, and that attending in a different way—adopting a different perspective—changes how we experience the world. Variations of this activity include drawing only the straight lines, or just the colors, or textures. The idea is to move away from autopilot by encouraging the children to look at the world in a different way. The post-activity inquiry can explore the idea that each of us sees the world differently and that no one perspective is the absolute truth. In this session, children are also introduced to "choice points," which arise in the moments between a trigger event and the response to that event. These are the points at which conscious choices can be made. The therapist also aims to elicit understanding of the distinctions between descriptive (objective) observations and thoughts and feelings (subjective).

Session 9. In this session, children practice bringing mindful awareness to the sense of touch. The activity we call *Being in Touch* involves an assortment of small objects that have a variety of textures. Examples might include a hairbrush, bubble wrap, cotton balls, pine cones, or Silly Putty. The children take turns being blindfolded and then one object is placed in his or her hand (or they may prefer to hold the object behind their back). Without seeing or naming it, the task is simply to describe the object. They can explore and describe the temperature, size, shape, texture, and weight of the object using only their sense of touch. The other children in the group can see the object, so they may add color, shading, and pattern to the description. All children help each other identify when a word becomes a subjective judgment rather than an objective description. Each child gets the opportunity to practice mindfully touching and describing one object. The inquiry should aim to bring out the idea that, although we often have little control over things that happen, we always have choices in how we respond to events. Learning to stay present with what is happening, separating the objective from the judgmental, we may discover different ways to respond—different "choice points" that may be more skillful or wholesome than their usual choices. The body awareness activity for this session is a 10-min body scan.

Session 10. Smells generally elicit rapid judgments, which are usually emotionally reactive "like" or "dislike" decisions. Because this is an instinctive, nearly automatic reaction, mindfulness of smells can be considered an advanced practice. In this session, children practice smelling and describing different scents—while observing their judgments about the scents. For this activity, the therapist offers five or six different scents, passing each scent, one at a time, around the circle to each child. The task is for the child to describe each scent using objective words, while bringing awareness to the thoughts, feelings, and body sensations that arise in response to the scent. They may discover how difficult it is to find descriptive words (e.g., strong, sour, or sweet) rather than judgmental words (e.g., nice, gross, or boring), particularly for strong scents. However, it can be easy to observe the emotional "like" or "dislike" response, particularly when the child moves closer to or away from the scent. This point can be brought out in the inquiry—that we frequently react to events by moving toward the ones we judge positively and by moving away from those we judge negatively. Neither reaction, often driven by emotional reactivity or quick judgments, reduces anxiety or increases happiness. The alternative is learning to use mindful awareness to see available "choice points," and then selecting skillful response choices. The second activity for this session is a mindful yoga movement activity.

Session 11. The last two sessions focus on maintaining and generalizing mindfulness as a way of life. Although practicing mindfulness in daily life is emphasized throughout the program, the aim of these last sessions is to reinforce the idea that practicing mindfulness can change how we relate to the world and to ourselves in significant ways. Practicing mindfulness can be done with formal breath or body practices, or informally, by remembering to bring mindful awareness to the things we see, hear, touch, smell, and taste. To promote decentering, the mindfulness activities for Session 11 again explore thoughts and feelings as internal events—mind stuff—not facts. Two visualization activities reinforce the idea that thoughts are not facts and that emotions are not facts either. Relationships between thoughts, emotions, and body sensations are also reexamined with a repeat of the same raisin activity that was done in Session 2. Each child then creates a drawing that becomes the back cover of his or her notebook. The drawing prompt is simply, "Mindfulness is ...." This activity gives children an opportunity to see how the meaning of mindfulness may have changed for them over the past months. We also take a few minutes to plan the Session 12 graduation party. The home practice activity for this week is to write a letter to themselves. Practicing mindfulness is not particularly difficult. Remembering to practice is. So, these letters will be mailed back to the children 3 months after the end of the program. The written home practice handout includes a number of prompts to focus the writing. Examples of these include the following:

- What does mindfulness mean to me now?
- Has mindfulness changed the way I interact with other people? If so, how?
- What skills have I learned that might be helpful to me?
- How would I like to continue cultivating mindful awareness in my everyday life?
- What could I do that will help me remember to practice?

Session 12. The first goal of Session 12 is to reflect and explore ways to continue the practice of mindfulness. Following the opening Three-Minute Breathing Space, the children are invited to share the *Letter to Myself* that they wrote after Session 11. By this time, the children tend to be very cohesive and supportive of each other, and although sharing is always optional, we have found that nearly all the children do choose to share what they wrote. This discussion is mostly guided by the children themselves. They discuss what influence the group may have had on their lives or their relationships; they explore everyday mindfulness and how it might bring changes to their lives; and share strategies for continuing the practices beyond the weekly group sessions. They may also share what they have learned from other group members or express appreciation for another child's presence in the group. The discussion ends with another Three-Minute Breathing Space. Then comes the graduation ceremony and farewell party. The graduation ceremony is a way of creating closure by marking the dissolution of the group, celebrating the shared journey, their achievements, and exchanging goodbyes. As therapists, we may choose to write a short farewell letter and prepare congratulatory "graduation from MBCT-C" certificates for each child. In addition, we share one more mindfulness activity in which we distribute a small remembrance of the program to each child (usually a brightly colored polished pebble). The children often bring cookies, drinks, balloons, or other party goods to the party. We generally bring pizza as our contribution.

6.5 When the Program Ends

Mindfulness as a way of life is emphasized throughout MBCT-C, and more so as the group nears its end. Children have considered ways to bring mindfulness into their everyday lives. Therapists have encouraged the children to explore the benefits of using mindfulness skills in daily activities and emphasized the importance of ongoing practice. Children, like adults, may find themselves skipping daily practices and moving back into old habits. The therapist has fostered motivation for ongoing practice and worked with the children to devise practical strategies for integrating mindfulness in everyday life. Three months after the program ends, letters the children wrote to themselves after Session 11 should be mailed back to them. In this way, we allow each child's own "words of wisdom" to motivate his or her ongoing practice.

6.6 Adapting MBCT-C for Individual Therapy

Individual therapy adaptations of MBCT-C interventions are not difficult and the treatment manual offers suggestions for adapting each session to this modality. The key considerations are the reduced time for individual sessions and the ways that the therapist engages with the child. Some activities will naturally take less time. Others can be abbreviated. Most do not require major modifications. We recommended shortening an activity rather than eliminating it altogether. Session review handouts may be taken home and read with the parent rather than reading them during the session. A few activities, for example, the expressive sounds activity (Session 6) do need significant modifications for an individual therapy format. The treatment manual suggests starting by prompting the child to identify thoughts, feelings, and body sensations associated with different situations that have triggered strong emotional responses (e.g., an argument with a sibling or not being invited to a party). Then, rather than assigning musical instruments to others, the child creates sounds that represent each identified emotion. The therapist participates by responding with sounds that represent his or her emotional response to the child's music.

More important is how the therapist engages with the child when a therapist–child dyadic relationship replaces the group interactions. The therapist will need to participate more fully in each of the activities, with mindful attention that their own contributions do not come across as being judgmental or critical of the child. Children are commonly corrected by adults, and are likely to be less familiar relating to them on an equal basis. Authenticity is important—you are on a shared journey and should expect to learn as much as your young client. Closure is important. Children in individual therapy often develop strong attachments to their therapists. Each child should have a small graduation, perhaps choosing to read the letter he or she wrote, reflect and discuss his or her experience of MBCT-C, and receive a small remembrance and graduation certificate from the therapist. Together, the therapist and child can plan a fun activity or share food to celebrate completing the program.

6.7   Adapting MBCT-C for Special Populations

No one treatment, including MBCT-C, suits the needs of all children. The treatment manual discusses adaptations as needed for children with special needs. Examples include eliminating the mindful tasting activities for children with food allergies or eating disorders and modifying the mindful movement activities for children with physical challenges. Children with severe anxiety frequently prefer to keep their eyes open during the breathing activities or body scan, and may feel more comfortable seeing the item they are touching rather than being blindfolded. Children with autism spectrum disorder or other developmental disabilities may have verbal, intellectual, or sensory issues that need to be addressed. Activities and language may need to be simplified, shortened, or the program extended to allow more frequent repetition to match the learning pace of the children. Children with reading difficulties may prefer not to read aloud to the group. Therapeutic sensitivity, mindful attention, and understanding the needs of each child in your group will allow you to modify the interventions appropriately.

7   Discussion

It is important to keep in mind that "mindfulness is not a set of techniques one can simply learn and then sit back and enjoy the fruits" [32]. There are a number of personal and clinical training considerations for therapists wishing to conduct mindfulness-based interventions. First, basic clinical competencies must be mastered. Given that MBCT-C is generally conducted in groups, therapists should be knowledgeable about the theory and practice of group facilitation. Group therapy adds the complexity of managing multiple interpersonal dynamics, while cultivating the wholesome therapeutic alliance that is essential to any therapy. When implementing any child therapy, some understanding of child development and family dynamics is helpful. Recognizing how depression, anxiety, trauma, and other issues manifest in children is essential, as well as understanding the precepts and practices of CBT.

In CBT, as in most traditional psychotherapies, effecting change is the primary aim. Therapists and clients tend to focus on what changes in thinking or behaviors might be beneficial. Seasoned therapists know that promoting change can be complex and difficult. Particularly in longer-term therapies, motivation and engagement can wax and wane. Alternatively, mindfulness-based interventions foster acceptance rather than change. Mindful therapists may become aware of desires in themselves to promote change in their clients. MBCT-C encourages observing rather than judging, participating with authenticity, accepting the client as he or she is, and letting go of the desire to push for change. Moving from a change-oriented model to an acceptance and awareness-based approach may be the biggest challenge for conventionally trained therapists.

As the field of child mindfulness continues to develop, mindful therapists and practitioners are evaluating what constitutes competency in teaching mindfulness. The use of mindfulness-based interventions in clinical settings has strengthened the call for unified standards and specialized clinical training, but no consensus has yet been reached. The UK Network for Mindfulness-Based Teacher Trainers has established *Good Practice Guidance for Teaching Mindfulness-Based Courses* [33]. These guidelines emphasize the need to be a trained mental health provider with experience providing evidence-based treatments or have a background in education, with experience in group-based teaching. Familiarity with the group population being served is recommended. Personal participation in the mindfulness-based curriculum one plans to facilitate and having a regular practice of mindfulness are also strongly encouraged. Other recommendations include completion of a mindfulness-based teacher training program, regular supervision, consultation, and collaboration with colleagues, and maintaining awareness of current empirical evidence. Saltzman's [34] *Qualifications for Teachers of Mindfulness-Based Stress Reduction Youth Programs* suggest qualifications for mindfulness group facilitation with children. These include five years of mindfulness experience, attendance at mindfulness retreats, experience teaching yoga or another movement practice in a mindfulness context, professional experience working with children, facilitating process groups, and creating a network of mindfulness colleagues.

Many mindfulness programs, including MBCT-C, emphasize the importance of cultivating a personal mindfulness practice. As noted in both sets of guidelines discussed above, we strongly encourage therapists interested in facilitating MBCT-C to learn and practice mindfulness in their own lives. Participating in an MBSR (Mindfulness-based Stress Reduction) [35] or MBCT program and developing a regular personal mindfulness practice can be a good beginning. Although MBCT-C does not yet have a formal therapist-training program, we agree with Saltzman's model, with one important addition. Since MBCT-C is intended as a clinical intervention, we believe that facilitators must be clinically trained to conduct child therapy and have an understanding of both CBT and child psychopathology.

Mindful therapists teach through their own embodiment of mindfulness. Embodying mindfulness includes cultivating present-focused awareness, acting and speaking with intention, and maintaining a nonjudgmental stance while incorporating acceptance and compassion [32]. Additional qualities include being able to establish a sense of safety in the group, convey genuine interest, open curiosity and compassion for all aspects of the human experience, "unshockability," and profound empathy [34]. However, it may not be realistic to expect oneself to embody all these traits at once. Crane [32] noted that becoming a mindful therapist requires "the development of awareness (clear seeing) cultivated during formal and informal practice in daily life; the intention to underpin one's practice and life with a particular attitudinal framework characterised by warm acceptance, non-striving and curiosity; and a willingness to engage in an alive exploration of what it is to be human and of how our suffering is caused" (p. 160). Development of these qualities takes practice, patience, and self-compassion. Remember that although this personal journey can be demanding, it also holds the possibility of a lifetime of enrichment.

References

1. Kessler RC et al (2012) Lifetime co-morbidity of DSM-IV disorders in the US national comorbidity survey replication adolescent supplement (NCS-A). Psychol Med 42:1997–2010
2. Merikangas KR et al (2010) Lifetime prevalence of mental disorders in US adolescents: results from the national comorbidity survey replication-adolescent supplement (NCS-A). J Am Acad Child Adolesc Psychiatry 49:980–989
3. Costello EJ et al (2003) Prevalence and development of psychiatric disorders in childhood and adolescence. Arch Gen Psychiatry 60:837–844
4. Kessler RC et al (1996) Comorbidity of DSM-III-R major depressive disorder in the general population: results from the US national comorbidity survey. Br J Psychiatry 168:17–30
5. Mojtabai R et al (2015) Long-term effects of mental disorders on educational attainment in the national comorbidity survey ten-year follow-up. Soc Psychiatry Psychiatr Epidemiol 50:1577–1591
6. Goodwin RD, Fergusson DM, Horwood LJ (2004) Association between anxiety disorders and substance use disorders among young persons: results of a 21-year longitudinal study. J Psychiatr Res 38:295–304
7. Connolly SD, Bernstein GA (2007) Practice parameter for the assessment and treatment of children and adolescents with anxiety disorders. J Am Acad Child Adolesc Psychiatry 46:267–283
8. Walkup JT et al (2008) Cognitive behavioral therapy, sertraline, or a combination in childhood anxiety. N Engl J Med 359:2753–2766
9. Flannery-Schroeder EC, Kendall PC (2000) Group and individual cognitive-behavioral treatments for youth with anxiety disorders: a randomized clinical trial. Cognitive Ther Res 24:251–278
10. James AC et al (2013) Cognitive behavioural therapy for anxiety disorders in children and adolescents. Cochrane Database Syst Rev 6
11. Williams JMG et al (2007) The mindful way through depression: freeing yourself from chronic unhappiness. Guilford, New York, NY
12. Semple RJ, Lee J (2011) Mindfulness-based cognitive therapy for anxious children: a manual for treating childhood anxiety. New Harbinger, Oakland, CA
13. Lenz AS, Hall J, Bailey SL (2016) Meta-analysis of group mindfulness-based cognitive therapy for decreasing symptoms of acute depression. J Spec Group Work 41:44–70
14. Kuyken W et al (2010) How does mindfulness-based cognitive therapy work? Behav Res Ther 48:1105–1112
15. Schoenberg PL, Speckens AE (2014) Modulation of induced frontocentral theta (Fm-theta) event-related (de-)synchronisation dynamics following mindfulness-based cognitive therapy in major depressive disorder. Cogn Neurodyn 8:373–388
16. Segal ZV, Williams JMG, Teasdale JD (2013) Mindfulness-based cognitive therapy for depression, 2nd edn. Guilford, New York, NY
17. Chiesa A, Serretti A (2011) Mindfulness based cognitive therapy for psychiatric disorders: a systematic review and meta-analysis. Psychiatry Res 187:441–453
18. Piet J, Hougaard E (2011) The effect of mindfulness-based cognitive therapy for prevention of relapse in recurrent major depressive disorder: a systematic review and meta-analysis. Clin Psychol Rev 31:1032–1040
19. Metcalf CA, Dimidjian S (2014) Extensions and mechanisms of mindfulness-based cognitive therapy: a review of the evidence. Aust Psychol 49:271–279
20. CEBC (2018) Mindfulness-based cognitive therapy for children. http://www.cebc4cw.org/program/mindfulness-based-cognitive-therapy-for-children-mbct-c/detailed
21. Semple RJ, Reid EF, Miller L (2005) Treating anxiety with mindfulness: an open trial of mindfulness training for anxious children. J Cogn Psychother 19:379–392
22. Lee J, Semple RJ, Rosa D, Miller L (2008) Mindfulness-based cognitive therapy for children: results of a pilot study. J Cogn Psychother 22:15–28
23. Semple RJ et al (2010) A randomized trial of mindfulness-based cognitive therapy for children: promoting mindful attention to enhance social-emotional resiliency in children. J Child Fam Stud 19:218–229
24. Cotton S et al (2016) Mindfulness-based cognitive therapy for youth with anxiety disorders at risk for bipolar disorder: a pilot trial. Early Interv Psychiatry 10:426–434
25. Esmailian N et al (2013) Effectiveness of mindfulness-based cognitive therapy on depression symptoms in children with divorced parents. J Clin Psychol 3:47–57
26. Ames CS et al (2014) Mindfulness-based cognitive therapy for depression in adolescents. Child Adolesc Ment Health 19:74–78
27. Haydicky J et al (2012) Evaluation of a mindfulness-based intervention for adolescents

with learning disabilities and co-occurring ADHD and anxiety. Mindfulness 3:151–164

28. Milani A, Nikmanesh Z, Farnam A (2013) Effectiveness of mindfulness-based cognitive therapy (MBCT) in reducing aggression of individuals at the juvenile correction and rehabilitation center. Int J High Risk Behav Addict 2:126

29. Haydicky J et al (2015) Evaluation of MBCT for adolescents with ADHD and their parents: impact on individual and family functioning. J Child Fam Stud 24:76–94

30. Strawn JR et al (2016) Neural function before and after mindfulness-based cognitive therapy in anxious adolescents at risk for developing bipolar disorder. J Child Adolesc Psychopharmacol 26:372–379

31. Abedini S, Habibi M, Abedini N, Achenbach TM, Semple RJ (under review). A randomized clinical trial of mindfulness-based cognitive therapy for children: managing internalizing and attention problems in children hospitalized with cancer. Mindfulness.

32. Crane R (2017) Mindfulness-based cognitive therapy: the CBT distinctive features series, 2nd edn. Routledge, Abingdon, Oxon, pp 159–160

33. UK Network for Mindfulness-Based Training Organisations (2015) UK network for mindfulness-based teachers: good practice guidelines for teaching mindfulness-based courses. Good practice guidance for teachers. https://www.mindfulnessteachersuk.org.uk/pdf/UK%20MB%20teacher%20GPG%202015%20final%202.pdf

34. Saltzman A (2014) A still quiet place: a mindfulness program for teaching children and adolescents to ease stress and difficult emotions. New Harbinger, Oakland, CA

35. Kabat-Zinn J (1994) Wherever you go there you are: mindfulness meditation in everyday life. Piatkus, London

Chapter 11

Cognitive Behavioral Therapy with Diverse Youth

Esteban V. Cardemil, Caitlin Straubel, and Jessica L. O'Leary

Abstract

Researchers and clinicians are increasingly recognizing the importance of attending to dimensions of diversity in clinical work given the changing demographics of the USA. There exists a growing evidence base demonstrating the effectiveness of cognitive behavioral therapy (CBT) with racial and ethnic minority youth, and there is some work examining CBT with sexual and gender minority youth. In this chapter, we review some of this literature and present a case example with a Latinx transgender girl and her parents. We focus on the importance of developing a strong working alliance within a CBT framework and use the literature on cultural competency to ground our recommendations.

   Key words Diversity, Cognitive behavioral therapy, Culture, Gender

1 Introduction

The demographics of the USA are changing rapidly and becoming increasingly diverse. The 2010 US Census noted that non-Hispanic White Americans represented only 63.7% of the population, a reduction from 69.1% in the 2000 Census [1]. Moreover, demographic projections suggest that by 2050, non-Hispanic White Americans will represent less than 50% of the overall US population [2]. These changing patterns have also been found with youth, where the percentage of non-Hispanic white youth under age 18 was 54% in 2010 and projected to drop to 50% by 2020 [3]. These demographic changes have already become evident in certain parts of the country. For example, the 2010 US Census reported that in California, the District of Columbia, Hawaii, New Mexico, and Texas, racial and ethnic "minorities" represented more than 50% of the total population [1].

Beyond race and ethnicity, there has also been growing awareness about the importance of other dimensions of diversity and identity, including sexual orientation/attraction and gender identity. In particular, according to the Williams Institute, approximately 4.5% of the US population identifies as lesbian, gay,

Robert D. Friedberg and Brad J. Nakamura (eds.), *Cognitive Behavioral Therapy in Youth: Tradition and Innovation*, Neuromethods, vol. 156, https://doi.org/10.1007/978-1-0716-0700-8_11, © Springer Science+Business Media, LLC, part of Springer Nature 2020

bisexual, or transgender (LGBT) [4], numbers that appear to be similar across racial and ethnic groups [5]. Although demographic data regarding the sexual and gender identity of youth is lacking, the Centers for Disease Control's (CDC) annual Youth Behavior Risk survey recently found that approximately 2.4% of youth under 18 identified as gay or lesbian, and 8.0% identified as bisexual [6]. There exist fewer estimates of the number of youth who identify as transgender or gender diverse[1]; however, one report suggests that approximately 0.7% of youth under age 18 identified as transgender [8].

Unfortunately, it is unclear if the mental health system is adequately meeting the needs of this increasingly diverse population. Considerable research has found that racial and ethnic minority (REM) youth are less likely to receive mental health services than White youth, despite similar need [9–12]. These disparities in receipt of mental healthcare have even been documented in school-based services, where the structural barriers to accessing care are less salient [9, 13, 14]. Although there is less research examining access and utilization of mental healthcare among sexual and gender minority (SGM) youth, the limited body of research to date has generally found that sexual minority youth utilize mental health services at higher rates than heterosexual youth [14, 15], but still demonstrate greater unmet need [16, 17].

Disparities in receipt or experience of mental healthcare are problematic for youth with marginalized identities. Among REM youth, researchers have documented elevated rates of depression and externalizing symptoms [18], and research has consistently documented that SGM youth are at elevated risk for a number of mental health problems, including depression and suicidality, as well as greater risk of victimization [19–21]. Although limited, research examining the experiences of individuals with multiple marginalized identities (e.g., both REM and SGM) suggests that they may have elevated risk of experiencing multiple forms of discrimination, including racism and heterosexism [22]. Given this elevated risk for mental health problems among youth with marginalized identities, finding ways to reduce the persistent disparities in mental healthcare is critical.

There are numerous explanations for these disparities [23]. Given that many of the explanations are structural in nature and therefore located at the level of society (e.g., funding for community health centers), communities (e.g., limited number of bilingual providers), and systems (e.g., insurance limitations), many of the changes needed to address these disparities take time to implement [24]. However, there is also ample evidence that

---

[1] The terms "transgender" and "gender-diverse" refer to individuals whose gender expressions and/or identities do not align with their sex assigned at birth and who may not conceptualize their identity along the gender binary [7].

important contributors to disparities exist at the level of service provision [24, 25]. In particular, the research examining the efficacy of psychotherapy interventions with racial and ethnic minority individuals is extremely limited [26, 27], raising important questions about the generalizability of standard interventions to diverse populations [28]. Moreover, research on therapist cultural competence has found it to be associated with the quality of provider–patient relationships, treatment engagement, and outcome [29–32]. Similarly, several studies have found that SGM are at risk for receiving treatment from providers who lack knowledge about LGBT issues, demonstrate heteronormative bias, or engage in discriminative and non-affirming behavior [33–38]. Thus, there is good reason to believe that mental healthcare disparities could be partially addressed through greater attention to the development, evaluation, and dissemination of culturally adapted interventions, as well as through increased therapist training in culturally sensitive techniques.

Cognitive behavioral therapy (CBT) has been at the forefront of research examining the generalizability of therapeutic interventions to diverse populations. Most of the relevant CBT research with youth has been with REM, but researchers have also begun to develop and evaluate approaches to working with SGM (e.g., [39]). However, very little research has examined the treatment experiences and outcomes of youth who identify as both REM and SGM. Attending to the needs of youth with multiple marginal identities will likely require interventions that not only draw on approaches that have been developed for REM youth and SGM youth but also recognize their unique intersectional needs. In particular, this work will likely require careful attention to the development of a strong working alliance that takes into consideration the specific needs of these youth and their families.

In this chapter, we describe a case example of a transgender Latina girl and her family and present recommendations for developing a strong working alliance. By way of background, we first briefly review the relevant research literature that has examined the efficacy of CBT with REM youth, SGM youth, and youth with multiple marginalized identities.

2 Theoretical and Empirical Foundations

2.1 CBT with Racial and Ethnic Minority Youth

Although limited, the research documenting the effectiveness of CBT with REM youth has grown, particularly with African American and Latinx[2] youth [27, 42]. Among African American youth, CBT has been found to be effective in treating anxiety-related

---

[2] Latinx is a panethnic term that refers to individuals of Latin American descent living in the US. It is increasingly being used in recognition that the gender binary of Spanish excludes many gender diverse individuals who do not subscribe to that binary [40, 41].

problems [43], conduct problems [44], and trauma-related problems [45]. Among Latinx youth, evidence supports the effectiveness of CBT with anxiety [46], depression [47, 48], and trauma-related problems [49]. In a meta-analysis examining the effectiveness of all evidence-based interventions for racial and ethnic minority youth, Huey and Polo [50] reported an overall medium effect size ($d = 0.44$), although they did not distinguish among specific treatment orientations.

Most of these interventions have been adapted in some way to make them more relevant and appealing to REM youth and their families [27]. Adaptations tend to be varied and can be included in the treatment format (e.g., group vs. individual format, inclusion of family sessions), treatment delivery (e.g., use of cultural values in delivery, ethnic match between the provider and the client) or treatment content (e.g., integrating culturally relevant material into manuals) [28]. Unfortunately, few of these adaptations have been evaluated in rigorous or systematic ways, despite the existence of various cultural adaptation frameworks (e.g., [51]). As such, there is no consensus in the literature as to whether (and which of) these adaptations increase the efficacy of the treatments (e.g., [50, 52]).

One example of a cognitive behavioral intervention with good evidence of effectiveness is the Cognitive Behavioral Intervention for Trauma in Schools (CBITS) [53]. This intervention was originally developed in collaboration with the Los Angeles Unified School District (LAUSD), an urban school district with a majority Latinx population. Using a standard 10-session group format, CBITS addresses trauma-related symptoms and related difficulties for school-age children and their families. Important elements include psychoeducation, relaxation training, stress reduction, trauma exposure, and problem-solving skills [49]. CBITS also incorporates a parental component that consists of psychoeducation and skill-building sessions. Several studies have found CBITS to be effective in reducing PTSD and depressive symptoms among Latinx children [49, 53]. Moreover, adaptations to make the intervention more relevant for immigrant Latinx families have been developed and preliminarily evaluated. These adaptations included delivering the intervention in Spanish, adjusting the content of the intervention to include material on family coping, and adding joint sessions with parents and children [49, 54]. Preliminary results from the research on these adaptations were promising and suggest that CBITS and its adaptations are effective with a range of Latinx youth.

In sum, the literature examining the effectiveness of CBT with REM youth is promising, particularly for African American and Latinx youth. There remain many gaps, however. In particular, there is almost no evidence base examining the efficacy of CBT, or any psychotherapy approach, for Asian American and American

Indian youth. Moreover, the majority of evidence demonstrating CBT's efficacy has been conducted by only one research team; replication by independent research teams is critical to more firmly establish the benefits of the interventions [27]. Finally, more research is needed to better evaluate the particular contributions made by cultural adaptations.

2.2 CBT with Gender and Sexual Minority Youth

In comparison to the research with REM youth, the literature examining CBT's effectiveness with SGM youth is more nascent. Although there are currently no published randomized controlled trials (RCT) in the literature, there exists a growing clinical literature describing affirmative approaches to treating SGM (e.g., [36, 55]). These approaches call for sexual and gender diversity to be treated as normal variants in human identity, and not viewed as pathological [55, 56]. They instead conceptualize the mental health problems experienced by SGM through the lens of minority stress theory [57], a perspective that identifies societal heterosexism, homophobia, and transphobia as the distal causes of the mental health problems that disproportionately affect SGM [36, 58]. These affirmative approaches are generally conceptualized transtheoretically, with many parallels to the literature on cultural competence and adaptations [36]. Indeed, the World Professional Association for Transgender Health (WPATH), the American Psychological Association (APA), as well as the American Academy of Pediatrics (AAP), have published guidelines on comprehensive care for transgender and gender diverse children that follow a gender affirmative approach to treatment [59–61]. These guidelines complement the existing guidelines on psychological practice with sexual minority clients (e.g., [62]).

Within this emerging literature, a few researchers have begun to develop and preliminarily evaluate cognitive behavioral approaches tailored for use with SGM youth [39, 63]. These interventions provide psychoeducation around gender and minority stress, validate positive expressions of SGM identities, and deliver cognitive behavioral content in an affirmative manner. Some of the cognitive behavioral interventions include challenging internalized feelings of heterosexism and marginalization and supporting the development of healthy, affirmative sources of support [39, 64, 65].

An exemplar of this work can be seen in the intervention, AFFIRM (http://www.projectyouthaffirm.org), an eight-session cognitive behavioral coping skills group for SGM youth [39, 66]. AFFIRM was developed through a community-based participatory approach that recognized the many identity-based stressors that SGM youth experience. With an affirming stance toward sexual and gender diversity, AFFIRM uses cognitive behavioral techniques to promote healthy coping and management of emotional distress that can result from victimization, family rejection, or internalization of homophobic and transphobic attitudes and beliefs [66]. One open-pilot trial of AFFIRM with thirty participants produced positive results: depression scores, as well as threat appraisals, improved significantly over the course of treatment and three months of follow-up. In addition, both quantitative and qualitative data indicated that participants found the intervention relevant and of value to their lives [39]. An analysis of the eight participants who identified as transgender yielded similar findings: significant reductions in depressive symptoms over the course of treatment and follow-up [66].

Although preliminary, the positive results from the AFFIRM trial are encouraging and suggest that adaptations for SGM youth are relevant and valuable, and could contribute to positive mental health outcomes. More research is clearly needed through the use of RCT methodology to evaluate the extent that AFFIRM (and other interventions) leads to positive outcomes. Moreover, evidence is needed that AFFIRM produces better outcomes than standard CBT, although finding evidence of better engagement rates may be as important as evidence of better outcomes [28].

2.3 CBT with Youth with Multiple Marginalized Identities

Given the limited research evaluating CBT with REM youth, and the even more limited work with SGM youth, it is not surprising we found only two empirical studies that reported on CBT approaches with youth who identify as both REM and SGM. Craig et al. [63] describe the implementation of a pilot group counseling program for high school sexual minority youth, the majority of whom identified as Latinx (74%) or Black (21%). The intervention consisted of 8–10 weekly group sessions that focused on promoting effective problem-solving and coping skills, particularly as they pertained to family rejection and peer victimization. Results were generally positive: over the course of the program, significant increases in coping skills and self-esteem were found.

In a case history, Duarté-Vélez, Bernal, and Bonilla [67] describe how they adapted CBT to work with a gay, Latino adolescent with depression and anxiety who was having difficulty accepting his sexual orientation. In their treatment approach, the authors identified problematic cognitions in three important domains of the client's life: sexuality, family, and spirituality. They worked with the negative cognitions to challenge their all-or-nothing quality, and they provided validation and affirmation for the very real challenge he would face in coming out to his parents. The therapy also focused on increasing the client's engagement with pleasurable activities by helping him identify personal life goals, which included support around navigating differences in career expectations between the client and his family. Throughout the treatment, the authors recognized the importance of several Latinx cultural values, including *familismo* (central role of the family in everyday life), *machismo* (traditional male gender role), and spirituality. By the end of treatment, the client reported reductions in depressive symptoms and negative cognitions, greater self-acceptance, and less internalized homophobia. Moreover, he was able to come out to his family, continue his involvement in his church, and move towards engaging more fully in his career goals.

In sum, although there is currently limited published literature on CBT with diverse youth, there is reason to be optimistic. The growing body of research examining adaptations of CBT with African American and Latinx youth has generally yielded positive outcomes, and the emerging literature on CBT with SGM youth is promising. To complement this growing literature, we present some recommendations for conducting CBT with youth who identify as both REM and SGM. We organize our recommendations around a case example that represents a compilation of CBT we have conducted with different youth and families. Importantly, we believe that focused training is needed to acquire the skills to work effectively with diverse youth and their families, particularly youth who identify as transgender or gender diverse [68]. And, yet, because there are insufficient numbers of clinicians with this expertise to meet the growing demand, it is likely that clinicians will at some point have the opportunity to work with these clients and their families. We present these recommendations with the hope that this chapter will encourage clinicians to seek out opportunities to acquire more skill in working with diverse youth.

3   Intervention[3]

María is an 8-year-old transgender, Mexican-American girl whose parents sought treatment for help and support in facilitating a healthy gender transition.[4] María presents as a kind, thoughtful, and playful child who enjoys spending time with her family and taking dance classes. For the past several years, María has been dressing in female-marketed clothing and requesting that her parents not use her birth name (Miguel) while at home. Recently, she has asked her parents to use her affirmed name and pronouns when interacting with extended family and community members. When presenting for initial assessment, without María present, her mother reports that María is transgender, and she and María's father are unsure how to navigate supporting their child's identity, as well as their own fears about how the family will be viewed by their community. María's father reports stress related to how his brothers will view his child being feminine, and he wonders if he is

---

[3] The case of María is confabulated from a combination of typical cases, as well as fictionalized information.

[4] Gender transition refers to the process of presenting one's true gender identity through either social or medical means. This true gender identity is often different from that associated with one's sex assigned at birth [61, 69].

at all to blame for her desire to be seen as female. In response, María's mother expresses frustration that María's father cannot accept their child for who she is.

María's parents were both born in Northern Mexico. Before María was born, they moved to the Southwestern USA, and then to a mid-sized city in New England. María's older sister and maternal grandmother, who only speaks Spanish, live with them. According to María's mother, both her sister and her grandmother are supportive of María's gender identity. Most of their extended family continues to live in Mexico; however, two of María's father's siblings also live within a short distance. Although María's parents are naturalized US citizens, not all of their extended family members have documentation that allows them to reside in the US. Church services play a large part in María's family's social network. Her parents believe that many congregants will be supportive, but they are nervous that some may not be. Similarly, her parents express concern with how the school will respond to María's transition. They live in a fairly politically progressive town, and María's parents feel many teachers will be supportive; however, they are unsure if the administration has supports in place to ensure María's emotional and physical safety. María has a few friends through dance class but they do not go to the same school and her parents worry that if she presents more feminine at school she will lose the few friends she has made there.

3.1 Recommendations for CBT with María and Her Family

In the following sections, we present recommendations for working with María and her family within a cognitive behavioral gender affirmative approach [65]. As noted earlier, this approach consists of applying a gender-affirming and minority-stress congruent stance to common CBT techniques, including assessment and case conceptualization, psychoeducation, and addressing maladaptive thinking and behaviors. Moreover, when including family members, this approach would support family members understanding and acceptance of their child's identity and help identify community resources to aid them during their transition [70, 71].

Establishing a strong working alliance would be critical in working with María and her family, and research suggests that the working alliance and cultural competence are interconnected [72, 73]. Therefore, our recommendations focus on integrating cognitive behavioral work with three aspects of cultural competence that have been defined in the literature: cultural awareness, cultural knowledge, and cultural skill [30, 74]. Cultural awareness refers to clinicians' understanding of the importance of culture on the client's lived experience and worldview, as well as the clinician's own biases and assumptions. Cultural knowledge refers to an understanding of the culturally salient life events, values, and traditions that are common to particular cultural groups, as well as an understanding of the disparities in mental health services that affect different groups. Finally, cultural skill refers to the ability of clinicians to actively engage their clients in the therapy process [31]. Although the literature on cultural competence has been primarily focused on REM, there are many ways in which the key concepts can also apply to SGM [75]. Therefore, our recommendations take into consideration aspects of María and her family's culture of origin, issues related to her gender identity, and the intersection of both. Although we present our recommendations in three sections (i.e., cultural awareness, cultural knowledge, and cultural skill), there is overlap among the three.

3.1.1 *Cultural Awareness*   In preparing to work with María and her family, clinicians should cultivate awareness of their own identity, privilege, and social location, given that unexamined assumptions and biases can affect the development of a healthy working alliance and adversely impact treatment [31, 59, 70]. For example, some therapists may hold ideas regarding gender identity, norms, and roles that may not match with traditional Latinx cultural values around gender. This area would be particularly important in working with María's family, as psychoeducation around gender and gender identity is a critical element of the treatment plan [64, 68, 69]. Relatedly, it would be important for clinicians to consider their own biases regarding definitions of family, and the extent to which their conceptions include only the nuclear family vs extended family. In many Latinx cultures, extended family members are understood as central members [76, 77]; in working with María's family, therapists would do well to consider the possibility that their own definition of family may not be shared by María and her family.

Beyond perspectives and biases shaped by culture, cultural awareness also includes understanding how identity, power, and privilege could shape interactions with María's family [78, 79]. For instance, it would be important for therapists to reflect on the impact of their own identity and expression of this identity on the development of a therapeutic relationship with María and her family. In particular, cisgender providers (i.e., providers whose gender identity and gender expression align with the sex they were assigned at birth) would benefit from considering how their gender identity and expression could impact how comfortable María or her parents feel about their ability to assess and understand María's lived experience. Conversely, clinicians who identify as gender diverse may want to consider that their gender identity and expression may raise concerns for María's parents about whether the provider can provide a neutral perspective in working with their child. Similarly, it may be important for therapists to consider the potential impact of their own racial and ethnic identity on María and her family's experience in treatment [78]. Therapists who identify as European American, or who phenotypically look White, may find it useful to consider María and her family's assumptions and ideas about their background, as well as their prior experiences with service providers, the majority of whom may have been White. Therapists who identify as a racial or ethnic minority—and particularly those who identify as Latinx—might do well to anticipate the possibility that María and her parents may assume a shared understanding of some cultural or lived experiences.

Unifying these various considerations is a recognition that María and her family may be experiencing considerable anxiety about both María's transition and the therapy process, and so a strong therapy alliance is critical to successful therapy. Awareness of how culture shapes both therapist and client worldviews and assumptions about normative functioning can attune therapists to subtle indications that the therapy alliance is not developing in positive ways, and create openings for openly addressing these concerns.

3.1.2 Cultural Knowledge

In addition to cultural awareness, clinicians should have some knowledge of the salient cultural life experiences, cultural values, and cultural worldviews that may be relevant for particular groups. Importantly, however, it is critical to consider both within-group and individual differences, and not use a stereotypic lens to understand the experiences of particular individuals [30, 80]. For example, the Latinx population in the USA is very heterogeneous, as Latinx have ancestry in over 20 countries or territories, with the largest groups having origins in Mexico (63.2%), Puerto Rico (9.5%), and Cuba (3.9%). Although there are important cultural commonalities across Latin America, there also exist important cultural and linguistic differences across the different regions [76]. In addition, although the majority of Latinx in the USA are American citizens (including those of Puerto Rican origin) and speak English [81], among immigrant Latinxs there exist considerable diversity in both the immigration experiences and the receiving contexts in which they find themselves (e.g., legal residents with a "green card," naturalized US citizens, visa holders, undocumented immigrants, refugees/asylum seekers). In the case of María's family, there is evidence of this diversity, as not all of her extended family has the necessary paperwork for long-term residence in the USA. This variability may have implications for how forthcoming María's parents may be in providing personal information about their living situation.

There is a considerable literature documenting important cultural values to which many Latinx adhere [82, 83]. Cultural values provide norms and guidance around interacting with the social world, which can include the role of the family, interpersonal interactions, and gender roles, as well as religious and spiritual values [77, 82]. Clinicians should understand, however, that although many Latinx families subscribe to traditional gender roles of masculinity and femininity, such as *machismo* (i.e., male gender role that emphasizes strength and power), *caballerismo* (i.e., male gender role that emphasizes protection and responsibility) or *marianismo* (i.e., female gender role that emphasizes nurturance and self-sacrifice), many Latinx families endorse more flexible approaches to gender roles [77]. Moreover, given the intergenerational variability in acculturation that often exists within families [84], it is quite likely that there will exist different perspectives on gender roles among individuals from the same family. Indeed, María's father has expressed more concern than María's mother about María's desire to express herself as a girl, and it is possible that he believes that María's gender identity reflects on his own sense of masculinity. Careful discussion of ideas about gender will be paramount to therapy. Beyond gender, several Latinx cultural values may also shape interactions with providers. For example, the cultural value of *respeto*, which refers to an expectation for a formal interaction style within hierarchical relationships [77, 82], may lead María's parents to show respect and deference to the clinician, and not openly acknowledge or raise any disagreement or ambivalence they might feel with the therapy.

With regards to the issues around gender identity and gender transitions, clinicians should be familiar with relevant and up-to-date information. For example, clinicians should have a basic knowledge of terminology (e.g., terms such as sex, gender identity, and gender expression), as well as some familiarity with the literature on developmental trajectories of gender identity, and the treatment guidelines regarding assessment and treatment of gender dysphoria [59, 61, 69, 85]. Moreover, therapists should know about the range of medical interventions available for transgender and gender diverse youth, as well as the landscape of state and federal laws that protect gender diverse youth and serve as obstacles to receiving the medical treatment they might desire [70, 85]. In María's case, her parents expressed uncertainty regarding the resources María's school might have to support her emotional and social well-being. Because the laws and protections vary by state, it would be important for clinicians to be familiar with the local regulations.

Clinicians should also be aware of the role that prejudice and discrimination play in the lives of many REM and SGM. For Latinx youth and families, the current political environment has raised the likelihood of experiencing personal discrimination [86]. And there is considerable research demonstrating that SGM youth are at markedly elevated risk for bullying and experiences of violence [34, 59, 87]. Clinicians should understand that although some of this discrimination is apparent and may be reported by their clients, some discrimination may not be understood as such by their clients without psychoeducation [69]. In María's case, there may be ways in which her family manages fear of racial or ethnic discrimination that contribute to them being reluctant to ask for support services from her school and church. Asking for support necessitates drawing attention to oneself, a particularly fraught exercise if any members of María's extended family or network of friends does not have legal documentation to support their residency in the USA.

3.1.3 Cultural Skill

Together with cultural awareness and knowledge, culturally competent therapy includes cultural skill, or the ability of clinicians to actively engage their clients in the therapy process. In addition to developing and maintaining a strong working alliance, this engagement can also be seen in the culturally sensitive application of specific cognitive behavioral techniques [88]. With María, clinicians will need to hold a nonjudgmental, affirmative perspective to create space for her to explore her gender identity and support her social transition [59, 69, 70, 79]. Moreover, clinicians will need to support the family in their adjustment to María's transition [71]. A strong working alliance with María and her family will be critical to help María safely explore and affirm her gender identity, as well as to increase the support María experiences from her family. In this regard, clinicians should consider María's preferred language, as well as that of her parents and any other family members brought into session. Research suggests that providing services in the client's language of preference is a powerful predictor of treatment outcomes [89]. Of note, even if a client is competent in English, and even uses English in other contexts such as work, the client's preferred language for therapy may still not be English [78]. Moreover, given the extensive use of worksheets and written material typical in CBT, translation of these materials may be important. Further, clinicians should understand that Spanish is a grammatically gendered language (i.e., many words make gendered assumptions), and thus care will be needed to use gender neutral language in line with a gender affirmative model [90]. Using gender neutral language may not come naturally to all members of María's family (e.g., parents, grandmother), and so finding a balance between affirming the client while also creating space for others to learn and accept the changes is important.

It can also be useful to assess for some culturally relevant constructs, such as adherence to traditional cultural values and level of acculturation, and gender relevant constructs, like gender dysphoria and identity [65, 77]. A clinical interview can be helpful for such assessments [75], but there also exist a number of formal measures that clinicians may want to use to assess for cultural constructs (*see* Table 1). Given the importance of working with María's family, these assessments may prove particularly useful in evaluating the diversity of perspective within María's family and may also offer clinicians some guidance in how to interact with

Table 1
Assessment instruments

Cultural-construct	Scale	Reference
Latinx relevant scales	Abbreviated Multidimensional Acculturation Scale	Zea et al. [91]
	Attitudinal Familism Scale	Lugo-Steidel et al. [92]
	Machismo Measure	Arciniega et al. [93]
	Marianismo Beliefs Scale	Castillo et al. [94].
	Latina/o Values Scale	Kim et al. [95]
	Mexican American Cultural Values Scale for Adolescents and Adults	Knight et al. [96]
Gender-relevant scales	Gender Identity/Gender Dysphoria Questionnaire for Adolescents and Adults	Deogracias et al. [97]
	Transgender Congruence Scale	Kozee et al. [98].
	Heterosexist Harassment, Rejection, and Discrimination Scale	Szymanski [99]
	Daily Heterosexist Experiences Questionnaire (DHEQ)	Balsam et al. [100]

María or her parents. For example, if María's parents or grandparents generally adhere to the cultural value *respeto*, they may prefer to be addressed with the formal "Usted" as opposed to the informal "Tú," if the therapy is being conducted in Spanish. Similarly, the cultural value of *personalismo*, a style of interpersonal interactions that is less formal and more warm and friendly, may encourage some clinicians to engage in small talk at the outset of a session, as well as engage in appropriate self-disclosure [77, 82].

This culturally sensitive work can facilitate the clinician's gender affirmative treatment approach with both María and her family. To actively engage María's parents in the therapeutic process, it will be essential for the clinician to create space for the parents to process their child's gender transition both individually and as a couple [71, 79]. Some of this work may consist of working in culturally sensitive ways with the normative mix of conflicting cognitions and feelings that María's parents may be experiencing. For example, they may report having a strong desire for María to be happy, while also expressing concerns and uncertainty about what their family members and church congregation think of María. Moreover, María's parents (particularly her father, in this case), may have negative cognitions about how María's gender identity is a reflection on their parenting. Working with them to reframe these cognitions in a balanced way can help them process and accept their child's SGM identity. This would be a critical aspect of therapy, as a balanced perspective can facilitate their understanding of the positive impacts of transitioning on María's mental health. The cognitive work of challenging the maladaptive or rigid cognitions experienced by María's parents would be more smoothly conducted in the context of a strong working alliance, wherein María's parents feel connected with and supported by the clinician. Thus, working with María's father's may necessitate validation of his feelings of confusion around gender identity and expression, prior to exploring and working with his unexamined assumptions about what it means to be a man. There may also be an opportunity to leverage the cultural values of *familismo* and *caballerismo*, whereby the role of family protector could motivate María's father to advocate for his daughter.

When working with María, the therapist should also be attuned to the possible expression of negative or rigid cognitions about what it means to be a particular gender. For example, María might state in session that "to be a girl I need to wear a dress." Although clinicians may rightly see this cognition as overly rigid and dichotomous, care should be taken when challenging these sorts of gender-based cognitions. It is possible that María might experience such a challenge as invalidating of her sense of self, especially if this particular thought represents an expression of María's desire to be seen as a girl. Instead of directly challenging these thoughts, clinicians working with María could help her develop more balanced thoughts, such as "wearing dresses doesn't make me a girl; telling the world with my voice makes me a girl." This example could help María experience her freedom in gender expression while also affirming her gender identity.

Beyond working with maladaptive cognitions, conducting gender affirmative therapy often means working as an advocate for clients [85]. Serving as an advocate is delicate, as too much clinician advocacy can undermine client autonomy. However, it is important for clinicians to understand that REM and GSM clients often have had many experiences of disempowerment when interacting with systems of care. Thus, clinicians may find that engaging with different systems with their clients will prove beneficial. In María's case, it may be important for clinicians to accompany María's parents as they meet with María's school principal and teachers. Offering to provide literature and information packets on gender diversity and gender transitioning to both the family and the schools may also prove useful, given that many schools do not have the necessary information to support transgender and gender diverse youth (*see* Table 2).

Finally, clinicians would do well to remember the many strengths that their clients and their families possess [85]. Some of these strengths can be found in the support that families can provide each other, while others can be found in their extended communities. In María's case, there is evidence that María has a significant amount of familial support from her sister, her grandmother, and her parents—her father's concerns notwithstanding.

Table 2
Culture- and gender-related resources

Resource type	Focus	Reference
Books	Latinx	Adames et al. [101]
		Arredondo et al. [102]
		Falicov [103]
	Clinician guides for working with gender	Pachankis and Safren [104]
		Krieger [56]
		Ehrensaft [105]
	Books for children on gender	Thorn [106]
		Pessin-Whedbee [107]
	Workbooks for youth on gender	Testa [108]
		Storck [109]
		Singh [110]
Websites	National Latinx Psychological Association	https://www.nlpa.ws
	Facts and information on Latinx populations	https://www.pewresearch.org/hispanic/
	World Professional Association for Transgender Health	https://www.wpath.org/
	Resources for clinicians and families	https://www.genderspectrum.org/
	National Association of School Psychologists	https://www.nasponline.org/resources-and-publications/resources-and-podcasts/diversity/lgbtq-youth/gender-inclusive-schools-faqs
	Transgender Law Center	https://transgenderlawcenter.org/

Moreover, given that church and religion are important parts of María's family life, it will be important for clinicians to help María's parents think through how to engage with their church around María's transition. Leveraging already existing community supports to help the family can offer a sense of security and stability for families who may feel adrift with the changes they are experiencing in their family.

4 Conclusion

Although CBT is at the forefront of evidence-based psychotherapies, there remain gaps in our understanding of the adaptations needed to work with diverse populations. These gaps are particularly pronounced when considering the particular needs of populations who identify with multiple marginalized identities. However, the cultural competency framework of cultural awareness, cultural knowledge, and cultural skills can offer clinicians a useful roadmap for conducting CBT with diverse youth and their families. In this case example, we presented recommendations for conducting culturally competent CBT within a gender affirmative approach when working with a young Latinx transgender girl and her family. Our emphasis was on developing a strong working alliance with María and her parents, as well how to apply some cognitive behavioral techniques in a culturally sensitive manner.

It is our hope that this case example and clinical recommendations can complement the emerging literature in this area and encourage more clinicians to engage with the challenging yet rewarding work of helping diverse clients in their journeys towards authentic self-expression, as well as helping their parents and families move towards acceptance, validation, and support. We are optimistic that greater engagement with the complexities of adapting CBT can help reduce the disparities in mental healthcare and increase service access and utilization for all.

References

1. Humes KR, Jones NA, Ramírez RR (2011) Overview of race and Hispanic origin: 2010. 2010 census briefs. United States Census Bureau, Washington, DC

2. Passel JS, Cohn D (2008) U.S. population projections: 2005–2050. Pew Research Center: Social and Demographic Trends, Washington, DC

3. Child Trends (2018) Racial and ethnic composition of the child population. https://www.childtrends.org/indicators/racial-and-ethnic-composition-of-the-child-population

4. Williams Institute (2019) Adult LGBT population in the United States. Williams Institute, UCLA School of Law, Los Angeles, CA

5. Gates GJ (2014) LGB/T demographics: comparisons among population-based surveys. Williams Institute, UCLA School of Law, Los Angeles, CA

6. Centers for Disease Control and Prevention (2018) Youth risk behavior surveillance — United States, 2017. MMWR Surveill Summ 67(8):1–114

7. Chen D, Hidalgo MA, Leibowitz S et al (2016) Multidisciplinary care for gender-diverse youth: a narrative review and unique model of gender-affirming care. Transgend Health 1(1):117–123

8. Herman JL, Flores AR, Brown TNT et al (2017) Age of individuals who identify as transgender in the United States. The Williams Institute, Los Angeles, CA

9. Barksdale CL, Azur M, Leaf PJ (2009) Differences in mental health service sector utilization among African American and Caucasian youth entering systems of care programs. J Behav Health Serv Res 37(3):363–373

10. Garland AF, Lau AS, Yeh M et al (2005) Racial and ethnic differences in utilization of mental health services among high-risk youths. Am J Psychiatry 162:1336–1343

11. Gudiño OG, Lau AS, Yeh M et al (2009) Understanding racial/ethnic disparities in youth mental health services. Do disparities vary by problem type? J Emot Behav Disord 17(1):3–16

12. Merikangas KR, He J, Burstein ME et al (2011) Service utilization for lifetime mental disorders in U.S. adolescents: results of the National Comorbidity Survey Adolescent Supplement (NCS-A). J Am Acad Child Adolesc Psychiatry 50(1):32–45

13. Thomas JF, Temple JR, Perez N et al (2011) Ethnic and gender disparities in needed adolescent mental health care. J Health Care Poor Underserv 22(1):101–110

14. Williams KA, Chapman MV (2015) Mental health service use among youth with mental health need: do school-based services make a difference for sexual minority youth? Sch Ment Health 7(2):120–131

15. McGuire J, Russell ST (2007) Health care utilization by sexual minority adolescents. J Adolesc Health 40:S28–S28

16. Williams KA, Chapman MV (2011) Comparing health and mental health needs, service use, and barriers to services among sexual minority youth and their peers. Health Soc Work 36:197–206

17. Williams KA, Chapman MV (2012) Unmet health and mental health need among adolescents: the roles of sexual minority status and child-parent connectedness. Am J Orthopsychiatry 82(4):473–481

18. Valdez CR, Rodgers CRR, Gudiño OG et al (2019) Translating research to support practitioners in addressing disparities in child and adolescent mental health and services in the United States. Cultur Divers Ethnic Minor Psychol 25(1):126–135

19. Marshal MP, Dietz LJ, Friedman MS et al (2011) Suicidality and depression disparities between sexual minority and heterosexual youth: a meta-analytic review. J Adolesc Health 49(2):115–123

20. Mustanski BS, Garofalo R, Emerson RM (2010) Mental health disorders, psychological distress, and suicidality in a diverse sample of lesbian, gay, bisexual, and transgender youths. Am J Public Health 100(12):2426–2432

21. Russell ST, Franz BT, Driscoll AK (2001) Same-sex romantic attraction and experiences of violence in adolescence. Am J Public Health 91:903–906

22. Velez BL, Moradi B, DeBlaere C (2015) Multiple oppressions and the mental health of sexual minority Latina/o individuals. Counsel Psychol 43(1):7–38

23. Cardemil EV, Nelson T, Keefe K (2015) Racial and ethnic disparities in depression treatment. Curr Opin Psychol 4:37–42

24. Gonzales JJ, Papadpoulos AS (2010) Mental health disparities. In: Levin BL, Hennessy KD, Petrilla J (eds) Mental health services: a public health perspective, 3rd edn. Oxford University Press, New York, NY, pp 443–464

25. Copeland V (2006) Disparities in mental health service utilization among low-income African American adolescents: closing the gap by enhancing practitioner's competence. Child Adolesc Soc Work J 23:407–431

26. Mak WWS, Law RW, Alvidrez J et al (2007) Gender and ethnic diversity in NIMH-funded clinical trials: review of a decade of published research. Admin Policy Ment Health Ment Health Serv Res 34:497–503

27. Huey SJ, Polo AJ (2018) Evidence-based psychotherapies with ethnic minority children and adolescents. In: Weisz JR, Kazdin AE (eds) Evidence-based psychotherapies for children and adolescents, 3rd edn. The Guilford Press, New York, NY, pp 361–378

28. Cardemil EV (2010) Cultural adaptations to empirically supported treatments: a research agenda. Sci Rev MentHealth Pract 7:8–21

29. Costantino G, Malgady RG, Primavera LH (2009) Congruence between culturally competent treatment and cultural needs of older Latinos. J Consult Clin Psychol 77(5):941–949

30. Sue S, Zane N, Nagayama Hall GC et al (2009) The case for cultural competency in psychotherapeutic interventions. Annu Rev Psychol 60:525–548

31. Soto A, Smith TB, Griner D et al (2018) Cultural adaptations and therapist multicultural competence: two meta-analytic reviews. J Clin Psychol 74:1907–1923

32. Tao KW, Owen J, Pace BT et al (2015) A meta-analysis of multicultural competencies and psychotherapy process and outcome. J Couns Psychol 62(3):337–350

33. Cochran SD, Mays VM (2013) Sexual orientation and mental health. In: Patterson CJ, D'Augelli AR (eds) Handbook of psychology and sexual orientation. Oxford University Press, New York, NY, pp 204–222

34. Grant JM, Mottet LA, Tanis J et al (2011) Injustice at every turn: a report of the National Transgender Discrimination Survey. National Center for Transgender Equality and National Gay and Lesbian Task Force, Washington, DC

35. Gridley SJ, Crouch JM, Evans Y et al (2016) Youth and caregiver perspectives on barriers to gender-affirming health care for transgender youth. J Adolesc Health 59(3):254–261

36. Pepping CA, Lyons A, Morris EMJ (2018) Affirmative LGBT psychotherapy: outcomes of a therapist training protocol. Psychotherapy 55(1):52–62

37. Shelton K, Delgado-Romero EA (2011) Sexual orientation microaggressions: the experience of lesbian, gay, bisexual, and queer clients in psychotherapy. J Couns Psychol 58:210–221

38. Shipherd J, Green KE, Abramovitz S (2010) Transgender clients: identifying and minimizing barriers to mental health treatment. J Gay Lesbian Ment Health 14:94–108

39. Craig SL, Austin A (2016) The AFFIRM open pilot feasibility study: a brief affirmative cognitive behavioral coping skills group intervention for sexual and gender minority youth. Child Youth Serv Rev 64:136–144

40. Cardemil EV, Millán F, Aranda E (2019) A new, more inclusive name: the Journal of Latinx Psychology. J Latinx Psychol 7(1):1–5

41. Santos C (2017) The history, struggles, and potential of the term Latinx. Latin Psychol Today 4(2):7–13

42. Williams MT, Chapman LK, Buckner EV et al (2016) Cognitive behavioral therapies. In: Brieland-Noble AM, Al-Mateen CS, Singh NN (eds) Handbook of mental health in

African American youth. Springer, New York, NY, pp 63–77

43. Ginsburg G, Drake K (2002) School-based treatment for anxious African American adolescents: a controlled pilot study. J Am Acad Child Adolesc Psychiatry 41(7):768–775

44. Lochman JE, Curry JF, Dane H et al (2001) The Anger Coping Program: an empirically-supported treatment for aggressive children. Resident Treat Child Youth 18:63–73

45. Cohen JA, Deblinger E, Mannarino AP et al (2004) A multisite, randomized controlled trial for children with sexual abuse-related PTSD symptoms. J Am Acad Child Adolesc Psychiatry 43:393–402

46. Silverman WK, Kurtines WM, Ginsburg GS et al (1999) Treating anxiety disorders in children with group cognitive–behavioral therapy: a randomized clinical trial. J Consult Clin Psychol 67:995–1003

47. Rosselló J, Bernal G (1999) The efficacy of cognitive-behavioral and interpersonal treatments for depression in Puerto Rican adolescents. J Consult Clin Psychol 67(5):734–745

48. Rosselló J, Bernal G, Rivera-Medina C (2009) Individual and group CBT and IPT for Puerto Rican adolescents with depressive symptoms. Cultur Divers Ethnic Minor Psychol 14 (3):234–245

49. Kataoka SH, Stein BD, Jaycox LH et al (2003) A school-based mental health program for traumatized Latino immigrant children. J Am Acad Child Adolesc Psychiatry 42 (3):311–318

50. Huey SJ, Polo AJ (2008) Evidence-based psychosocial treatments for ethnic minority youth: a review and meta-analysis. J Clin Child Adolesc Psychol 37:262–301

51. Bernal G, Domenech Rodriguez MM (eds) (2012) Cultural adaptations: tools for evidence-based practice with diverse populations. American Psychological Association, Washington, DC

52. Benish SG, Quintana S, Wampold BE (2011) Culturally adapted psychotherapy and the legitimacy of myth: a direct-comparison meta-analysis. J Couns Psychol 58 (3):279–289

53. Stein BD, Jaycox LH, Kataoka SH et al (2003) A mental health intervention for schoolchildren exposed to violence: a randomized controlled trial. JAMA 290 (5):603–611

54. Santiago CD, Lennon JM, Fuller AK et al (2014) Examining the impact of a family treatment component for CBITS: when and for whom is it helpful? J Fam Psychol 28 (4):560–570

55. Hidalgo MA, Ehrensaft D, Tishelman AC et al (2013) The Gender Affirmative Model: what we know and what we aim to learn. Hum Dev 56:285–290

56. Krieger I (2017) Counseling transgender and non-binary youth: the essential guide. Jessica Kingsley Publishers, Philadelphia, PA

57. Meyer IH (2003) Prejudice, social stress, and mental health in lesbian, gay, and bisexual populations: conceptual issues and research evidence. Psychol Bull 129(5):674–697

58. Pachankis JE, Goldfried MR (2004) Clinical issues in working with lesbian, gay, and bisexual clients. Psychother Theor Res Pract Train 41(3):227–246

59. American Psychological Association (2015) Guidelines for psychological practice with transgender and gender nonconforming people. Am Psychol 70(9):832–864

60. Rafter J, AAP Committee on Psychosocial Aspects of Child and Family Health, AAP Committee on Adolescence, and AAP Section on Lesbian, Gay, Bisexual, and Transgender Health and Wellness (2018) Ensuring comprehensive care and support for transgender and gender diverse children and adolescents. Pediatrics 142(4):e20182162

61. World Professional Association for Transgender Health (2012) Standards of care for the health of transsexual, transgender, and gender-nonconforming people (7th version). Int J Transgend 13(4):165–232

62. American Psychological Association (2012) Guidelines for psychological practice with lesbian, gay, and bisexual clients. Am Psychol 67 (1):10–42

63. Craig SL, Austin A, McInroy LB (2014) School-based groups to support multiethnic sexual minority youth resiliency: preliminary effectiveness. Child Adolesc Soc Work J 31:87–106

64. Austin A, Craig SL (2015) Transgender affirmative cognitive behavioral therapy: clinical considerations and applications. Prof Psychol Res Pract 46(1):21–29

65. Austin A, Craig SL, Alessi EJ (2017) Affirmative cognitive behavior therapy with transgender and gender nonconforming adults. Psychiatr Clin North Am 40(1):141–156

66. Austin A, Craig SL, D'Souza SA (2017) An AFFIRMative cognitive behavioral intervention for transgender youth: preliminary effectiveness. Prof Psychol Res Pract 49(1):1–8

67. Duarté-Vélez Y, Bernal G, Bonilla K (2010) Culturally adapted cognitive-behavior

therapy: integrating sexual, spiritual, and family identities in an evidence-based treatment of a depressed Latino adolescent. J Clin Psychol 66(8):895–906

68. Woodward EN, Willoughby B (2014) Family therapy with sexual minority youths: a systematic review. J GLBT Fam Stud 10(4):380–403

69. Austin A (2018) Transgender and gender diverse children: consideration for affirmative social work practice. Child Adolesc Soc Work J 35:73–84

70. Edwards-Leeper L, Leibowitz S, Sangganjanavanich VF (2016) Affirmative practive with transgender and gender nonconforming youth: expanding the model. Psychol Sex Orientat Gend Divers 3(2):165–172

71. Tishelman AC, Kaufman R, Edwards-Leeper L et al (2015) Serving transgender youth: challenges, dilemmas, and clinical examples. Prof Psychol Res Pract 46(1):37–45

72. Anderson KN, Bautista CL, Hope DA (2018) Therapeutic alliance, cultural competence, and minority status in premature termination of psychotherapy. Am J Orthopsychiatry 89 (1):104–114

73. Ishikawa RZ, Cardemil EV, Alegría M et al (2014) Uptake of depression treatment recommendations among Latino primary care patients. Psychol Serv 11(4):421–432

74. Whaley AL, Davis KE (2007) Cultural competence and evidence-based practice in mental health services. Am Psychol 62:563–574

75. Heck NC, Flentje A, Cochran BN (2012) Intake interviewing with lesbian, gay, bisexual, and transgender clients: starting from a place of affirmation. J Contemp Psychother 43(1):23–32

76. Arredondo P, Gallardo-Cooper M, Delgado-Romero EA et al (2014) Culturally responsive counseling with Latinas/os. American Counseling Association, Alexandria, VA

77. Edwards LM, Cardemil EV (2015) Clinical approaches to assessing cultural values in Latinos. In: Geisinger K (ed) Psychological testing of Hispanics: clinical and intellectual issues, 2nd edn. American Psychological Association, Washington, DC, pp 215–236

78. Hays PA (2016) Addressing cultural complexities in practice: assessment, diagnosis, and therapy, 3rd edn. American Psychological Association, Washington, DC

79. Butler C (2009) Sexual and gender minority therapy and systemic practice. J Fam Ther 31:338–358

80. Lakes K, López SR, Garro LC (2006) Cultural competence and psychotherapy: applying anthropologically informed conceptions of culture. Psychother Theor Res Pract Train 43(4):380–396

81. Flores A, López G, Radford J (2017) Facts on U.S. Latinos, 2015: statistical portrait of Hispanics in the United States. Pew Research Center, Washington, DC

82. Añez LM, Silva MA, Paris M et al (2008) Engaging Latinos through the integration of cultural values and motivational interviewing principles. Prof Psychol Res Pract 39:153–159

83. González-Prendes AA, Hindo C, Pardo YF (2011) Cultural values integration in cognitive-behavioral therapy for a Latino with depression. Clin Case Stud 10 (5):376–394

84. Liu P (2015) Intergenerational cultural conflict, mental health, and educational outcomes among Asian and Latino/a Americans: qualitative and meta-analytic review. Psychol Bull 141(2):404–446

85. Collazo A, Austin A, Craig SL (2013) Facilitation transition among transgender clients: components of effective clinical practice. Clin Soc Work 41:228–237

86. Cerezo A (2016) The impact of discrimination on mental health symptomatology in sexual minority immigrant Latinas. Psychol Sex Orientat Gend Divers 3(3):283–292

87. Kosciw JG, Greytak EA, Giga NM et al (2016) The 2015 National School Climate Survey: the experiences of lesbian, gay, bisexual, transgender, and queer youth in our nation's schools. GLSEN, New York, NY

88. Hays PA (2009) Integrating evidence-based practice, cognitive-behavior therapy, and multicultural therapy: ten steps for culturally competent practice. Prof Psychol Res Pract 40 (4):354–360

89. Griner D, Smith TB (2006) Culturally adapted mental health interventions: a meta-analytic review. Psychother Theor Res Pract Train 43:531–548

90. Keo-Meier C, Ehrensaft D (2018) The Gender Affirmative Model: an interdisciplinary approach to supporting transgender and gender expansive children. American Psychological Association, Washington, DC

91. Zea MC, Asner-Self KK, Birman D, Buki LP (2003) The Abbreviated Multidimensional Acculturation Scale: empirical validation with two Latino/Latina samples. Cultur Divers Ethnic Minor Psychol 9(2):107–126

92. Lugo-Steidel AG, Contreras JM (2003) A new familism scale for use with Latino populations. Hispanic J Behav Sci 25:312–330

93. Arciniega MC, Anderson TC, Tovar-Blank Z, Tracy TC (2008) Toward a fuller conception of machismo: development of a traditional machismo and caballerismo scale. J Couns Psychol 55:19–33

94. Castillo LG, Perez FV, Castillo R, Ghosheh MR (2010) Construction and initial validation of the Marianismo Beliefs Scale. Counsel Psychol Q 23:163–175

95. Kim BSK, Soliz A, Orellana B, Alamilla S (2009) Latina/o Values Scale: development, reliability, and validity. Measur Eval Counsel Dev 42:71–91

96. Knight GP, Gonzales NA, Saenz DS, Bonds DD, Germán M, Deardorff J et al (2010) The Mexican American cultural values scale for adolescents and adults. J Early Adolesc 30:444–481

97. Deogracias JJ, Johnson LL, Meyer-Bahiburg HF et al (2007) The gender identity/gender dysphoria questionnaire for adolescents and adults. J Sex Roles 44(4):370–379

98. Kozee HB, Tylka TL, Bauerband LA (2012) Measuring transgender individuals' comfort with gender identity and appearance: development and validation of the Transgender Congruence Scale. Psychol Women Q 36 (2):179–196

99. Szymanski DM (2006) Does internalized heterosexism moderate the link between heterosexist events and lesbians' psychological distress? Sex Roles 54(3-4):227–234

100. Balsam KF, Beadnell B, Molina Y (2013) The daily heterosexist experiences questionnaire. Measur Eval Counsel Dev 46(1):3–25

101. Adames HY, Chavez-Dueñas NY (2017) Cultural foundations and interventions in Latino/a mental health: history, theory, and within group differences. Routledge Press, New York, NY

102. Arredondo P, Gallardo-Cooper M, Delgado-Romero EA, Zapata AL (2014) Culturally responsive counseling with Latinas/os. American Counseling Association, Alexandria, VA

103. Falicov C (2014) Latino families in therapy, 2nd edn. The Guilford Press, New York, NY

104. Pachankis, J.E., & Safren, S.A. (2019). Handbook of evidence-based mental health practice with sexual and gender minorities.

105. Ehrensaft D (2016) The gender creative child: pathways for nurturing and supporting children who live outside gender boxes. The Experiment Publishing, New York, NY

106. Thorn T (2019) It feels good to be yourself: a book about gender identity. Henry Holt and Company, New York, NY

107. Pessin-Whedbee B (2017) Who are you? The kids' guide to gender identity. Jessica Kingsley Publishers, London

108. Testa RJ (2015) The gender quest workbook: a guide for teens and young adults exploring gender identity. New Harbinger Publications, Oakland, CA

109. Storck K (2018) The gender identity workbook for kids: a guide to exploring who you are. New Harbinger Publications, Oakland, CA

110. Singh AA (2018) The queer and transgender resilience workbook. New Harbinger Publications, Oakland

Chapter 12

Modular CBT for Youth: Principles and Guides

Maya Boustani, Jennifer Regan, and Cameo Stanick

Abstract

The past decade has seen a proliferation of modular interventions. Modular interventions differ from traditional cognitive behavioral therapy (CBT) manuals by formalizing clinical decision-making via a system of rules and algorithms that clinicians can use to identify their clients' needs and best strategies to address those needs. Depending on the intervention, this typically allows clinicians to personalize treatment (e.g., shuffling the order of modules, extending or shortening treatment duration). This chapter provides clarification about what constitutes a modular treatment by reviewing the four concepts of modularity and how these can help address common limitations of traditional CBT manuals. The chapter also reviews currently available modular CBT interventions for youth, common strategies for implementing modular treatments, and two case examples to help further clarify how to use modular design principles in clinical practice.

Key words Modular, CBT, Youth, Treatment, Transdiagnostic

1 Introduction

The popularity of manualized cognitive behavioral therapy (CBT) approaches has led to a proliferation of CBT manuals. Traditionally, these treatment manuals focus on a single disorder or cluster of disorders and follow a fixed sequence for all youth regardless of presentation, age, or comorbidity [1]. This structure has allowed the field to scientifically test CBT and provide evidence of its effectiveness in controlled efficacy trials. However, this increased rigor has also led to a decrease in providers' ability to individualize treatment delivery. Youth in community care tend to present with high comorbidity [2] or show evidence of a new problem during the course of treatment [1, 3]. Furthermore, youth in community care often present with numerous psychosocial stressors such as family problems, poverty, and academic difficulties, that may impede treatment and therapeutic alliance if left unaddressed [1, 4, 5]. Traditional single disorder and fixed session-by-session treatment manuals do not typically provide guidance about how to address such issues that naturally arise in therapy (e.g., comorbidity,

Robert D. Friedberg and Brad J. Nakamura (eds.), *Cognitive Behavioral Therapy in Youth: Tradition and Innovation*, Neuromethods, vol. 156, https://doi.org/10.1007/978-1-0716-0700-8_12, © Springer Science+Business Media, LLC, part of Springer Nature 2020

crises, changes to family structure, poor engagement). In addition, providers reflect that they dislike the rigidity of prescribed treatments [6], which limits their capacity to make evidence-informed clinical decisions in the moment, based on emerging concerns. To move from efficacy trials to true implementation, adoption, and sustainability of evidence-based treatments (EBT) into usual care settings, providers must be able to meet the needs of populations in community care, without discounting the evidence from years of research findings. Modular treatment designs attempt to address these challenges by being dynamic while retaining structure and evidence-based content [7, 8]. In this chapter, we review the concept of modularity in treatment design and explore how modularity can help address some of the challenges encountered by traditional CBT manuals. We also review some common strategies to use when delivering modular treatments, examples of modular treatments for youth, and provide two case examples to illustrate how to treat a child using a modular treatment design.

2   Theoretical and Empirical Foundations

Modular-based CBT formalizes clinical decision-making via a system of rules and algorithms that clinicians can use to identify their clients' needs and best strategies to address those needs [9]. Formalized research on modular treatment approaches to CBT has increased in the past 10 years [10–12]; targeting populations with depression, anxiety, conduct problems, trauma, and comorbid autism spectrum disorder with anxiety (see Sect. 3.2 for a review of modular interventions). Modular treatment designs organize practice elements (discrete clinical techniques or strategies, such as "self-monitoring" or "psychoeducation" [13]) into self-contained "content modules," which can be used repeatedly, combined with other modules, or not used at all [14]. Modules can be designed for a single diagnosis, multiple diagnoses (i.e., the same module can be used to treat different diagnoses), or to treat different diagnoses sequentially. Algorithms or "coordination modules" [13] help guide decision-making around which modules to use and in which order for a particular case. This could mean that any two children may experience very different treatments, even when receiving the same modular treatment approach for the same problem [14]. Although it may seem that modularity is synonymous with flexibility—that is not always the case. Some flexible treatments are not modular (e.g., they may be highly flexible but not made up of independent modules) while some treatments can be modular but not flexible (some modular treatments have rigid coordination modules that dictate how the modules should be delivered).

2.1 Four Criteria for Modularity

To be truly considered modular, a treatment needs to meet four criteria [13, 15]. First, the treatment must be *partially decomposable*, meaning that it can be divided into meaningful independent units or "modules." Each module can be delivered in any given amount of time (part of a session, a full session, or multiple sessions). Each module may contain one or more practice elements. Second, each module must have *a specific purpose*. This means that a module is not simply a subdivision of a treatment, but rather has a specified goal and result. A "relaxation training" module, for example, has its own goal and can be delivered independently of other modules. Third, each module must have a *standardized interface*, meaning that it is able to connect with other modules in a standardized way, so that independent modules may be combined to build a treatment, without problems in how they interact. For example, the "relaxation training" module may be combined with other modules such as "exposure" or "psychoeducation about anxiety" to create a treatment plan for an anxious person. Using a similar layout for each module makes it easier for a clinician to navigate between modules. Fourth, each module should be *self-contained*, such that it includes all the information needed to be presented on its own, without the need for information from a different module. In a purely modular design, none of the content modules would have information dictating their order of delivery in the treatment [16], although the coordination modules may dictate sequencing.

Indeed, it is important to note that modularity does not imply that any module can be administered in any order without concern for logic. On the contrary, part of modular therapy includes having structure. Modular treatments often have proposed coordination modules [13] that can help providers select the appropriate module to administer, providing a logic to the ordering of modules based on factors such as the presenting problem, comorbidity, treatment progress, and interfering problems (e.g., crises, poor engagement). These coordination modules [1, 13] are an essential part of guiding the treatment process.

2.2 Advantages of Modular Treatment Design

There are multiple advantages to modular treatment designs over traditional designs. (1) *Reduced training burden*: by virtue of modules being stand-alone, one can capitalize on prior training to learn how to administer a module. For example, a clinician already trained in Coping Cat (an evidence-based CBT manual for anxiety [17]) would not need to retrain on the administration of the exposure module of a modular treatment, and could instead focus on modules with which he or she is less familiar. One can even envision a training protocol in which providers earn "credits" for learning certain modules, so that they can avoid repeating them when learning new interventions. In addition, for a group of clinicians who have reached a certain level of competency on specific practices, training could primarily focus on the process tools and logic models of modular treatment, rather than the content modules themselves (e.g., focus on learning how to measure progress rather than relearning a familiar module such as behavioral activation). (2) *Parsimony*: clinicians may omit irrelevant modules, or end treatment when clients meet their clinical goals, potentially making therapy more efficient. For example, if clients are able to learn the skills rapidly and their symptoms remit, they could end treatment weeks earlier compared to a traditional manual in which a minimum number of sessions is required. (3) *Personalized care*: modular designs allow care to be personalized to the client, by selecting modules that are matched to the unique needs of the client, including addressing issues such as comorbidity, crises, lack of engagement, and even pace of treatment. For example, if a child has a learning disability and needs more time to learn a skill, the modular approach may allow the clinician to extend the amount of time spent on a module. (4) *Likeability*: finally, it appears that clinicians like modular approaches, particularly for their involvement in the "design" of an episode of treatment. Once they learn how to use a modular approach, they are more inclined to use it again with other cases [18, 19].

The main disadvantage associated with modular treatment designs is some loss of step-by-step guidance and streamlining, which is common in traditional manuals. It may initially feel uncomfortable and challenging for inexperienced providers to transition to a modular approach.

3   Interventions

3.1   Strategies for Modular Treatment Delivery

We recommend specifying S.M.A.R.T. goals (Specific, Measurable, Attainable, Realistic, and Time-constrained) at the outset of treatment to help guide treatment decisions. You should decide on goals collaboratively with the youth and the family at the beginning of treatment. Goals should be realistic, achievable in a relatively short amount of time, and reflect improvement for the youth. Treatment goals can look very different for two youth with the same presenting problem based on the family's primary concerns and the level of impairment. For example, a youth with severe depression might have a primary goal of increasing school attendance from one time per week to three times per week, as school attendance is incredibly important for a youth's well-being and might take precedence over other depressive symptoms or concerns (notwithstanding any safety concerns). A youth with more mild depression might have a goal of increasing pleasant activities from three times a day to seven times a day because this youth is already engaging in some pleasant activities and may need to increase this number or specifically plan for pleasant activities around low mood times of the day.

We also recommend using these goals to monitor treatment progress. Modular treatments vary in terms of the extensiveness of their clinical progress monitoring recommendations. Some treatments do not incorporate it at all, while others use it as an integral part of treatment planning. Although progress monitoring is by no means a staple of modularity, it can help with clinical decision-making, which is often a part of modular treatment design. Progress monitoring can help identify next steps in treatment, such as whether a module needs to be repeated, or if the youth is ready for termination [20, 21]. There are a number of ways to track a client's progress. You can use standardized and validated tools, such as the Strengths and Difficulties Questionnaire or the Youth Outcomes Questionnaire (*see* a Becker-Haimes and colleagues [22] for a list of other free measurement tools). You can also track behaviors and/or symptoms linked to your treatment goals using an idiographic/individualized measure such as tracking the frequency of certain behaviors, the severity of certain experiences (mood, anxiety, etc.), among other metrics. For example, you may ask a parent to track how many tantrums his or her child had in the past week. Ideally, you would include BOTH a standardized measure and an idiographic/individualized measure at minimum. Typically, you would administer the standardized questionnaires less frequently and the individualized measures more frequently, such as weekly or daily.

The decision to move from one module to the next will vary depending on the modular treatment you use. Some treatments will have coordination modules to guide that decision; others will recommend that the decision be made based on feedback from the client. That is, rather than moving on because you have covered all the content in the module, you will need to make sure that the client understands the content and has mastered the skills within that module. If the client has successfully mastered the skills, you may move on. If your client is not progressing, you must decide whether the client needs to spend more time on that module or whether there are interfering factors that need to be addressed (e.g., poor engagement, literacy concerns, psychosocial stressors that are distracting from the core content). In addition, you should ask yourself if the module is a good fit for the client at this time and if you are delivering the module as intended. For instance, it would be a false negative to assume that *Exposure* was not working for an anxious client if you are not doing *Exposure* as it was meant to be delivered in the evidence base. That is, are the exposure exercises appropriately targeting that particular client's fears? Did the client actually experience anxiety during the exposure exercises and/or habituation to the feared stimulus? Did you repeat exposure exercises frequently enough for the client to experience reduction in anxiety? As an example, the Modular Approach to Treatment for Children (*MATCH* [23]) uses clinical dashboards to monitor client change on standardized and individualized measures during treatment, while simultaneously tracking the administration of modules (*see* Fig. 2 and case examples). Such dashboards allow providers to explore whether the use of certain modules is associated with symptom reduction or improvement in functioning [21]. You do not need a "premade" dashboard to track your clients' treatment progress. You can create a simple spreadsheet or Word document with dates and scores or ratings of the outcomes you are measuring. You can even keep track using pencil and paper and graph your results manually. For example, if you are tracking the number of temper tantrums a child is having per week, you can keep track in a table (on paper and pencil or in a word document) and manually draw a graph to help your client visualize his or her progress. If you feel comfortable with Excel, you can use formulas and embedded tools to draw graphs.

Unlike in traditional treatment manuals, modular treatment is not considered completed simply because one has covered all the content in the manual. The decision to terminate needs to be made based on the algorithms for each treatment. For each youth, the end of treatment might look very different and collaborative goals can help to determine if the desired outcome has been achieved. Once a client is no longer exhibiting symptoms or impairment, or these experiences have become sufficiently manageable based on an agreed upon level, termination may be considered.

3.2 Modular Interventions Review

There are multiple modular interventions for youth, most of which are CBT-based. It is important to note that not all of the interventions reviewed here are fully modular, based on the criteria previously outlined. Some are partially modular, meaning that only part of the intervention is modular or only some characteristics of modularity are integrated. However, all of them include some elements of modularity and strive to meet the unique needs of youth. In determining which modular therapies to use, you may want to consider whether you tend to be a generalist or a specialist, the types of populations with whom you typically work, and your professional development goals. A generalist who works with individuals with a wide range of mental health problems may benefit from learning a modular therapy that targets multiple problem areas, such as *MATCH*, or the evidence-based framework of *MAP*. These interventions allow providers to flexibly adapt treatment to target multiple problem areas and/or diagnoses and to address emergent treatment concerns. A specialist who focuses on a specific problem area or set of disorders may benefit from learning and using a more targeted modular therapy whose flexibility primarily lies in the arrangement of a specific set of core practices. The decision of whether to generalize or specialize might be determined by the population that comes to your setting. For example, a provider working in primary care is likely to see any number of mental health concerns and cannot be limited to a particular type of problem whereas a provider in a specialty clinic can arrange to see individuals within a particular area. The number of available mental health providers in your area may also affect your decision. For example, a provider who is located in a rural area with a limited number of providers may benefit from receiving training that can serve a greater range of youths. Modular interventions may also help to develop your skills in certain areas in which you would like to receive more training (e.g., progress monitoring, decision-making).

There are a number of modular treatments for internalizing disorders. The first clearly identified modular therapy for children—*Modular CBT* [24]—was developed to address anxiety disorders in children. This treatment is based on CBT work with adults [25, 26] and with children [27, 28]. It includes a minimum of four core modules (self-monitoring, psychoeducation about anxiety, exposure, and maintenance), along with optional cognitive modules and content for addressing disruptive behaviors, lack of motivation, and poor social skills. An algorithm in the form of a flowchart is used to guide decision-making. *Building Confidence* is another, more recent, modular treatment for anxiety disorders that was adapted from a manualized youth CBT treatment [29]. Providers select and sequence modules based on the client's anxiety disorder and administer treatment for one to sixteen sessions, depending on the child's needs and ongoing symptomatology. Session order and treatment duration are not predetermined, but rather reflect the needs of the child. Modules are selected using an algorithm adapted from Chorpita and colleagues [30]. Supplementary modules are available based on barriers to treatment (e.g., using rewards to increase motivation to complete exposure exercises, teaching social skills to youth who struggle to connect with peers). Providers may terminate treatment once clinical anxiety problems have reduced to nonclinical levels. Another modular treatment for anxious children, *Behavioral Interventions for Anxiety in Children with Autism* (BIACA [31]), is a CBT approach for youth with comorbid autism spectrum disorder (ASD). It is based on a CBT manual for typically developing anxious youth and uses a modular treatment approach that addresses problematic anxiety and non-anxiety-based symptoms, by accounting for barriers that may occur when working with children on the autism spectrum, such as low motivation, difficulty relating to peers, and other comorbidities. Modules are selected based on the child's ongoing needs and a treatment algorithm. Caregiver modules focus on supporting parents in completing homework such as exposure tasks, parent training, and support in working with schools. Core child modules focus on addressing anxiety by developing coping skills and conducting exposure. Additional modules include social skills, perspective taking, and communication skills. In addition,

ASD specific modules include social and daily living skills, and suppression of restricted interests and repetitive behaviors. The *Unified Protocol for the Treatment of Emotional Disorders—Youth* [32] is a partially modular treatment that can address various emotional disorders (i.e., depression and anxiety). Chapter 13 of this volume is dedicated to the Unified Protocol. Five treatment sections are required in a fixed order: (1) psychoeducation regarding emotions, (2) emotion regulation, (3) cognitive skills, (4) exposure, and (5) relapse prevention. The number of sessions associated with each module is flexible. Three optional modules are available that involve skills covering motivational enhancement, crisis management, and parent training.

We are aware of two modular treatments for externalizing problems. The *Modular treatment for children with Oppositional Defiant Disorder or Conduct Disorder* [33] is a modular treatment approach for children with conduct disorder or oppositional defiant disorder. Seven treatment modules derived from evidence-based treatment families such as CBT, parent management training, family therapy, and case management are combined according to an algorithm that helps to determine which modules to deliver in which order. The child CBT/Skills Training modules include anger-control and relaxation, problem solving, social skills training, and the inclusion of a medication consult to address comorbid Attention Deficit Hyperactive Disorder. The parent management training modules include psychoeducation specific to social learning theory and behavioral tools to improve parenting skills, as well as discipline practices such as contingency management procedures. Family-focused modules include communication and problem-solving practices, and teacher consultation and social skills modules can be included to address ecological concerns. Finally, crisis management modules exist in case families have to deal with an emergency, including safety planning, and referrals to the psychiatrist and emergency department if necessary. A nurse-administered version of the protocol, *Protocol for On-site Nurse-administered Intervention* (*PONI*; [34]), for delivery in primary care settings is also available. *Modular Intervention for Foster Parents* [35] employs a highly structured modular design. Some modules are prescribed, while others are optional and to be used only if necessary. Modules that are required cover practices involving positive reinforcement, enhancing structure such as giving effective instructions and limit setting, and final modules focused on maintenance and termination. An additional set of three modules focused on problem-solving, self-monitoring, and autonomy skills are mandatory but can be offered at any time. A limit-setting module focused on negative consequences such as ignoring, losing privileges and timeout along with two modules dedicated to avoiding escalations and evaluating parenting behaviors are optional.

Some modular therapies can address both internalizing and externalizing problems, allowing the clinician to address the majority of presenting child concerns. *CBT+* [36] can address problems related to depression, anxiety, disruptive behaviors, and trauma. For anxiety and trauma, modules include emotion regulation skills, exposure, and cognitive reprocessing. For depression, modules include emotion regulation, behavioral activation, problem-solving, and cognitive reprocessing. For disruptive behaviors, parent management techniques such as praise, rewards, and consequences are recommended. *CBT+* recommends a sequence according to the problem area but allows for tailoring as needed to address comorbidity. The *Common Elements Treatment Approach—Youth* (*CETA*, [37]) is based on the common elements research of Chorpita and colleagues [10] and has been adapted for delivery by a lay workforce in underresourced international settings. Most of the research on *CETA* was conducted with adult populations [37, 38], although a more recent study highlights findings from a trial of *CETA-Youth* [39]. *CETA-Youth* is delivered in 6–12 sessions based on the child's needs, with caregiver involvement if possible. To facilitate decision-making, lay counselors can use default "flows" (or a coordination module that suggests which modules to use in a recommended order) based on the evidence-based treatments for the problem area. *CETA-Youth* includes the following modules (referred to as "elements" by Murray and colleagues): engagement, psychoeducation, parenting skills, anxiety management (relaxation), behavioral activation, cognitive coping, exposure, problem solving, and safety planning that can be ordered in various flows depending on the presenting problem.

Another comprehensive modular approach that can address multiple problems is *Managing and Adapting Practice* (*MAP*); [40]). *MAP* modules are not grouped by problem area into a treatment manual, with worksheets and scripts. Rather, it is a system of resources that clinicians use to design, deliver and evaluate their own treatment [41]. First, *MAP* provides clinicians access to a database of research on evidence-based treatments. From this searchable database, practitioners can identify treatments that have been found to be effective in clinical trials for certain populations with specific problems. Furthermore, practitioners can discover the most common practice elements found within those treatments, along with practice and process guides (modules) to facilitate implementation of those practice elements. Finally, clinicians can develop a clinical dashboard for each of their cases to track progress and practices delivered into a visual summary [21]. *MAP* functions as a tool kit with which clinicians can select and use content to adhere to an existing treatment or build a new, individualized one from a set of knowledge resources. *MAP* is highly scalable, well-liked by providers, and yields large effect sizes [7, 40]. *MATCH-ADTC* (*Modular Approach to Treatment for Children with Anxiety,*

Depression, Trauma and Conduct problems [23]) is made up of 33 CBT-based modules similar to *MAP*, that can address common youth problems such as anxiety (e.g., exposure, relaxation), depression (e.g., behavioral activation, problem-solving), trauma (e.g., trauma narrative) and conduct problems (e.g., time-out, planned ignoring). The modules are administered independently to respond to each individual client's needs. Unlike *MAP*, it is a full treatment manual and includes scripts, worksheets, along with flowcharts to guide treatment decisions and treatment planning depending on symptoms and interferences. While a default flow chart suggests a sequencing of modules based on a target or "focus" problem, the provider is free to pick which modules to deliver based on progress monitoring and feedback information, including current needs, emerging interferences, and responses to prior modules. In order to illustrate how a modular treatment such as *MATCH* can allow clinicians to address comorbid presentations by changing treatment focus during the course of treatment, we provide two case examples inspired by real *MATCH* cases seen at a community mental health agency.

Case 1: *Sabrina* (pseudonym) is a 15-year-old-female, presenting with a dramatic mood change in the middle of her sophomore year and a history of separation anxiety.:
Precipitating factor: Sabrina lied to her parents about attending a weekend event for her soccer club, and instead went to an out-of-town party with her friends. After finding out about the party, Sabrina's parents expressed extreme disappointment with her behavior. They implemented consequences by not letting her go out on the weekends without a parent attending, and limiting her contact with her friends during weeknights (e.g., no phone calls over 15 min) for 1 month.

History: During the intake evaluation (about 6 weeks after the party), Sabrina reported having persistent low mood for about 1 month. She reported losing interest in activities that had previously made her happy, such as playing soccer or going to the local mall with friends. She also reported a diminished sense of joy when she pushed herself to participate in these activities. She described having low energy, decreased appetite to the point where her friends or parents often reminded or encouraged her to eat, and difficulty focusing in class. She reported having passive thoughts about death, such as, "What would happen if I died?" or "Would my family be happier if I was not around?" about 4–5 times a week. She denied any current or past active thoughts of suicide or suicidal behaviors. Sabrina reported that she had never previously experienced such a dramatic change in her mood. She described losing her parents' trust after the party as very difficult for her because they were a close-knit family, and the incident had created some distance between them. She expressed guilt about upsetting her parents and said that it felt like they no longer had the same relationship. Sabrina's parents described her as a very happy child who had no trouble making friends from an early age and who thoroughly enjoyed playing sports and being active. They sought treatment because they were very concerned by the shift in her mood. They believed her mood would improve once they lifted the consequences, and were surprised to see that she remained quiet and withdrawn.

Sabrina's parents also reported that Sabrina had a history of separation anxiety, primarily related to being away from her father, and current worries about her performance at school. Sabrina agreed that she focuses on doing well in school and hates to make mistakes. She reported spending hours on her homework, often checking it multiple times for errors, and constantly worrying about how her teachers feel about her work.

Assessment: At intake, Sabrina received a total score of 17 on the Patient Health Questionnaire (PHQ-9; [42]), which is in the moderately severe range, and a t-score of 67 on the Total Anxiety scale of the Revised Child Anxiety and Depression Scale (RCADS; [43]). Her *t*-scores on the panic, generalized anxiety, and separation anxiety subscales were 68, 73, and 62, respectively. Her mother's scores on the parent version of the RCADS followed a similar pattern. Sabrina reported that her average mood rating (from 0 to 10, with 10 representing the best she has ever felt) was at a "4" during the week of the intake. She also reported spending about 70% of her time at home alone in her room.

Treatment plan: Following the intake, the therapist concluded that Sabrina met the diagnostic criteria for Major Depressive Disorder, single episode moderate, and Generalized Anxiety Disorder. Due to the dramatic nature of her change in mood and its impairment on Sabrina's functioning, the clinician and the family decided to focus first on depression as the primary target of treatment. The clinician and the family also agreed to track Sabrina's average mood rating and percentage of free time spent alone in her room on a weekly basis. They also planned to administer the PHQ-9 every 2 weeks, and the RCADS total anxiety score every 3 months over the course of treatment.

Treatment course: The clinician proceeded with administering depression-focused modules in the order recommended by the *MATCH* depression flow chart (*see* Fig. 1). After completion of the initial modules (*Getting Acquainted, Psychoeducation for Depression for Child and Caregiver*), the family expressed that they were most concerned about Sabrina's increased time spent isolating in her room. Therefore, the clinician made the decision to prioritize a session on *Activity Selection* (behavioral activation) to engage her in mood-boosting activities. Sabrina generated a list of possible activities she could do to improve her mood and then

Depression

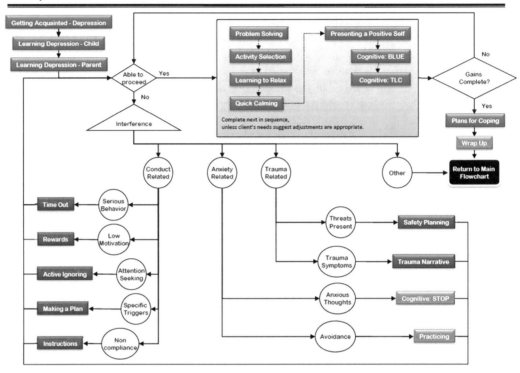

Fig. 1 Depression flowchart from MATCH manual (Reproduced with permission from PracticeWise, LLC)

sorted them into categories of activities that she would consider implementing versus activities that she would definitely implement. The clinician repeated a few sessions of this module, including practicing pleasant activities in session (e.g., walking to the park, playing a game) and reviewing Sabrina's ratings before and after participating. Sabrina also included multiple pleasant activities that she could do by herself, with friends, and with her parents (e.g., going to the movies, playing "Risk") and practiced those between sessions. As she incorporated these activities into her life on a regular basis again, Sabrina began reporting higher average mood ratings. The clinician then moved on to the *Problem Solving* module and worked with Sabrina first on hypothetical problems and then on problems that Sabrina generated (e.g., not being able to see her friends alone on the weekend, spending too much time on homework).

Measurement and tracking: After completion of the first few modules of the depression flowchart, Sabrina's mood improved dramatically (*see* Sabrina's dashboard, Fig. 2). Her scores on the PHQ-9 dropped to the mild range and her average mood rating moved up the scale to consistent "7" and "8" ratings compared to "4" at intake. She also reported spending an average of 40% of her

Progress and Practice Monitoring Tool   **Case ID: "Sabrina"**

Age (in years): 14.8   Gender: Female
Primary Diagnosis: Major Depressive Disorder, single episode, moderate;   Ethnicity: Asian American
Generalized Anxiety Disorder

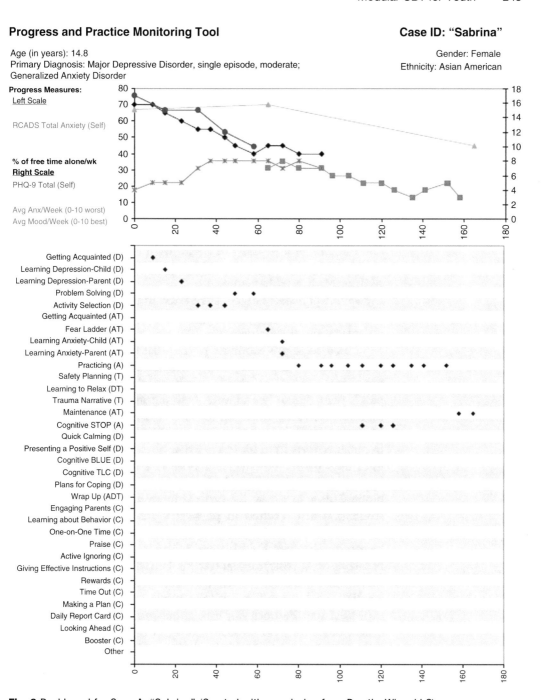

Fig. 2 Dashboard for Case A, "Sabrina" (Created with permission from PracticeWise, LLC)

free time in her room by herself, mainly when she needed to work, versus 70% at intake.

Change in treatment focus. Once the depression stabilized, as indicated by her presentation in therapy, the results of her PHQ-9, her mood ratings, and reported time spent isolating, Sabrina became more concerned with her anxiety about school performance. She reported that the excessive time she was spending on her homework felt more impairing than her currently improved mood. Thus, the clinician decided to change the focus of treatment to anxiety. The therapist consulted the anxiety flowchart and used the *Fear Ladder* module to start building a fear ladder around performance concerns. From that point, the clinician administered the *Psychoeducation for Anxiety for Child and Caregiver, Exposure* and *Cognitive Anxiety: Stop* modules. The clinician and Sabrina identified anxiety measures she could track on her own (i.e., time spent looking over homework per school night, average fear rating per week) and continued to track depression symptoms to ensure that Sabrina maintained her gains in this area.

Case 2: *Hugo*, a 9-year-old male, with a history of attention deficit hyperactivity disorder (ADHD) and presenting with oppositional behaviors.:

Precipitating factor: Hugo was referred to treatment by his school counselor because he had been getting into physical fights with his classmates and was frequently arguing with his brother at home.

History: Hugo's mother reported that he had been diagnosed with ADHD by a primary care physician when he was 7 and was prescribed a stimulant, that he has been taking on and off for the past 2 years due to her admittedly forgetting to administer it. She reported that Hugo is more focused when taking his medication but has still had difficulty keeping up with his schoolwork. She gets frequent reports from his teacher that he is off task during the day, interrupts other students, and does not follow along with her instructions. Right before the intake evaluation, he had been sent home 2 days in a row for provoking an argument with another student, which resulted in physical fights. Hugo's mother also reported that he does not follow instructions at home and often blames his brother for his own behavior. She sought treatment because she is concerned about his behaviors escalating and she wants him to do well in school. Hugo reported that he is easily irritated by his brother and the other children at school. He expressed that he does not care about getting along better with others but does want to stop getting in trouble. Hugo's mother also noted concerns with his level of anxiety when separating from her. She reported trying to minimize their separations to avoid tantrums. She expressed that Hugo often complained of stomachaches or muscle pains when having to go to school and that she has taken him to the emergency room several times for physical concerns. During these visits, no underlying physical concern had been identified.

Assessment: Hugo received a t-score of 80 on the oppositional defiant problems subscale and a *t*-score of 85 on the Attention Deficit/Hyperactivity Problems subscale of the Child Behavior

Check List-Parent Report Form (CBCL; Achenbach, 2009). His score was also elevated (t-score $= 73$) on the anxiety problems subscale. His scores were equally high on the oppositional defiant problems subscale (t-score $= 78$) and the ADHD scales (t-score $= 82$) of the Teacher Report Form (TRF) teacher version. The clinician, Hugo, and his mother also agreed to track his behavior over the week with a 0–10 scale in which 10 was reflective of very good behavior.

Treatment plan: Hugo met criteria for Oppositional Defiant Disorder and had an existing diagnosis of ADHD, combined presentation. He also met criteria for Separation Anxiety Disorder. Hugo's mother expressed that she was less concerned about his separation anxiety because she did not mind him sticking close to her at home and on errands. The initial plan was to focus on parent management skills for a disruptive behavior target of treatment as this target area would address both the ODD and ADHD diagnoses.

Treatment course: The clinician consulted the *MATCH* conduct problems flowchart and started to work with Hugo's mother by going through the *Engaging Parents* module to introduce the concept of treatment and address any potential barriers to regularly attending appointments. Following that, the clinician covered the *Psychoeducation about Disruptive Behavior* module, and a few sessions of the *Attending*, *Praise*, and *Rewards* modules. Hugo responded well to rewards and was highly motivated to earn airplane models for completing his tasks (e.g., being kind to his brother, following through on his mother's instructions, completing homework). He and his mother also reported enjoying the quality time they were spending together building the models.

Measurement and tracking: When the reward module was introduced, the clinician started tracking how many days out of the week that Hugo earned his reward for that day. As it was a topic that was coming up more frequently during sessions, the clinician had also started to track the number of times that the family had gone to the emergency room for physical concerns during the week. Over the course of treatment, Hugo's ratings on the oppositional defiant problems subscale of the CBCL and TRF decreased by about one standard deviation (t-score $= 68$ and 70, respectively) and his ratings on the anxiety problems subscale had increased (t-score $= 75$). An RCADS parent form revealed that separation anxiety was his most elevated subscale (t-score $= 78$).

Change in treatment focus: As the visits to the emergency room were time-consuming for the family and were getting in the way of being able to adhere to a regular routine, the clinician and the family made the decision to change the treatment focus to anxiety. The clinician started by going through the *Fear Ladder* and *Psychoeducation about Anxiety* modules for both Hugo and his mother. The remainder of the treatment episode focused on the

Exposure module, practicing exercises in which Hugo separated from his mother as well as *Relaxation* to teach Hugo other coping strategies when experiencing physical anxiety symptoms. The clinician also worked with Hugo's mother to decrease her modeling of her own anxiety and to identify situations that require a higher level of care and those that could be better managed at home. With practice, the visits to the emergency room eventually decreased and Hugo's scores on the anxiety problems and oppositional defiant problems subscale moved into the non-clinically significant range.

4   Conclusion

Though it is a relatively new area of study in the evidence-based practice literature, the field of modular treatment design is rapidly expanding and offers exciting treatment alternatives for children and families. Modular treatments provide clinicians the ability to tailor treatment to their clients based on certain guiding principles. Clinicians report preferring them to standard approaches [19], and it stands to reason that the ability for therapy to progress in an individualized manner where efficiency can also be maximized is a client preference as well.

Nevertheless, despite the effectiveness of evidence-based treatments (including modular treatments), close to 50% of youth drop out of treatment prematurely [44, 45]. Modular treatments do not necessarily address engagement challenges a family might experience (stigma, lack of understanding how therapy works, transportation problems, poor relationship with clinician, etc.) Modular treatments have the capacity to seamlessly integrate engagement-focused modules into their structures, allowing providers to address engagement concerns as they occur in treatment. Becker and colleagues [46] have piloted a modular engagement protocol that includes 12 modules specifically designed to address the most common engagement problems encountered by families in therapy. In the pilot study, these modules were integrated into a variety of treatments (modular and traditional manualized EBTs), with some preliminary evidence of feasibility and success. When considering the advantages of modular approaches, specific modules for engagement (when dictated by progress monitoring data or algorithms inherent to modular designs) appear to be a promising avenue for further maximizing treatment access and outcomes.

In addition to the challenges posed by poor engagement, addressing the high frequency of crises encountered by community mental health clinicians [4] is a challenge that modular treatments are well-positioned to address. Modular treatment developers have yet to integrate proposed, structured responses to in-session crises or emergent life events. Guan and colleagues [47] explored how existing modules from *MATCH* can be used to address in-session crises, while allowing for simultaneous reinforcement of skills learned in therapy, capitalizing on resources already available in the treatment manual, and maintaining the focus on treatment goals. This is an important area for future research and perhaps a next phase in the evolution of modular treatments.

Finally, there is a need to train graduate students and clinicians to use assessment and progress monitoring data adequately and consistently. Measurement and progress monitoring can easily be integrated into modular therapies. Measurement facilitates informed decision-making about which modules to implement, in what order, or when to terminate treatment. For instance, *MAP* providers are trained to use clinical dashboards as part of treatment planning and monitoring. In addition, clients are taught how to self-monitor and parents are taught how to monitor the behavior of their children, in order to track progress. Other modular therapies and evidence-based treatments could benefit from integrating and training in the use of measurement and progress monitoring.

In conclusion, modular therapies can provide helpful alternatives to standardized evidence-based treatment for clinicians. Although modular therapies are by no means a "silver bullet" to solve all youth mental health problems, they provide strategies that are responsive, efficient, and potentially efficacious. They may represent a "new wave" for how to apply the evidence base in a manner that is more parsimonious and efficient. This is especially important considering the mental health burden and mental health workforce shortage in the USA, especially in rural areas, and around the world [48–50]. Additional work is needed to ensure that modular therapies address even more of the most common concerns that are prominent in community care, including comorbidity, in-session crises, and poor treatment engagement. We look forward to seeing what the future of modular therapies holds, including the application of evidence in new and exciting ways.

References

1. Weisz JR, Chorpita BF (2012) "Mod squad" for youth psychotherapy: restructuring evidence-based treatment for clinical practice. In: Kendall PC (ed) Child and adolescent therapy: cognitive–behavioral procedures. American Psychological Association, Washington, DC

2. Southam-Gerow MA, Weisz JR, Kendall PC (2003) Youth with anxiety disorders in research and service clinics: examining client differences and similarities. J Clin Child Adolesc Psychol 32(3):375–385

3. Weisz JR, Southam-Gerow MA, Gordis EB, Connor-Smith JK, Chu BC, Langer DA, McLeod BD, Jensen-Doss A, Updegraff A, Weiss B (2009) Cognitive–behavioral therapy versus usual clinical care for youth depression: an initial test of transportability to community clinics and clinicians. J Consult Clin Psychol 77 (3):383

4. Chorpita BF, Korathu-Larson P, Knowles LM, Guan K (2014) Emergent life events and their impact on service delivery: should we expect the unexpected? Prof Psychol Res Pract 45 (5):387

5. Guan K, Park AL, Chorpita BF (2017) Emergent life events during youth evidence-based treatment: impact on future provider

adherence and clinical progress. J Clin Child Adolesc Psychol 48:S202–S218

6. Addis ME, Krasnow AD (2000) A national survey of practicing psychologists' attitudes toward psychotherapy treatment manuals. J Consult Clin Psychol 68(2):331

7. Bruns EJ, Walker JS, Bernstein A, Daleiden E, Pullmann MD, Chorpita BF (2014) Family voice with informed choice: coordinating wraparound with research-based treatment for children and adolescents. J Clin Child Adolesc Psychol 43(2):256–269

8. Rith-Najarian LR, Daleiden EL, Chorpita BF (2016) Evidence-based decision making in youth mental health prevention. Am J Prev Med 51(4):S132–S139

9. Chu BC, Merson RA, Zandberg LJ, Areizaga M (2012) Calibrating for comorbidity: clinical decision-making in youth depression and anxiety. Cogn Behav Pract 19(1):5–16

10. Chorpita BF, Daleiden EL, Weisz JR (2005) Identifying and selecting the common elements of evidence based interventions: a distillation and matching model. Ment Health Serv Res 7(1):5–20

11. Weisz JR, Chorpita BF, Palinkas LA, Schoenwald SK, Miranda J, Bearman SK, Daleiden EL, Ugueto AM, Ho A, Martin J (2012) Testing standard and modular designs for psychotherapy treating depression, anxiety, and conduct problems in youth: a randomized effectiveness trial. Arch Gen Psychiatry 69(3):274–282

12. Lyon AR, Ludwig K, Romano E, Koltracht J, Vander Stoep A, McCauley E (2014) Using modular psychotherapy in school mental health: provider perspectives on intervention-setting fit. J Clin Child Adolesc Psychol 43 (6):890–901

13. Chorpita BF, Daleiden EL, Weisz JR (2005) Modularity in the design and application of therapeutic interventions. Appl Prev Psychol 11(3):141–156

14. Ng MY, Weisz JR (2016) Annual research review: building a science of personalized intervention for youth mental health. J Child Psychol Psychiatry 57(3):216–236

15. Boustani MM, Gellatly R, Westman JG, Chorpita B (2017) Advances in cognitive behavioral treatment design: time for a glossary. Behav Ther 40(6):199–207

16. Lyon AR, Lau AS, McCauley E, Vander Stoep A, Chorpita BF (2014) A case for modular design: implications for implementing evidence-based interventions with culturally diverse youth. Prof Psychol Res Pract 45(1):57

17. Kendall PC (2006) Coping cat workbook. Workbook Publishing, Ardmore, PA

18. Borntrager CF, Chorpita BF, Higa-McMillanC, Weisz JR (2009) Provider attitudes toward evidence-based practices: are the concerns with the evidence or with the manuals? Psychiatr Serv 60(5):677–681

19. Palinkas LA, Weisz JR, Chorpita BF, Levine B, Garland AF, Hoagwood KE, Landsverk J (2013) Continued use of evidence-based treatments after a randomized controlled effectiveness trial: a qualitative study. Psychiatr Serv 64 (11):1110–1118

20. Bickman L, Rog DJ (2008) The SAGE handbook of applied social research methods. Sage Publications, Thousand Oaks, CA

21. Chorpita BF, Bernstein A, Daleiden EL (2008) Health RNoYM: driving with roadmaps and dashboards: using information resources to structure the decision models in service organizations. Adm Policy Ment Health Ment Health Serv Res 35(1–2):114–123

22. Becker-Haimes EM, Tabachnick AR, Last BS, Stewart RE, Hasan-Granier A, Beidas RS (2020) Evidence base update for brief, free, and accessible youth mental health measures. J Clin Child Adolesc Psychol 49(1):1–17. https://doi.org/10.1080/15374416.2019. 1689824

23. Chorpita B, Weisz J (2009) Modular approach to therapy for children with anxiety, depression, trauma, or conduct problems (MATCH-ADTC). Practice Wise LLC, Satellite Beach, FL

24. Chorpita BF (1998) Modular cognitive behavior therapy for child and adolescent anxiety disorders: therapist manual. Unpublished manuscript

25. Chorpita BF, Barlow DH (1998) The development of anxiety: the role of control in the early environment. Psychol Bull 124(1):3

26. Beck A, Rush A, Shaw B, Emery G (1979) Cognitive therapy of depression. Guilford, New York, NY

27. Kendall PC, Hedtke KA (2006) Cognitive-behavioral therapy for anxious children: therapist manual. Workbook Publishing, Ardmore, PA

28. Silverman WK, Kurtines WM, Ginsburg GS, Weems CF, Lumpkin PW, Carmichael DH (1999) Treating anxiety disorders in children with group cognitive-behavioral therapy: a randomized clinical trial. J Consult Clin Psychol 67(6):995

29. Chiu AW, Langer DA, McLeod BD, Har K, Drahota A, Galla BM, Jacobs J, Ifekwunigwe M, Wood JJ (2013) Effectiveness of modular CBT for child anxiety in elementary schools. Sch Psychol Q 28(2):141

30. Chorpita BF, Taylor AA, Francis SE, Moffitt C, Austin AA (2004) Efficacy of modular cognitive behavior therapy for childhood anxiety disorders. Behav Ther 35(2):263–287

31. Storch EA, Arnold EB, Lewin AB, Nadeau JM, Jones AM, De Nadai AS, Mutch PJ, Selles RR, Ung D, Murphy TK (2013) The effect of cognitive-behavioral therapy versus treatment as usual for anxiety in children with autism spectrum disorders: a randomized, controlled trial. J Am Acad Child Adolesc Psychiatry 52 (2):132–142

32. Ehrenreich JT, Goldstein CR, Wright LR, Barlow DH (2009) Development of a unified protocol for the treatment of emotional disorders in youth. Child Fam Behav Ther 31(1):20–37

33. Kolko DJ, Dorn LD, Bukstein OG, Pardini D, Holden EA, Hart J (2009) Community vs. clinic-based modular treatment of children with early-onset ODD or CD: a clinical trial with 3-year follow-up. J Abnorm Child Psychol 37(5):591–609

34. Kolko DJ, Campo JV, Kelleher K, Cheng Y (2010) Improving access to care and clinical outcome for pediatric behavioral problems: a randomized trial of a nurse-administered intervention in primary care. J Dev Behav Pediatr 31 (5):393

35. Vanschoonlandt F, Vanderfaeillie J, Van Holen F, De Maeyer S (2012) Development of an intervention for foster parents of young foster children with externalizing behavior: theoretical basis and program description. Clin Child Fam Psychol Rev 15(4):330–344

36. Dorsey S, Berliner L, Lyon AR, Pullmann MD, Murray LK (2016) A statewide common elements initiative for children's mental health. J Behav Health Serv Res 43(2):246–261

37. Bolton P, Lee C, Haroz EE, Murray L, Dorsey S, Robinson C, Ugueto AM, Bass J (2014) A transdiagnostic community-based mental health treatment for comorbid disorders: development and outcomes of a randomized controlled trial among Burmese refugees in Thailand. PLoS Med 11(11): e1001757

38. Murray LK, Dorsey S, Haroz E, Lee C, Alsiary MM, Haydary A, Weiss WM, Bolton P (2014) A common elements treatment approach for adult mental health problems in low-and middle-income countries. Cogn Behav Pract 21 (2):111–123

39. Murray L, Hall B, Dorsey S, Ugueto A, Puffer E, Sim A, Ismael A, Bass J, Akiba C, Lucid L (2018) An evaluation of a common elements treatment approach for youth in Somali refugee camps. Global Ment Health 5: e16

40. Southam-Gerow MA, Daleiden EL, Chorpita BF, Bae C, Mitchell C, Faye M, Alba M (2014) MAPping Los Angeles County: taking an evidence-informed model of mental health care to scale. J Clin Child Adolesc Psychol 43 (2):190–200

41. Chorpita BF, Daleiden EL (2014) Structuring the collaboration of science and service in pursuit of a shared vision. J Clin Child Adolesc Psychol 43(2):323–338

42. Kroenke K, Spitzer RL, Williams JB (2001) The PHQ-9: validity of a brief depression severity measure. J Gen Intern Med 16 (9):606–613

43. Chorpita BF, Ebesutani C, Spence SH (2011) Revised Children's Anxiety and Depression Scale. http://www.childfirst.ucla.edu/Resources.html. Accessed 12 Dec 2011

44. Pellerin KA, Costa NM, Weems CF, Dalton RF (2010) An examination of treatment completers and non-completers at a child and adolescent community mental health clinic. Commun Ment Health J 46(3):273–281

45. Nock MK, Ferriter C (2005) Parent management of attendance and adherence in child and adolescent therapy: a conceptual and empirical review. Clin Child Fam Psychol Rev 8 (2):149–166

46. Becker KD, Park A, Boustani M, Chorpita BF (2019) A pilot study to examine the feasibility and acceptability of a coordinated intervention design to address treatment engagement challenges in school mental health services. J Sch Psychol 76:78–88

47. Guan K, Boustani MM, Chorpita BF (2019) "Teaching moments" in psychotherapy: addressing emergent life events using strategies from a modular evidence-based treatment. Behav Ther 50(1):101–114

48. Thomas KC, Ellis AR, Konrad TR, Holzer CE, Morrissey JP (2009) County-level estimates of mental health professional shortage in the United States. Psychiatr Serv 60 (10):1323–1328

49. Merwin E, Hinton I, Dembling B, Stern S (2003) Shortages of rural mental health professionals. Arch Psychiatr Nurs 17(1):42–51

50. Bruckner TA, Scheffler RM, Shen G, Yoon J, Chisholm D, Morris J, Fulton BD, Dal Poz MR, Saxena S (2011) The mental health workforce gap in low-and middle-income countries: a needs-based approach. Bull WHO 89:184–194

Chapter 13

Unified Protocol for Transdiagnostic Treatment of Emotional Disorders in Children and Adolescents

Elizabeth R. Halliday and Jill Ehrenreich-May

Abstract

Treatment for emotional disorders in children and adolescents is increasingly being applied in a transdiagnostic manner to address the need for more broadly applicable, yet mechanism-focused psychotherapy. The Unified Protocols for Transdiagnostic Treatment of Emotional Disorders in Children and Adolescents (UP-C/A) are modular, flexible, evidence-based treatment manuals that utilize empirically supported, largely cognitive and behavioral strategies to treat anxiety, mood, and other emotional disorders. In this chapter, we describe the rationale and empirical support for the use of the UP-C/A, provide a pragmatic overview of its contents, and briefly discuss application of core intervention components.

Key words Transdiagnostic, Unified protocol, Children, Adolescents, Anxiety, Depression, Emotional disorders, Internalizing disorders, Cognitive behavioral therapy

1 Introduction

The Unified Protocols for Transdiagnostic Treatment of Emotional Disorders in Children and Adolescents (UP-C/A [1–3]) are evidence-based interventions designed purposefully to be applied to a wide variety and/or combination of emotional disorders. Originally developed in concert with the Unified Protocol (UP [4, 5]) for adults, the UP-C/A describes how a clinician may use an array of cognitive and behavioral strategies in a general emotion-focused language to promote change across transdiagnostic targets. Utilizing elements of emotion education, traditional cognitive techniques, a full range of behaviorally driven activation and exposure strategies, and a strong focus on mindful awareness techniques, the UP-C/A aims to address common underlying mechanisms that are theorized to predispose and maintain emotional disorders [6]. The UP-C and UP-A can be delivered to children and adolescents, approximately 6–12 and 12–17 years old, respectively, struggling with any emotional disorder, including any combination of anxiety disorders (i.e., generalized anxiety,

Robert D. Friedberg and Brad J. Nakamura (eds.), *Cognitive Behavioral Therapy in Youth: Tradition and Innovation*, Neuromethods, vol. 156, https://doi.org/10.1007/978-1-0716-0700-8_13, © Springer Science+Business Media, LLC, part of Springer Nature 2020

social anxiety, specific phobias, separation anxiety, panic disorder, illness anxiety disorder, agoraphobia) and depressive disorders (persistent depressive disorder and major depressive disorder). In theory, this approach may also be used for trauma and stress-related disorders, somatic disorders, tic disorders, and obsessive-compulsive disorders, although the evidence base regarding UP-C/A for non-anxiety and depressive disorders is more limited by comparison at present. Applications to borderline personality disorder [7], non-suicidal self-injury (NSSI [8]), eating disorders [9, 10], early psychosis and bipolar disorder [11], along with irritable presentations of emotional disorders are also being explored [12]. Being that comorbidity is the rule rather than the exception [13, 14], and that youth are likely to progress between emotional disorder conditions over time without effective intervention [15, 16], the UP-C/A is written with examples applicable to the experiences of anger, sadness, and anxiety/fear for most techniques, making it particularly useful for children presenting with multiple clinical and/or subclinical emotional disorders.

The movement toward a transdiagnostic treatment approach is supported, in part, by the notion that emotional disorders co-occur at high rates, that children who experience anxiety early in life are likely to experience anxiety and depression in the future, and that anxiety and depressive disorders share common genetic, neurobiological, and environmental risk factors [17–20]. Transdiagnostic approaches can also achieve multi-problem symptom reduction simultaneously and may offset some treatment barriers and burden for client and clinicians alike, such as accessibility, training time and costs, and treatment length, especially in low- and middle-income populations [21].

Neuroticism, postulated as a core dysfunction across emotional disorders [6, 22], is a specific focus of the UP and its youth adaptations. Individuals high in neuroticism demonstrate high levels of negative affect causing them to experience strong emotions (i.e., fear, anxiety, sadness, anger) and in response to these emotions, they become distressed, anxious, and uncomfortable so much so that they take actions to suppress, avoid, escape, distract, or otherwise control these unwanted feelings. These behaviors are negatively reinforced over time because when the child avoids or escapes strong emotions and the situations that elicit them, the discomfort appears to reduce more quickly. Over the long term, however, using avoidant strategies prevents an individual from learning more helpful behaviors and being positively reinforced for using such adaptive behaviors. For youth, this can have far-reaching implications for the development of one's self-efficacy to deal with the situations and experiences that elicit strong emotions [23]. With this in mind, the UP-C/A's overarching intervention principles include: (a) increasing emotional awareness; (b) preventing emotional avoidance by engaging in present-focused awareness during intense emotions; (c) increasing cognitive flexibility; and (d) modifying maladaptive action tendencies (i.e., "emotional behaviors") through personalized exposure and activation techniques. The mechanism of change across these strategies in the UP has generally been framed as extinction of distress and anxiety associated with the experience of strong emotions. By applying these UP principles in a flexible manner to youth with a range of emotional disorders, distress relative to the experience of strong emotion is lessened (or tolerated more easily) through new experience, and negative reinforcement cycles associated with avoidance and other problematic action tendencies (e.g., aggression) are reduced or eliminated.

2 Theoretical and Empirical Foundations

The UP has been studied extensively, in group [24] and individual formats [25], for a variety of disorders [26] and has been shown to target theoretically linked factors, such as maladaptive emotion regulation strategies [27], negative affect, fear of negative emotions, and anxiety sensitivity [28]. Similar results have been found for children and adolescents, supporting the use of a transdiagnostic approach for youth instead of a single emotional disorder approach to evidence-based treatment. Some of the existent research is highlighted in Table 1 [25–35].

The adolescent version of the UP (UP-A), in particular, may be considered a modular approach to treatment. Modules are "containers" of research-validated practice elements, decision-making tools, and delivery techniques [36]. These treatments can improve both externalizing [37–40] and internalizing symptoms [32, 41]. Modular treatments may be better accepted by practitioners than standard manuals [42] because they are more flexible, allow for parsimonious implementation of several evidence-based skills, and permit clinician personalization across a wider variety of problems or foci [43]. Modular treatments may also be more effective than standard manuals [44].

The UP-A consists of eight core modules that can be further expanded upon or minimized depending on the needs of the adolescent. Though the UP-A is written and described in this chapter as an individual treatment, it is increasingly being used in a group format [11]. The duration of treatment varies by client need; however, the UP-A can often be delivered in its entirety in 12–16 sessions [33]. Recommendations for potential module length is outlined in Table 2. The parent portion of the UP-A manual, Module P, is used as needed and varies greatly depending on the parent's involvement and influence on the adolescent's symptoms and treatment.

Table 1
Evidence supporting the UP, UP-A, and UP-C

Design	Citation	Comparison condition	Outcome
Randomized Controlled Trial (RCT)	Farchione et al. [27]; Bullis et al. [29]	UP vs. waitlist	In the UP condition, there was a large effect for reductions in primary and co-occurring disorders post-treatment, and treatment gains were maintained at the 6-month follow-up
RCT	Barlow et al. [25]	UP vs. single emotional disorder treatment	Results showed comparable improvements in treating symptoms There was less treatment dropout in the UP condition than that in the single disorder treatment condition
Open trial and multiple baseline	Ehrenreich et al. [30] Trosper et al. [31]	UP-A	There was significant improvement inanxiety and depression symptoms from pre- to post-treatment in a clinical sample of adolescents
Waitlist-controlled RCT	Ehrenreich-May et al. [32] Queen et al. [33]	UP-A vs. waitlist	Compared to the waitlist condition, those who were treated with the UP-A showed greater improvements in anxiety, depression, and overall global severity at post-treatment and at the 6-month follow-up
Open trial	Ehrenreich-May and Bilek [34]	UP-C	The UP-C conferred improvements in symptoms of anxiety, depression, and related disorders from pre- to post-treatment
RCT	Kennedy et al. [35]	UP-C vs. anxiety-focused CBT	Participants in both conditions showed significant and equal changes in anxiety symptoms Parents in the UP-C condition reported lower depression symptoms than the anxiety-focused CBT condition The UP-C condition conferred greater improvements in sadness dysregulation and cognitive reappraisal

The version of the UP for children under 13 (UP-C) is published as a 15-session group treatment, with separate child and parent group curriculum available. Most commonly, the parent and child components are administered in concert with one another. Group sizes typically range from five to seven children with one or both parents participating, though this can vary depending on clinician availability and group members. The UP-C is also frequently modified to be delivered as a more flexibly administered individual treatment [45]. Further guidelines for individual treatment are described in the Therapist Guide [1].

Table 2
UP-A and UP-C content overview

Module	UP-A module title	Session count	Content	Corresponding UP-C session	Parent content for UP-C session
1.	Building and Keeping Motivation	1 or 2	Build rapport Set Top Problems and goals. Determine what motivates the adolescent to change	Session 1: Intro to UP-C (C)	Introduce parents to treatment structure and CLUES skills Introduce the three-component model of emotions
2.	Getting to Know Your Emotions and Behaviors	2 or 3	Teach psychoeducation about emotions, including the purpose of emotions and the three parts of emotions Introduce the cycle of avoidance and other emotional behaviors.	Session 2: Getting to Know Your Emotions (C)	Discuss the cycle of emotional behaviors. Introduce parents to Double Before, During, and After Introduce "emotional parenting behaviors" and their "opposite parenting behaviors" Discuss positive reinforcement as the opposite parenting behavior for criticism
3.	Introduction to Emotion-Focused Behavioral Experiments	1 or 2	Introduce opposite action and emotion-focused behavioral experiments Teach the adolescent how to track their emotions and activity levels Engage the adolescent in emotion-focused behavioral experiments for sadness (and other emotions)	Session 3: Using Science Experiments to Change our Emotions and Behavior (C)	Introduce science experiments Discuss how parents can support their children in completing science experiments Discuss ways to reinforce children
4.	Awareness of Physical Sensations	1 or 2	Review the connection between physical feelings and strong emotions Develop adolescent's awareness of their physical feelings	Session 4: Our Body Clues (C)	Introduce somatization Teach parents bodyscanning Introduce sensational exposures and practice as a group Teach parents how to express empathy Conduct sensational exposures to help them tolerate uncomfortable physical feelings

(continued)

Table 2
(continued)

Module	UP-A module title	Session count	Content	Corresponding UP-C session	Parent content for UP-C session
5.	Being Flexible in Your Thinking	2 or 3	Develop adolescent's ability to think flexibly about emotional situations Introduce thinking traps Link thoughts to actions through Detective Thinking and Problem-Solving skills	Sessions 5, 6, and 7: Look at my Thoughts (L), Use Detective Thinking (U), Problem Solving and Conflict Management (U)	Introduce cognitive flexibility. Introduce the four common "thinking traps" Discuss inconsistency and its opposite parenting behaviors consistent reinforcement and discipline Introduce and practice Detective Thinking Introduce emotional parenting behavior of overcontrol/overprotection and its opposite parenting behavior, healthy independence granting Introduce and practice Problem Solving for interpersonal conflicts Discuss reassurance-seeking and accommodation
6.	Awareness of Emotional Experiences	1 or 2	Introduce and practice present-moment awareness and nonjudgmentalawareness Conduct generalized emotion exposures by asking the adolescent to practice awareness skills when exposed to emotional triggers	Session 8: Awareness of Emotion Exposure (E)	Discuss importance of learning to experience emotions instead of avoiding them Introduce and practice present-moment awareness and nonjudgmental awareness Begin completing the *Emotional Behavior Form*

7.	Situational Emotion Exposure	2+	Review skills learned so far Discuss rationale for situational emotion exposures, or "behavioral experiments"	Sessions 9–14: Introduction to Emotion Exposure (E), Experience our Emotions	Introduce parents to situational emotion exposures Explain parents' role in practicing exposures at home Introduce parents to the emotional parenting behavior of excessive modeling Conduct situational emotion exposures in session and assign exposures for home learning Parts 1 and 2 (E) of intense emotions and avoidance, and its opposite parenting behavior, healthy emotional modeling Continue to develop the *Emotional Behaviors Form* in preparation for upcoming exposures Review situational emotion exposures and discuss application to different symptoms Introduce and discuss safety behaviors Explain how parents can support exposures at home Introduce the *Emotion Ladder* for exposures and assist parents in finalizing *Emotional Behavior Form*
8.	Reviewing Accomplishments and Looking Ahead	1	Review skills and progress toward goals Create relapse prevention plan	Session 15: Wrap up and Relapse Prevention (S)	Review Emotion Detective skills and "opposite parenting behaviors" Discuss and celebrate children's progress Create a plan for sustaining and furthering progress Help parents distinguish lapses from relapses

(continued)

Table 2
(continued)

Module	UP-A module title	Session count	Content	Corresponding UP-C session	Parent content for UP-C session
P.	Parenting the Emotional Adolescent	1–3	Build parents' awareness of their responses to their adolescent's distress Introduce emotional parenting behaviors and their opposite parenting behaviors	NA	NA

Table 3
Overview of UP-C/A weekly sessions

UP-C/A weekly session format
1. Client and parent will give their weekly ratings of Top Problems
2. Address any pressing issues from the past week and/or engage in some conversation to build rapport
3. Review the home learning assignment to reinforce the previously learned skill
4. Introduce the new skills of current module/session
5. Practice the new skill first with a neutral example
6. Practice the new skill in the context of the client's personal emotional experience
7. Assign a home learning assignment

The UP-C is delivered in the context of an "Emotion Detectives" metaphor. Through the course of the intervention, children in UP-C learn CLUES Skills (Consider How I feel, Look at my Thoughts, Use detective thinking & problem solving, Experience my feelings, and Staying healthy and happy) to help "solve the mystery of their (strong) emotions." When administered similarly to how it has been implemented in clinical trials, both UP-C and UP-A sessions follow the same basic format, outlined in Table 3.

The UP-C/A recognize that parenting behaviors and practices also (often inadvertently) reinforce youths' emotional disorder symptoms, and these parenting behaviors are therefore targeted directly in treatment. The Unified Protocols hold that most parents and caregivers interacting with youth exhibiting strong emotions will normatively act to lessen or minimize the experience of that strong emotion [46]. These actions often arise out of adaptive parenting practices designed to soothe a distressed child. However, in the case of chronically distressed youth, these parenting behaviors may backfire, potentially furthering the notion that situations and experiences associated with strong emotions should be avoided, are frustrating to others or may be dangerous in some capacity. With this in mind, four "emotional" parenting behaviors addressed in the manuals are: excessive criticism, overcontrol/overprotection, modeling of avoidance, and inconsistent behavior patterns. These behaviors are then countered using a series of "opposite" parenting behaviors that can be employed by parents. In the UP-C, there is parent-directed content specified for each of its 15 group sessions. In the UP-A, these materials are more flexibly applied as needed. In Session 1 of the UP-A, the teen and clinician should discuss their parents' involvement, and the most practical and helpful use of Module P with parents. Suggested guidelines for use of Module-P materials in the UP-A manual are outlined in the Therapist Guide [1].

3   Intervention

The UP-C and UP-A share two halves of the same published therapist guide [1], with the UP-C and UP-A each having separate published workbooks [2, 3]. The UP-C workbook contains sections directed at the child client and parent/caregiver separately, whereas the UP-A workbook is directed almost solely at the adolescent client. A series of "Module Summary Forms" are available as parent-directed handouts for the UP-A at the end of each corresponding chapter in the UP-A portion of the therapist guide. Additional materials that may be helpful in administering the UP-C groups include a roll of butcher paper, markers, small plastic containers to make "Clues Kits," and reinforcement items such as stickers, small prizes, and tokens.

In the section to follow, each of the eight core modules of the UP-A are described first, followed by the corresponding UP-C and parent session content. A summary of the UP-A modules, UP-C sessions and their child and parent-directed content is also found in Table 2. Importantly, as you review these materials, please keep in mind that the UP-C/A is meant to be used with a good deal of flexibility and attendance to the individual personalized needs of each youth and their family. The modules themselves are presented in a recommended order, but there are many instances in which the clinician may wish to move through the modules or sessions with greater (or lesser) rapidity, skip forward to, for example, more saliently exposure-focused content, or eliminate an activity that appears less relevant to a particular client. These types of personalization decisions are encouraged in the UP-C/A. However, given all the potential options, personalization decisions may be easiest to implement following a more consistent practice with the materials in the order and amounts initially specified in the therapist guide.

3.1   Module 1: Building and Keeping Motivation

The goal of Module 1 is to orient the adolescent to the treatment concepts and structure of each session, including how their parent (s) will be involved. We recommend beginning the session together to introduce both adolescent and parent to what the UP-A is, how what is learned and practiced in therapy will be applied in real life, the length of treatment, and the use of the workbook. Then the parent can be excused and the clinician can address the goal of building rapport with their new adolescent client. For a more reluctant adolescent, more time needs to be spent on rapport building or personal perspectives on treatment needs to create a safe space and get to know them. Several directions for client-centered discussion consistent with principles of motivational enhancement are provided in the therapist guide. If appropriate, you can spend less time on this goal and move straight into eliciting the adolescent's experiences with strong emotions and their impression of their main or top problems at present. At minimum, three **Top Problems** should be determined [47]. Because these problems are subsequently rated weekly by the youth and parent as a personalized change measure, it is often useful to obtain the most precise grouping of top problems and provide feedback on those that are more or less likely to change during the course of this particular intervention. Some examples of Top Problems consistent with the UP include isolating self when upset, feeling like there is nothing he/she enjoys, engaging in specific compulsions, and avoiding interacting with others. If time permits, a **SMART goal** may be determined for each Top Problem. SMART goals are specific, measurable, attainable, relevant, and time-bound. Like Top Problems, SMART goals should be as behaviorally focused as possible, so as to create clear indicators of change over time in treatment.

If there is time, or if adolescent is showing low motivation for change, identifying initial steps to achieve the adolescent's SMART goals to help build commitment to treatment and self-efficacy for change may be useful. Before the end of the first session, the parent should be brought in to help finalize and rate Top Problems, as well. Parents' motivation for treatment should also be addressed if this is perceived to be a potential issue for the youth's treatment. By doing so, potential barriers to treatment attendance and engagement can be identified and discussed, as needed. The therapist may utilize problem-solving techniques from Module 5 to generate solutions to barriers.

Session 1 of the UP-C aims to be fun and enjoyable to reinforce children's attendance and participation. Through introducing the theme of Emotion Detectives and doing hands-on activities, including some crafting, a positive atmosphere is formed. Clinicians should begin with an icebreaker to introduce families to each other. Next, as parent and child dyads, clinicians should establish Top Problems and related SMART Goals. After separating, the child group is introduced to the Emotion Detectives metaphor/theme of the treatment and then "Detective Rules," including a review of confidentiality, are established. Children make a CLUES Kit using small plastic containers, stickers, letters, and other crafts, in which they will store tokens earned through good behavior and puzzle pieces earned for completing home-learning assignments. Tokens may be exchanged for a small prize of the child's choosing, while puzzle pieces are collected in Session 15 to allow children to identify a group prize for their end of treatment celebration (this is typically a food item like cookies or pizza, as appropriate for each group). Children are assigned the *My Emotions at Home* worksheet to begin working on identifying emotions [3].

In the first session, parents are encouraged to introduce themselves and explain why they are seeking treatment for their child. Parents often enjoy sharing and finding commonalities among their parenting experiences; however, it is most beneficial to keep this part of the group brief and structured on a weekly basis so that time is sufficient to deliver group content. In this first meeting, parents are told about the structure of group, including recommended attendance, their role as "coaches" to practice skills at home, and emphasis on parenting strategies. They also learn about the CLUES skills and the three parts of an emotion—thoughts, body feelings, and behaviors. Parents are introduced to the concept of the cycle of avoidance, including safety behaviors, and how angry behaviors can also be reinforced through negative reinforcement patterns. To apply this content, parents are assigned a worksheet requiring them to identify the trigger and three parts of an emotional experience their child has that during the week.

3.2   Module 2: Getting to Know Your Emotions and Behaviors

Module 2 broadly consists of education about emotions and aids the adolescent in better labeling and identifying the components of their emotional experience with this new knowledge. Thus, the goals of Module 2 are for the adolescent to begin learning emotion identification skills, receive information on the functions of their emotions, learn how emotions impact behavior, discuss how reinforcement principles maintain certain less adaptive or helpful behaviors, and learn how to break down an emotional experience using a functional assessment tool that references the antecedents, responses, and short- vs. long-term consequences of actions that occur during emotional experiences (i.e., the "*Before, During, and After*" [*B/D/A*]; [1–3]). To learn emotion identification skills, the therapist introduces the teen to the concept of an **emotional behavior**, or an emotionally driven behavior, which is any behavior that is motivated by an emotional experience—some positive like smiling because one is happy, some necessary and appropriate like reaching out to a teacher because of bullying, and some maladaptive, the focus of the UP-C/A, like avoiding a party due to fear of interacting with peers. The adolescent also learns that emotions are comprised of three parts: the feelings in their body, their actions, and their cognitions. Using the *Emotion Twister*, the therapist explains that these three parts of an emotional experience whirl around together, building up quickly and intensely, blurring together making it difficult to figure out how to deal with the emotional experience. The adolescent is prompted to describe an emotional experience as fully as possible and then attempts to break down the experience. When introducing the concept of reinforcement as a maintenance mechanism, it is important to consider that some of the adolescent's avoidance or other maladaptive behaviors may be subtle. Sometimes these are safety behaviors like carrying an object with them, like their cell phone, and reassurance seeking when worrying.

Applying the concept of negative reinforcement, the adolescent learns how their emotional behaviors result in short-term relief of their strong emotions, but may end up intensifying emotional experiences over the longer-term. More specifically, avoiding situations that provoke strong emotions lessens the opportunities for positive experiences, adaptive learning, and effective problem-solving in emotional situations. The adolescent is then stuck in this **cycle of avoidance**, which maintains their symptoms. This discussion can be adapted to refer to aggressive, compulsive or more overtly withdrawal-oriented behaviors, as needed. Importantly, the clinician wants to maintain a positive attitude about this discussion. Awareness of and labeling one's emotional experience are key to motivating more helpful behavioral choices. This is only the start of this discussion with the adolescent.

This is often a good time to bring parents into the discussion of emotional behaviors by introducing the concept of the "*Double B/D/A*" [1]. The *Double B/D/A* is an exercise in functional assessment to examine how parental responses to their adolescent's distress moderates their child's emotional experience. The exercise uses the same monitoring tool their child is being asked to use weekly but is applied to both their child's emotional behaviors and any emotional behaviors that a parent provides in response to their child. Parents can be introduced to the four emotional parenting behaviors and each can be discussed, along with an opposite parenting behavior, as needed. However, given time constraints, the therapist is encouraged to consider which of the four parenting behaviors might be most relevant to the adolescent's current functioning and social environment to better focus this discussion in subsequent sessions.

In Session 2 of the UP-C, Getting to Know Your Emotions, the same concepts as in UP-A Module 2 are introduced in a developmentally appropriate manner. Children use the "Emotion Thermometer" to take the "temperature" of their emotions in varying situations. The therapist uses the activity to facilitate discussion about normative emotional experiences and differences and similarities of the group's emotional reactions. Children play the "Alarm Game" to illustrate the idea that although emotions are nature's way of encouraging us to take action because something dangerous or harmful may be happening, for some kids this "alarm" goes off even when no danger is present, or harm is unlikely. Children can better differentiate true and false alarms by identifying the three parts of emotional experiences, and their emotional behaviors. Further, the idea of short- and long-term consequences is explained. To further elucidate this idea, children are assigned the *B/D/A* worksheet for home learning [3]. From this point on, they will be responsible for tracking the before (trigger), during (thoughts, body clues, behaviors), and after (short- and long-term consequences) of an emotional experience every week for the remainder of treatment.

UP-C Session 2 parent group addresses emotional parenting behaviors in a supportive way that normalizes parents' struggles. Parents learn about the *Double B/D/A* and independently work to expand their *C for Parents* home learning assignment [3] to empathetically point out how certain "emotional parenting" responses may unintentionally reinforce less adaptive child and parent behaviors and encourage observation of this interaction. Clinicians then facilitate a discussion about four particular emotional parenting behaviors that may reinforce less helpful child responding to emotions (criticism, overprotection/overcontrol, inconsistent reinforcement and discipline, and modeling of emotional behaviors). In this session, clinicians focus initially on criticism specifically, and one of its **opposite parenting behaviors**: positive reinforcement. Parents learn about the importance of using neutral, descriptive language and focusing on things that are going well in light of their children's propensity for sensitivity to criticism. Parents are encouraged to generate simple reinforcement ideas, such as smiling, nodding, and verbal praise, and small rewards. For home learning, clinicians ask parents to complete at least one *Double B/D/A* for a time their child evidenced an emotional behavior that week. Session ends by rejoining with the children and completing the *Rewards List* [3]. Importantly, some parents may be suspect of the idea of encouraging new, more adaptive behaviors using positive reinforcers. This could be due to potential expenses or concerns about whether behaviors that have some natural reinforcement capacity (talking to peers, doing schoolwork, etc.) should be motivated by extrinsic rewards. In navigating this discussion, it is important to relay that rewards in such situations are meant to "kick start" behaviors that are functionally important to healthy adaptation and can be faded over time once such behaviors occur with greater consistency.

3.3 Module 3: Introduction to Emotion-Focused Behavioral Experiments

The primary focus of Module 3 is to introduce and reinforce "opposite action" or **acting opposite**, which is explained as doing the opposite of (or something different than) what the teen's emotions seemingly want them to do, in situations where current actions are unhelpful or maladaptive to long-term adaptation. This is practiced in the context of emotion-focused behavioral experiments, to test out beliefs about the outcomes of changing one's actions when experiencing a strong emotion. The concept of conducting such experiments and observing their outcome is an essential metaphor throughout the UP-C/A. It can be illustrated in this module through a relatively straightforward application of behavioral activation, although other experiments related to exposure, exposure with response prevention, or opposite actions for anger and aggression may be undertaken as alternative options. In terms of activation, the therapist will initially help identify activities the adolescent enjoys, explain the idea of tracking their emotions and activity levels, and engaging in an "experiment" to see if the adolescent's mood, in fact, changes with increases in personally valued and/or functionally impactful activity.

For some adolescents, behavioral activation will be the primary opposite action engaged as you continue moving onward through the modules. Such behavioral experiments for sadness/withdrawal may need to be modified to include more basic vocational behaviors, focus on other emotions, and/or account for barriers in the environment, as time progresses.

In Session 3 of the UP-C, Using Science Experiments to Change our Emotions and Behavior, the therapist(s) introduces acting opposite and conducting *science experiments* for emotional behaviors. To illustrate the concept, *Nina's Mood and Activity Diary* is discussed and children set up their own experiments. In the parent group, they are similarly introduced to the concepts and how to support their child in acting opposite. Additionally, ways to reinforce a child during or after conduct of a science experiment are offered. Together, the information is reviewed, and home learning is assigned. Through the week, children will complete *My Emotion and Activity Diary* and parents will complete the *Double B/D/A* and implement their reinforcement plan [3].

As noted above, Module 3 may offer important flexibility for clients necessitating earlier introduction of exposure work, such as children with OCD, challenging exposure content (e.g., disgust, irritability) or those who may need more shaping and time with exposures, such as children with ASD. Many therapists use the "opposite action" and "science experiment" frameworks as a jumping off point for the introduction and implementation of exposures prior to Module 7/Session 9.

3.4 Module 4: Awareness of Physical Sensations

Module 4 broadly focuses on helping the client become more aware of their physiological sensations and how feelings in their body influence their emotional experiences. The goals of this module are to teach the adolescent about the concept of physiological sensations and their relationship to intense emotions, work with them to identify and reinterpret their feelings during emotional experiences using a body scanning exercise, and finally to use interoceptive exposure to help normalize and increase awareness of these sensations.

By becoming more aware of physical sensations that occur during emotional experiences, teens can consider how to best respond to these "body feelings" (as described in Module 2's B/D/A). Because the sensations can feel very intense, make us feel uncomfortable, and try to prompt us to do something to make the sensations go away quickly, it is especially important to understand such sensations and learn to better tolerate them to more effectively manage them. Body scanning is a key mindful awareness technique that helps the adolescent achieve this tolerance through practice in session and over time. The goal of body scanning is for the adolescent to notice the sensations, say something about them, and to experience them fully [48], allowing themselves to identify any distracting cognitions that occur about sensations in their body or otherwise and return to the present-moment by anchoring back in the sensations they are experiencing at present. By continuously practicing attentiveness to their bodily sensations, they can "watch" as the intensity changes and naturally reduces over time. Adolescents should practice rating the intensity of sensations, verbally or mentally describing the sensations, and intentionally staying in the present-moment until the feelings are manageable. For some teens, especially those who experience panic attacks, high levels of anxiety sensitivity, or unwarranted hyperarousal, it may be helpful to first objectively review the fight-or-flight response and the bodily symptoms that occur or the rationale for exposure (see Module 7) more so in presenting sensational exposure and body scanning skills.

Similar to individual adolescent clients, in UP-C Session 4, children learn how to recognize "body clues" and how they relate to their emotions. Children learn the idea of *Becoming a Body Detective* and engage in a body drawing and body scanning [3]. The final goal of this session is for children to practice at least three interoceptive (or as they are referred to in the UP-C/A, *sensational*) exposures. Examples are provided in the Therapist Guide [1]. By enduring the exposure until stated tolerance of any discomfort, children can see for themselves that the feelings, which are often associated with strong emotions, will lessen or become more tolerable over time. In some cases, the clinician would likely use a larger number of sensational exposures with their youth clients. For children with high anxiety sensitivity or very reactive symptoms to certain body signals (e.g., cases of emetophobia or Avoidant Restrictive Food Intake Disorder) it is especially important to comprehensively illustrate the relationship between thoughts and body clues and reinforce the use of body scanning and mindful awareness as a means to decatastrophize any negative interpretations of these experiences.

In this session, parents are familiarized with somatic responses to emotional experiences and help to detect these in their children. Parents also learn the concept of body scanning, how to assist their children in using the skill, and when it might be helpful. To convey the purpose and encourage empathy for uncomfortable feelings, parents will also engage in a sensational exposure. Finally, the importance of expressing empathy is discussed and practiced with a worksheet [3]. At the end of Session 4, children earn their C badge for their CLUES Kit.

3.5 Module 5: Being Flexible in Your Thinking

Module 5 includes several traditional cognitive and behavioral strategies; however, there is stronger emphasis on utilizing cognitive reappraisal in the anticipation of an emotion reaction or difficult situation (**antecedent cognitive reappraisal**), so to preemptively address the difficulty of engaging in this technique in the midst of an intensely emotional situation, when it may be less effective or unhelpfully distracting. Additionally, the problem-solving techniques in Module 5 are designed to increase flexibility when thinking about solutions in or following emotionally intense situations. Thus, whether referring to reappraisal or problem-solving, the overarching idea of Module 5 is to increase **flexible thinking**. The clinician discusses with the adolescent that people may make automatic, emotionally relevant interpretations of stimuli or situations around or within them. To illustrate this concept, the clinician can first use one of a number of possible optical illusions. The clinician introduces the adolescent to an optical illusion image and invites them to just tell the clinician what they see first, which may include any number of interpretations. Discuss with the adolescent that there may be several possible interpretations for the pictures, there can be many right answers, and for whatever reason, people will see different things. Typically, the first idea that comes to mind is the one that is most familiar or requires the least mental effort to identify; whereas, additional interpretations may require more time, assistance or practice to identify. The main idea to convey through this example is that there may also be more than one interpretation of situations causing strong emotions around or within us. By only considering one possible explanation for a situation or for our emotions, one may ignore other possibilities, which may negatively influence an emotional experience.

This experience is normalized by discussing "thinking traps," classically known as cognitive distortions, which may lead youth to more typically identify negative or threat-related cognitions before more neutral or positive interpretations. Common thinking traps are introduced: Jumping to Conclusions (Probability Overestimation), Thinking the Worst (Catastrophizing), and Ignoring the Positive, although others may be discussed. Once the adolescent is more adept at identifying thinking traps, the therapist can explain how the adolescent can use **Detective Thinking**, or cognitive reappraisal, to be more flexible in these interpretations. The steps of Detective Thinking include relatively traditional cognitive restructuring techniques to encourage looking for clues to see if one's thoughts are realistic and what other interpretations may be possible. Steps of Detective Thinking are summarized below:

3.5.1 Identify the Interpretation

Use one of the adolescent's own examples of a time their automatic interpretation probably was not accurate or realistic and then see if they can identify the trap. For example, you can discuss the example of a thought like, "If I don't get straight A's on my report card, I'll never get into a good college." In this case, the thinking traps might be Thinking the Worst (assigning chances of something bad happening are greater than they actually are) and Ignoring the Positive (by implying that one could not cope well with such an outcome).

3.5.2 *Evaluate the Evidence*

This includes using dispute handles (or "detective questioning") to clarify the accuracy and likelihood of one's automatic interpretation, as well as the adolescent's ability to cope with an intense emotional situation. Have the adolescent answer at least two of the questions on the *Being a Detective—Steps for Detective Thinking* worksheet in the UP-A workbook to accomplish this [2]. Some examples include (but are not limited to) the following:

- Am I 100% sure that ____ will happen?
- Can I really tell what someone else is thinking?
- How many times has ____ happened in the past?

After completing Detective Thinking steps, identify the most realistic outcome(s)—even if this includes the "thinking trap" or what may appear to be a negative interpretation—and how they would cope if outcome were to occur. Explain to the adolescent why Detective Thinking is most useful before entering situations and remind them of this as needed. Remind the youth that this skill is about *realistic thinking*, not just thinking more positively, as needed.

The remainder of Module 5 focuses on **Problem Solving**. This skill can be used to help the adolescent get out of traps where they may feel "stuck" or already enacted an unhelpful emotional behavior. With adolescents, it can be helpful to apply this skill to interpersonal conflicts, but problem solving can be useful anytime in treatment for determining helpful behavioral solutions, including more crisis-oriented situations. The steps of problem solving in the UP-A are summarized in Table 4.

The UP-C/A Therapist Guide provides an example of a possible script to practice this skill in a neutral case and a case where the adolescent is experiencing a strong emotion [1]. For home learning, the teen is asked to complete the *B/D/A* and Worksheet 5.4 *Getting Unstuck—Steps for Solving Problem* (*see* Fig. 1).

The concepts covered in Module 5 of the UP-A are presented over the course of three sessions in UP-C: Sessions 5—Look at My Thoughts, Session 6—Use Detective Thinking, and Session 7—Problem Solving and Conflict Management. Similar to the UP-A, the concept of flexible thinking is first introduced with optical illusions, followed by some non-personal examples, and then younger clients can begin to challenge their **snap judgments**, or automatic interpretations. Thinking traps are then presented as illustrated characters in the workbook: Mind reading or *Psychic Suki*, jumping to conclusions or *Jumping Jack*, thinking the worst or *Disaster Darrell*, and ignoring the positive or *Negative Nina*. To practice recognizing the thinking traps, children can complete the

Table 4
Steps of problem solving

Steps of problem solving
1. Define the problem in the simplest and most straightforward terms possible
2. Identify some solutions, as many as possible, making sure not to judge any options too soon
3. List the good and bad about each solution (at least one good and one bad for each)
4. Pick one to try, with a set date or time specified
5. Evaluate how the solution worked
6. If it did not work, try a second or go back to step one (redefine problem or think of other solutions)

Worksheet 5.4: Getting Unstuck—Steps for Solving a Problem

When trying to figure out how to solve a problem there is a set of steps we can use every time. In fact, these may be steps that you are sometimes using without even realizing it.

The first step is simple, **Define the Problem**. Be careful; the way you define the problem will influence the solutions you arrive at. Try to keep the problem as simple as possible.

What is the problem you are trying to solve?

Now, try to determine all the possible solutions or all the things you could do to solve the problem. Remember not to judge your options right now, just list out as many options as you can.

What are all the possible things you could do in this situation?

1. _____
2. _____
3. _____
4. _____
5. _____

	What are the **good** things about each solution?	What are the **bad** things about each solution?
Solution 1.		
Solution 2.		
Solution 3.		
Solution 4.		
Solution 5.		

Now, circle the solution you think is the best one and try it out!

Pick a specific time that you plan to try out your solution:

If the solution you choose does not work, go through the process again. With the information you now have, reevaluate your options and pick another solution to try.

Now that you have listed out some possible solutions, let us start to think about what is good and what is bad about each of the options. What are the likely outcomes of each solution?

Using the chart on the next page, write down the good and bad things about each of the solutions you listed above in the appropriate column.

Fig. 1 The adolescent should practice the Problem Solving steps from Module 5 at home using this worksheet (Adapted from *Unified Protocol for Transdiagnostic Treatment of Emotional Disorders in Adolescents: Workbook* (pp. 64–65), by J. Ehrenreich-May, S.M. Kennedy, J.A. Sherman, S.M. Bennett, and D.H. Barlow, 2018, New York: Oxford University Press. Copyright 2018 by Oxford University Press. Reprinted with permission form)

Match the Thought to the Trap! worksheet [3]. In parent group, parents are familiarized with cognitive flexibility so that they can aid their children in recognizing the Thinking Traps. They also discuss the parenting behavior of inconsistency, along with consistent reinforcement and discipline techniques. Parents learn the basics about positive and negative reinforcement and positive and negative punishment. With this knowledge, the therapist can help parents become more consistent in appropriately praising and punishing their child through consistent behavior management, even in the case of more intense emotional displays. Page 129 of the UP-C Workbook supplies guidelines for such a system ([3]; *Guidelines for Creating an Effective Behavior Management System at Home*). Parents' home learning assignment will be to implement this system and complete Session 5 worksheets [3].

Session 6 of the UP-C focuses on Detective Thinking. Children are encouraged to participate in a *Mystery Game* in which they attempt to solve a non-emotional mystery. Throughout the game, children can learn that sometimes they must look for clues to figure out what is going on around them, that there may be more than one right answer, and that they can discover clues to solve such mysteries (alone or in a team) in many different ways. Using the Worksheet *U for Children* [3], the group then learns the Detective Thinking steps to get out of thinking traps—what are referred to as the "Stop, Slow, Go steps" [3]. Then, after going through an example, the group can use their new skill to solve the mysteries in the *Detective Thinking Practice* worksheet [3]. They will practice Detective Thinking for their own experiences at home using the *U for Children at Home* worksheet [3]. In this session, parents will complete Detective Thinking steps using an example from their own lives, followed by a child-focused example. The parenting behavior of overcontrol/overprotection is discussed, along with healthy independence granting and shaping of new behaviors through reinforcement, planned ignoring, and differential reinforcement. By reviewing the origins and consequences of this parenting behavior and how they can engage in opposite parenting behaviors, parents can help their child more independently cope with distress and build self-efficacy.

Session 7 of the UP-C focuses on Problem Solving, which mirrors the adolescent version of the skill. By the end of Session 7, children will also earn their L and U badges. For home learning, children complete the *BDA* and the *Problem Solving at Home with Others* worksheets [3]. In the parent portion of the group, parents are introduced to problem solving as another skill to develop cognitive flexibility. They discuss applications of Problem Solving in their own lives, specifically to resolve interpersonal conflicts. Finally, the group reviews parents' efforts at healthy independence granting using the *Encouraging Independent Behaviors* worksheet [3], and connects this opposite parenting behavior to the skills learned in L and U. Parents are aided in viewing Detective Thinking and Problem Solving as alternatives to reassurance-seeking and the provision of accommodation. Parents have learned how providing reassurance, either explicitly or less obviously, is a form of overprotection that while quick and easy, may also facilitate children's dependence on their parents to evaluate emotional situations. Because all parents want to protect their children from distress, parents often try to tell their child which solution will work best to help their child avoid choosing the less helpful or less efficient solution. As a result, children do not get as much practice problem solving, become more inflexible, and may experience feelings of low self-efficacy. Clinicians point out the lay idea that making mistakes often helps children learn best and that such mistakes may not be catastrophic, but rather essential for growth and adaptive development. The group explores what **shaping** would look like when helping their children use these skills, and suggestions for simple substitutes for reassurance (i.e., "What do you think?") or prompts to use their skills are provided [1]. For home learning, parents are assigned *Shaping Detective Thinking and Problem Solving at Home* [3].

3.6 Module 6: Awareness of Emotional Experiences

Module 6 builds on and enhances the behavioral and cognitive skills learned and practiced thus far by incorporating present-moment awareness and nonjudgmental awareness during emotional experiences. For some teens, these skills can be thought of as an opposite action to suppression and rumination that may occur during emotionally intense scenarios. Clinicians should provide adolescents with the rationale for learning these awareness skills. They can explain that in contrast to Module 5's cognitive techniques, present-moment awareness can be practiced before, during, or after an emotional experience or situation to clarify and deepen the experience itself (as opposed to escape from or avoid it). Clinicians may further introduce the concept by saying:

> Present-moment awareness means we are fully engaged in the 'here and now'. We are not thinking about the past or the future (although thoughts about these may pass through your mind), but about what is happening in the moment. We are focusing on one thing at a time and letting go of distractions or distracting thoughts. I'd like to suggest that it is okay to just notice and say something to yourself about what you are experiencing and leave it at that, so you can participate in the present moment. In the end, this may help you slow down the emotion twisters you experience, which will help you better manage your emotions in the future [1].

Clinicians should describe a situation where this skill would make sense for their client to use, individualizing and relating this concept to the client's specific experience. Overall, adolescents should understand that practicing present-moment awareness will help them slow down and better utilize their skills to approach problems and manage strong emotions. Similar to the steps of body scanning, adolescents will also use the *Notice it, Say something about it, and Experience it* worksheet as needed for this exercise [48]. *See* Fig. 2.

Worksheet 6.1: Notice It, Say Something About It, Experience It

Noticing it, saying something about it, and experiencing it can help you practice present-moment awareness.

Notice it:

What did you notice? Name the object, food, location, person, event, or emotion that you observed using present-moment awareness:

Say something about it:

Describe what you noticed in as much detail as you can (e.g., what colors, textures, tastes, temperatures, sensations, smells, and/or people do you notice?). Remember not to make any judgments or interpretations, just describe what is there:

Experience it:

How are you staying in the present moment?

Remember: When you notice a distracting thought or judgment, gently bring yourself back to the present moment.

Fig. 2 The adolescent can use this worksheet to facilitate the practice present-moment awareness (Adapted from *Unified Protocol for Transdiagnostic Treatment of Emotional Disorders in Adolescents: Workbook* (p. 73), by J. Ehrenreich-May, S.M. Kennedy, J.A. Sherman, S.M. Bennett, and D.H. Barlow, 2018, New York: Oxford University Press. Copyright 2018 by Oxford University Press. Reprinted with permission form)

Adolescents then put this skill to use in session by practicing at least one present-moment awareness exercise. Some suggestions include General Breathing Awareness, Exploring a Candy Exercise, Mindful Walking, Play-Doh Exercise, and Guessing the Flavor or Scent. *See* Box 1 for the script included in the therapist guide for General Breath Awareness.

Box 1 General Breathing Awareness Exercise

Instruct the adolescent to do the following:

1. Assume a comfortable posture lying on your back or sitting. If you are sitting, keep the spine straight and let your shoulders drop.

2. Close your eyes if it feels comfortable.

3. Bring your attention to your belly, feeling it rise or expand gently on the in-breath and fall or recede on the out-breath.

4. Keep your focus on your breathing, "being with" each in-breath for its full duration and with each out-breath for its full duration, as if you were riding the waves of your own breathing.

5. Every time you notice that your mind has wandered off the breath, notice what it was that took you away and then gently bring your attention back to your belly and the feeling of the breath coming in and out.

6. If your mind wanders away from the breath a thousand times, then your job is simply to bring it back to the breath every time, no matter what it becomes preoccupied with.

7. Practice this exercise for 5–15 min at a convenient time every day, whether you feel like it or not, for one week and see how it feels to incorporate present-moment practice into your life. Be aware of how it feels to spend time each day just being with your breath without having to do anything.

Adapted from *Unified Protocol for Transdiagnostic Treatment of Emotional Disorders in Children and Adolescents: Therapist Guide* (pp. 107–108), by J. Ehrenreich-May, S.M. Kennedy, J.A. Sherman, S.M. Bennett, and D.H. Barlow, 2018, New York: Oxford University Press. Copyright 2018 by Oxford University Press. Reprinted with permission form.

In this presentation of skills, **nonjudgmental awareness** is framed as a more purposefully compassionate and accepting stance on the notion of present-moment awareness. Instead of evaluating one's experiences or thoughts as good, bad, or wrong as they present in the moment, adolescents are encouraged to approach their emotions with empathy and understanding, as they would with a close friend. Instead of being critical or trying to distract one's self from uncomfortable self-judgment, which can make one avoid situations or feel worse, adolescents can learn to pay attention to their emotions, let go of the judgment, reengage in the present moment, and acknowledge and accept the experience for what it is. The three steps (Notice it, Say Something About it, Experience it) in present-moment awareness should be applied here as well. For example, when an adolescent is talking to someone, she might notice that her face gets red, and might think, "I'm so stupid! What's wrong with me? This person will never like me if I don't calm down." Using nonjudgmental awareness, this teen might say to herself, "This is how it is right now (notice it), there go my thoughts again (say something about it)," and then gently bring her attention back to the person and conversation (experience it [in the present moment]). Empathize with the adolescent in the sense that doing this may be harder than it sounds, but it is not impossible and does not need to be done perfectly to be helpful. The more we practice this skill in session and then at home, the more natural and routine it becomes. Lastly, in preparing for Module 7, teens are introduced to the concept of a **generalized emotion exposure** and practicing nonjudgmental and present-moment awareness in context. Teens are encouraged to participate in activities that may induce some broad and general emotional experiences, like listening or watching a sad movie or reading or writing a narrative. The goal in engaging in a generalized emotion exposure is to help the adolescent learn that they do not need to avoid experiencing uncomfortable emotions and to help encourage practice of their Module 6 skills. This practice can be introduced using a sample script [1].

Family media rules, rapport, and clinical severity should be considered when coming up with stimuli needed for conducting general emotion exposures. Clinicians may choose to progress from exposures that induce positive, to mixed, to negative emotions, or with significantly depressed youth only induce pleasant emotions, remembering that there is more time to conduct more personalized exposures in Module 7 to come.

In UP-C Session 8, Awareness of Emotion Exposure, children are also introduced to the concept of being more fully aware of their emotions and taught the steps of present-moment awareness. The therapist can provide examples of how sitting with an emotion, similar to how they sat with uncomfortable feelings in Session 4, can help them learn that they are not dangerous and help stop the cycle of avoidance. After a neutral exercise, the group plays a "Using My Five Senses" game. The therapist and group can choose between options, one being the "Exploring a Candy Exercise" [3, 49]. Parents discuss the importance of learning to experience emotions instead of avoiding them, practice present-moment and nonjudgmental awareness for situations in parents' lives, and begin to prepare for the upcoming exposure therapy activities [3].

3.7 Module 7: Situational Emotion Exposure

Before moving into exposures, skills from previous modules should be reviewed, along with any remaining emotional behaviors and their triggers, as the therapist works with the adolescent and their parent(s) to create the adolescent's list of exposure situations. Module 7 capitalizes on the principles of exposure therapy in order to help the adolescent engage in increasingly uncomfortable situations they will ideally endure, progressively perceive as harmless, and ultimately alter their cognitions and behaviors to allow for continued engagement without emotional behaviors. **Exposure** is included as a treatment component in almost all treatment research studies that show positive treatment effects of CBT for youth anxiety. Further, much of the improvement in anxiety symptoms in CBT treatment takes place after exposures are initiated [50] and exposures are associated with few side effects [51]; therefore, careful consideration should be made in this flexible, modular application in terms of how soon one can conceivably and comfortably utilize Module 7 with their adolescent client.

Exposure or further opposite action exercises can occupy many sessions and may have already been introduced in treatment at this stage. Some adolescents with depression symptoms may have already been practicing exposure through opposite action. For some irritable or anger-prone adolescents, a combination of irritability/frustration exposures, mindful awareness and problem-solving steps may be combined to promote more effective engagement with triggering situations and environments. For youth with OCD, combining exposure with response prevention in this module is emphasized, whereas the narrative form in Module 6 may be used as a starting point for exposure to trauma-related cues that may be expanded as needed for trauma-exposed youth. Finally, for those with social skill deficits, instructions are provided in this module for how to combine some basic social skill building with graduated exposure steps. Across triggers and situations, Module 7 is designed to directly reinforce and realistically engage with extinguishing or tolerating distress associated with situations in which emotional behaviors continue to occur. Clinics may vary in their ability to accommodate varying types of exposures or exposure locations other than the office. Regardless, exposures often require creativity in using existing therapy space when limited to such, and creative use of immediately surrounding areas, if possible, such as retail and food stores, schools or classrooms, along with access to confederates, and props. For these reasons, those new to exposure provision are encouraged to seek support and consultation from supervisors and peers to increase their confidence levels in exposure activities. In addition to review and resources about exposure basics in the UP-C/A, other adjunctive materials [52] may be useful.

Some examples of emotional behaviors to target in UP-A exposures include: avoidance, escape, withdrawal/isolation, bringing along other people, distraction, emotion suppression, worry, rituals, anger outbursts, and physical aggression. Situations and behaviors are listed pragmatically on the *Emotional Behavior Form*. Some clinicians may complete the form hierarchically, with entries listed from lowest degree of emotion elicited to the strongest; however, this not necessary. Most importantly, the clinician should be mindful of where they might start and ask questions that would allow for new safety learning to be easily achieved in exposure (e.g., what would make this harder, easier, etc.).

Exposure may help adolescent learn that nothing truly dangerous or terrible is likely happening when they do not engage in emotional behaviors while in feared situations [53]. There are certainly many theories as to why exposure is effective, including extinction/habituation, learning through increased practice between sessions, disconfirmation of beliefs about potential threats, and inhibitory learning [54–56]. Inhibitory learning principles suggest that exposure creates new memories to teach youth that their previously feared situations are safe. Repeated practice facing feared situations may also lead to habituation, or extinction of distress and anxiety in previously fear-evoking situations. Clinicians can use the *Emotion Curve—Habituation* figure to illustrate habituation-related reductions in distress during exposure to the adolescent, as this rationale may not be readily understood by adolescents [2]. However, clinicians may wish to clarify that it is not always necessary to habituate during an exposure. While traditional models posit that habituation is essential for fear extinction, more recent research shows this may not be valid [57]. An inhibitory learning approach posits that fear tolerance and learning that the situation is safe is central to fear extinction. Habituation is less important in this model, while memory and learning, retrieval, and influences of time and context are key [58–60]. Yet if habituation does not occur during an in-session exposure, the clinician should process with parents, as well as the youth, following exposure as to not alarm parents or bring up concerns about teen's ability to handle the exposure activity.

Before exposures commence, the clinician should obtain general consent for planned activity but avoid excess reassurance provision and focus on labeling of any affect and affirmation of the youth's self-efficacy to engage in exposure. The clinician should also pay close attentive to any use of safety behaviors, such as having a phone, another person, food or drink present during subsequent exposure practices. Dealing with subtle avoidance and/or safety behaviors can be imperative for the efficacy of the exposure practice

[61]. Adolescents should continue to monitor their progress and outcomes of the exposure through use of awareness skills throughout each exposure "experiment." The adolescent can rate their distress before, during, and after exposure using a Subjective Units of Distress (SUDS) scale [62]. It is also important to leave enough time to debrief after the exposure activity. Issues to address are as follows (Box 2):

Box 2 Debriefing After Exposures

- Was there anything that surprised the adolescent? Did what they expect to happen occur?

- Were the adolescent's hypotheses supported?

- Did the adolescent learn anything new about the situation or about their behaviors?

- Did the adolescent's ability to observe the strong emotion experienced improve?

- Was the adolescent's emotion level highest before the exposure started? What level was their emotion at when the exposure was over?

Additionally, after each exposure, you may want to point out any noticeable patterns in how the adolescent's self-ratings changed over the course of the exposure and discuss if the adolescent experienced habituation, referencing the emotion curve to show how in the past their emotion decreased likely at the time they decided to, or took measures to, avoid the feared situation. Discuss how this response has changed since practicing exposure. Encourage the client to verbally express anything new they learned about the feared situation or about their ability to cope with strong emotions if habituation did not occur, understanding that this is not always a necessary precursor to enhanced safety learning.

Home learning should expand upon exposures or other opposition action practice, varying the type and scope. Assign approximately two situational exposure activities or opposite action practices each week between the remaining sessions, and instruct the adolescent to monitor reactions during the experiments using the *Tracking the B/D/A* form [2]. Module P can be important in this stage of treatment in order to plan and conduct exposures. For some adolescents, this may necessitate resources such as materials/props, transportation, supervised activities, and support. If unavailable or limited, exposure situations should be modified.

In UP-C Session 9, skills are reviewed, opposite action experiments are reviewed, and the concept of a new type of science experiment, "exposure," is introduced. The therapist completes a

Form 7.1: Emotional Behavior Form

Use this form to identify and describe situations that cause you to feel strong emotions, as well as the emotional behaviors you use in these situations. Using the Emotions Scale thermometer below, rate how much uncomfortable <u>emotion</u> you experience in each situation. When creating this list, think of behaviors like avoidance, escape, or other undesired actions (like aggression) which you may wish to change during treatment. As time goes on in treatment, you can use the last column (Did you work on it?) to see how much progress you've made on these behaviors over time.

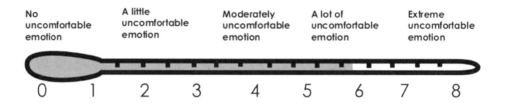

Situation	Emotional Behavior	Emotion (0–8)	Did you work on it? (Y/N)
		▼	▼
		▼	▼
		▼	▼
		▼	▼
		▼	▼
		▼	▼
		▼	▼
		▼	▼
		▼	▼
		▼	▼

Fig. 3 The clinician and adolescent will complete this form decide which emotional behaviors to concentrate on next (Adapted from *Unified Protocol for Transdiagnostic Treatment of Emotional Disorders in Adolescents: Workbook* (p. 87), by J. Ehrenreich-May, S.M. Kennedy, J.A. Sherman, S.M. Bennett, and D.H. Barlow, 2018, New York: Oxford University Press. Copyright 2018 by Oxford University Press. Reprinted with permission form)

demonstration of an exposure [3]. Together, families complete the *Emotional Behavior Form* (*see* Fig. 3). In the parent only portion of the session, parents are provided with thorough psychoeducation of exposure therapy, as it requires their cooperation to implement exposures, reinforce, and help their child continuously engage in this type of "experiment." When the group returns for Session 10, the concept and rationale for exposures is reviewed and the group learns about safety behaviors. They practice an exposure and plans are made for individualized situational emotion exposures. In Sessions 11–14, individualized exposures are conducted. Children then earn their E badge.

3.8 Module 8: Reviewing Accomplishments and Looking Ahead

The end of treatment is a time to celebrate, reinforce the hard work that has been done, ensure the changes endure, and skills continue to be used. The last session of treatment is devoted to such celebration and a review the progress made and skills that have been most useful. Changes in Top Problems are reviewed with parents. Items on the *Emotional Behavior Form* should also be re-rated. The clinician and client should also formulate a plan for the future, which may include ongoing exposures. The difference between lapses and relapses is also instructed, so that clients and their parents can appropriately and adaptively respond to either. In the UP-C, children earn the S badge and use their earned pieces to complete the puzzle. Parents, clinicians and children then acknowledge youth accomplishments in becoming an "Emotion Detectives" with a party and receipt of a certificate acknowledging this achievement.

4   Conclusion

The UP-C/A offer a unique, flexible approach to the transdiagnostic treatment of emotional disorders in youth. They contain both "common elements" of change for these conditions and present such materials in a singular manner that reflects attendance to potential core dysfunction across these conditions and an overarching theory of change. By presenting common change strategies associated with cognitive behavior therapy and third-wave behavior therapy in a flexible manner that allows clinicians to utilize a range of emotion states (fear, anxiety, anger, and sadness in particular) and progress monitoring tools (Top Problems, Emotional Behaviors Form, continued use of the Before/During/After) to help personalize their treatment approach, we believe the UP-C/A offer a comprehensive and elegant set of tools for treating a range of emotional disorders and related conditions. However, there is certainly more work to be done on the horizon. While initial efficacy trials have shown the adult UP, along with their adolescent and child versions, to be efficacious in research settings, effectiveness and implementation efforts have just begun. It is very likely that some refinement of the protocol to aid clinicians in abbreviating or optimally tailoring the approach to various circumstances and settings will likely result from these efforts, as aspects of the UP-C/A may be considered too lengthy, require too much a priori knowledge of cognitive or behavior therapy principles or appear cumbersome or nonspecific to some community clinicians initially approaching them at present. More so, there has been substantial interest in adapting the UP-C/A for use in other settings (pediatric and school settings, for example), and for youth with more severe mental illness, disruptive behavior conditions with strong mood features, borderline disorder features, and eating disorders. However, this work is in its nascent stages and needs to be furthered in order to better understand how well this approach works with other settings and populations. Although largely developed in South Florida, with predominantly Hispanic/Latinx samples, there may be needs to also further linguistically or culturally tailor the intervention or its parenting components, in particular, as time goes on. Nevertheless, the UP-C/A share an important focus on transdiagnostic presentation of skills, a strong parenting component, a fundamental fusion of cognitive behavioral and third-wave behavioral principles to promote change [63] and a flexibly intended set of treatment elements that, at least in research settings, promote significant improvement in emotional disorders among youth.

Acknowledgement

Disclosure statement: Dr. Jill Ehrenreich-May receives royalties from sale of the Unified Protocols for Transdiagnostic Treatment of Emotional Disorders in Children and Adolescents (UP-C and UP-A). She also receives payments for UP-C and UP-A clinical trainings, consultation and implementation support services.

References

1. Ehrenreich-May J, Kennedy SM, Sherman JA, Bennett SM, Barlow DH (2018) Unified protocol for transdiagnostic treatment of emotional disorders in children and adolescents: therapist guide. Oxford University Press, New York, NY
2. Ehrenreich-May J, Kennedy SM, Sherman JA, Bilek EL, Barlow DH (2018) Unified protocol for transdiagnostic treatment of emotional disorders in adolescents: workbook. Oxford University Press, New York, NY
3. Ehrenreich-May J, Kennedy SM, Sherman JA, Bilek EL, Barlow DH (2018) Unified protocol for transdiagnostic treatment of emotional disorders in children: workbook. Oxford University Press, New York, NY
4. Barlow DH, Farchione TJ, Fairholme CP, Ellard KK, Boisseau CL, Allen LB, Ehrenreich-May J (2011) Unified protocol for transdiagnostic treatment of emotional disorders: therapist guide. Oxford University Press, New York, NY
5. Barlow DH, Farchione TJ, Sauer-Zavala S, Murray Latin H, Ellard KK, Bullis JR, Bentley KH, Boettcher HT, Cassiello-Robbins C (2018) Unified protocol for transdiagnostic treatment of emotional disorders: therapist guide (Rev. edn). Oxford University Press, New York, NY
6. Barlow DH, Ellard KK, Sauer-Zavala S, Bullis JR, Carl JR (2014) The origins of neuroticism.

Perspect Psychol Sci 9(5):481–496. https://doi.org/10.1177/1745691614544528

7. Lopez ME, Stoddard JA, Noorollah A, Zerbi G, Payne LA, Hitchcock CA et al (2015) Examining the efficacy of the unified protocol for transdiagnostic treatment of emotional disorders in the treatment of individuals with borderline personality disorder. Cogn Behav Pract 22(4):522–533. https://doi.org/10.1016/j.cbpra.2014.06.006

8. Bentley KH (2017) Applying the unified protocol transdiagnostic treatment to nonsuicidal self-injury and co-occurring emotional disorders: a case illustration. J Clin Psychol 73(5):547–558. https://doi.org/10.1002/jclp.22452

9. Thompson-Brenner H, Boswell JF, Espel-Huynh H, Brooks G, Lowe MR (2018) Implementation of transdiagnostic treatment for emotional disorders in residential eating disorder programs: a preliminary pre-post evaluation. Psychother Res 2018:1–17. https://doi.org/10.1080/10503307.2018.1446563

10. Boisseau CL, Boswell JF (2018) The unified protocol for eating disorders. In: Barlow DH, Farchione TJ (eds) Applications of the unified protocol for transdiagnostic treatment of emotional disorder. Oxford University Press, New York, NY, pp 150–163

11. Weintraub MJ, Zinberg J, Bearden CE, Miklowitz DJ (2019) Applying a transdiagnostic unified treatment to adolescents at high risk for serious mental illness: rationale and initial case studies. https://doi.org/10.1016/j.cbpra.2019.07.007

12. Malmberg J, Kennedy SM, Holzman J, Ehrenreich-May J (2020) Development and application of an innovative transdiagnostic treatment approach for pediatric irritability. Behav Ther 51(2):334–349

13. Angold A, Costello EJ, Erkanli A (1999) Comorbidity. J Child Psychol Psychiatry 40:57–87. https://doi.org/10.1111/1469-7610.00424

14. Brady EU, Kendall PC (1992) Comorbidity of anxiety and depression in children and adolescents. Psychol Bull 111(2):244–255. https://doi.org/10.1037/0033-2909.111.2.24

15. Cummings CM, Caporino NE, Kendall PC (2014) Comorbidity of anxiety and depression in children and adolescents: 20 years after. Psychol Bull 140(3):816–845. https://doi.org/10.1037/a0034733

16. Leyfer O, Gallo KP, Cooper-Vince C, Pincus DB (2013) Patterns and predictors of comorbidity of DSM-IV anxiety disorders in a clinical sample of children and adolescents. J Anxiety Disord 27(3):306–311. https://doi.org/10.1016/j.janxdis.2013.01.010

17. Boomsma DI, Van Beijsterveldt CEM, Hudziak JJ (2005) Genetic and environmental influences on anxious/depression during childhood: a study from the Netherlands Twin Register. Genes Brain Behav 4(8):466–481. https://doi.org/10.1111/j.1601-183X.2005.00141.x

18. Eley TC, Bolton D, O'Connor TG, Perrin S, Smith P, Plomin R (2003) A twin study of anxiety-related behaviours in pre-school children. J Child Psychol Psychiatry 44:945–960

19. Middeldorp CM, Cath DC, Van Dyck R, Boomsma DI (2005) The co-morbidity of anxiety and depression in the perspective of genetic epidemiology. A review of twin and family studies. Psychol Med 35(5):611–624. https://doi.org/10.1017/s003329170400412x

20. Wilamowska ZA, Thompson-Hollands J, Fairholme CP, Ellard KK, Farchione TJ, Barlow DH (2010) Conceptual background, development, and preliminary data from the unified protocol for transdiagnostic treatment of emotional disorders. Depress Anxiety 27:882–890. https://doi.org/10.1002/da.20735

21. Martin P, Murray LK, Darnell D, Dorsey S (2018) Transdiagnostic treatment approaches for greater public health impact: Implementing principles of evidence-based mental health interventions. Clin Psychol Sci Pract. https://doi.org/10.1111/cpsp.12270

22. Marchette L, Weisz JR (2017) Practitioner review: empirical evolution of youth psychotherapy toward transdiagnostic approaches. J Child Psychol Psychiatry. https://doi.org/10.1111/jcpp.12747

23. Chorpita BF, Barlow DH (1998) The development of anxiety: the role of control in the early environment. Psychol Bull 124(1):3–21. https://doi.org/10.1037/0033-2909.124.1.3

24. Bullis JR, Sauer-Zavala S, Bentley KH, Thompson-Hollands J, Carl JR, Barlow DH (2015) The unified protocol for transdiagnostic treatment of emotional disorders: preliminary exploration of effectiveness for group delivery. Behav Modif 39(2):295–321. https://doi.org/10.1177/0145445514553094

25. Barlow DH, Farchione TJ, Bullis JR, Gallagher MW, Murray-Latin H, Sauer-Zavala S et al (2017) The unified protocol for transdiagnostic treatment of emotional disorders compared with diagnosis-specific protocols for anxiety disorders: a randomized clinical trial. AMA

Psychiatry.   https://doi.org/10.1001/
jamapsychiatry.2017.2164

26. Conklin LR, Cassiello-Robbins C, Brake CA, Sauer-Zavala S, Farchione TJ, Ciraulo DA, Barlow DH (2015) Relationships among adaptive and maladaptive emotion regulation strategies and psychopathology during the treatment of comorbid anxiety and alcohol use disorders. Behav Res Ther 73:124–130. https://doi.org/10.1016/j.brat.2015.08

27. Farchione TJ, Fairholme CP, Ellard KK, Boisseau CL, Thompson-Hollands J, Carl JR, Barlow DH (2012) Unified protocol for transdiagnostic treatment of emotional disorders: a randomized controlled trial. Behav Ther 43(3):666–678. https://doi.org/10.1016/j.beth.2012.01.001

28. Sauer-Zavala S, Boswell F, Gallagher MW, Bentley KH, Ametaj A, Barlow DH (2012) The role of negative affectivity and negative reactivity to emotions in predicting outcomes in the unified protocol for the transdiagnostic treatment of emotional disorders. Behav Res Ther 50(9):551–557. https://doi.org/10.1016/j.brat.2012.05.005

29. Bullis JR, Fortune MR, Farchione TJ, Barlow DH (2014) A preliminary investigation of the long-term outcome of the unified protocol for transdiagnostic treatment of emotional disorders. Compr Psychiatry 55(8):1920–1927. https://doi.org/10.1016/j.comppsych.2014.07.016

30. Ehrenreich JT, Goldstein CR, Wright LR, Barlow DH (2009) Development of a unified protocol for the treatment of emotional disorders in youth. Child Fam Behav Ther 31(1):20–37

31. Trosper SE, Buzzella BA, Bennett SM, Ehrenreich JT (2009) Emotion regulation in youth with emotional disorders: implications for a unified treatment approach. Clin Child Fam Psychol Rev 12:234–254

32. Ehrenreich-May J, Rosenfield D, Queen AH, Kennedy SM, Remmes CS, Barlow DH (2017) An initial waitlist-controlled trial of the unified protocol for the treatment of emotional disorders in adolescents. J Anxiety Disord 46:46–55.   https://doi.org/10.1016/j.janxdis.2016.10.006

33. Queen AH, Barlow DH, Ehrenreich-May J (2014) The trajectories of adolescent anxiety and depressive symptoms over the course of a transdiagnostic treatment. J Anxiety Disord 28(6):511–521.   https://doi.org/10.1016/j.janxdis.2014.05.007

34. Ehrenreich-May J, Bilek EL (2012) The development of a transdiagnostic cognitive behavioral group intervention for childhood anxiety disorders and co-occurring depression

symptoms. Cogn Behav Pract 19(1):41–55. https://doi.org/10.1016/j.cbpra.2011.02.003

35. Kennedy SM, Bilek EL, Ehrenreich-May J (2018) A randomized controlled pilot trial of the unified protocol for transdiagnostic treatment of emotional disorders in children. Behav Modif Adv. https://doi.org/10.1177/0145445517753940

36. Weisz JR, Chorpita BF (2012) "Mod squad" for youth psychotherapy: Restructuring evidence-based treatment for clinical practice. In: Kendall PC (ed) Child and adolescent therapy: cognitive-behavioral procedures, 4th edn. Guilford Press, New York, NY, pp 379–397

37. Kolko DJ, Baumann BL, Bukstein OG, Brown EJ (2007) Internalizing symptoms and affective reactivity in relation to the severity of aggression in clinically referred, behavior-disordered children. J Child Fam Stud 16(6):745–759.   https://doi.org/10.1007/s10826-006-9120-3

38. Kolko DJ, Dorn LD, Bukstein OG, Pardini D, Holden EA, Hart J (2009) Community vs clinic-based modular treatment of children with early-onset ODD or CD: a clinical trial with 3-year follow-up. J Abnorm Child Psychol 37(5):591–609.   https://doi.org/10.1007/s10802-009-9303-7

39. Kolko DJ, Pardini DA (2010) ODD dimensions, ADHD, and callous–unemotional traits as predictors of treatment response in children with disruptive behavior disorders. J Abnorm Psychol 119(4):713–725. https://doi.org/10.1037/a0020910

40. Shaffer A, Lindhiem O, Kolko DJ, Trentacosta CJ (2013) Bidirectional relations between parenting practices and child externalizing behavior: a cross-lagged panel analysis in the context of a psychosocial treatment and 3-year follow-up. J Abnorm Child Psychol 41(2):199–210. https://doi.org/10.1007/s10802-012-9670-3

41. Chorpita BF (2007) Modular cognitive-behavioral therapy for childhood anxiety disorders. Guilford Press, New York, NY

42. Borntrager CF, Chorpita BF, Higa-McMillan C, Weisz JR (2009) Provider attitudes toward evidence-based practices: are the concerns with the evidence or with the manuals? Psychiatr Serv 60(5):677–681. https://doi.org/10.1176/appi.ps.60.5.677

43. Wiltsey Stirman S, Comer JS (2018) What are we even trying to implement? Considering the relative merits of promoting evidence-based protocols, principles, practices, or policies. Clin Psychol Sci Pract. https://doi.org/10.1111/cpsp.12269

44. Weisz JR, Chorpita BF, Palinkas LA, Schoenwald SK, Miranda J, Bearman SK et al (2012) Testing standard and modular designs for psychotherapy treating depression, anxiety, and conduct problems in youth: a randomized effectiveness trial. Arch Gen Psychiatry 69 (3):274–282. https://doi.org/10.1001/archgenpsychiatry.2011.147

45. Tonarely NA, Halliday ER, Ehrenreich-May J (2019, November) Comparing individual and group administration of the Unified Protocol for Transdiagnostic Treatment of Emotional Disorders in Children (UP-C). Poster presented at the meeting of the Association for Behavioral and Cognitive Therapies Convention, Atlanta, G.A.

46. Hudson JL, Rapee RM (2002) Parent-child interactions in clinically anxious children and their siblings. J Clin Child Adolesc Psychol 31 (4):548–555. https://doi.org/10.1207/153744202320802214

47. Weisz JR, Chorpita BF, Frye A, Ng MY, Lau N, Bearman SK, Hoagwood KE (2011) Youth top problems: using idiographic, consumer-guided assessment to identify treatment needs and to track change during psychotherapy. J Consult Clin Psychol 9(3):369–380. https://doi.org/10.1037/a0023307

48. Linehan MM, Korslund KE, Harned MS, Gallop RJ, Lungu A, Neacsiu AD et al (2015) Dialectical behavior therapy for high suicide risk in individuals with borderline personality disorder: a randomized clinical trial and component analysis. JAMA Psychiat 72 (5):475–482. https://doi.org/10.1001/jamapsychiatry.2014.3039

49. Williams JM, Teasdale J, Segal Z, Kabai-Zinn J (2007) The mindful way through depression: freeing yourself from chronic unhappiness. Guilford Press, New York, NY

50. Peris TS, Compton SN, Kendall PC, Birmaher B, Sherrill J, March J et al (2015) Trajectories of change in youth anxiety during cognitive—behavior therapy. J Consult Clin Psychol 83(2):239–252. https://doi.org/10.1037/a0038402.supp

51. Foa EB, Liebowitz MR, Kozak MJ, Davies S, Campeas R, Franklin ME et al (2005) Randomized, placebo-controlled trial of exposure and ritual prevention, clomipramine, and their combination in the treatment of obsessive-compulsive disorder. Am J Psychiatry 162 (1):151–161. https://doi.org/10.1176/appi.ajp.162.1.151

52. Raggi VL, Samson JG, Felton JW, Loffredo HR, Berghorst LH (2018) Exposure therapy for treating anxiety in children and adolescents: a comprehensive guide. New Harbinger Publications, Oakland, CA

53. Barlow DH, Conklin LR, Bentley KH (2015) Psychological treatments for panic disorders, phobias, and social and generalized anxiety disorders. In: Nathan PE, Gorman JM (eds) A guide to treatments that work, 4th edn. Oxford University Press, New York, NY, pp 409–461

54. Bouton ME (1993) Context, time, and memory retrieval in the interference paradigms of Pavlovian learning. Psychol Bull 114:80–99. https://doi.org/10.1037/0033-2909.114.1.80

55. Miller RR, Matzel LD (1988) Contingency and relative associative strength. In: Klein SB, Mowrer RR (eds) Contemporary learning theories: Pavlovian conditioning and the status of traditional learning theory. Lawrence Erlbaum Associates, Hillsdale, NJ, pp 61–84

56. Wagner AR (1981) SOP: a model of automatic memory processing in animal behavior. In: Spear NE, Miller RR (eds) Information processing in animals: memory mechanisms. Erlbaum, Hillsdale, NJ

57. Baker A, Mystkowski J, Culver N, Yi R, Mortazavi A, Craske MG (2010) Does habituation matter? Emotional processing theory and exposure therapy for acrophobia. Behav Res Ther 48:1139–1143. https://doi.org/10.1016/j.brat.2010.07.009

58. Abramowitz JS, Arch JJ (2014) Strategies for improving long-term outcomes in cognitive behavioral therapy for obsessive-compulsive disorder: Insights from learning theory. Cogn Behav Pract 21(1):20–31. https://doi.org/10.1016/j.cbpra.2013.06.004

59. Craske MG, Kircanski K, Zelikowsky M, Mystkowski J, Chowdhury N, Baker A (2008) Optimizing inhibitory learning during exposure therapy. Behav Res Ther 46(1):5–27. https://doi.org/10.1016/j.brat.2007.10.003

60. Craske MG, Treanor M, Conway CC, Zbozinek T, Vervliet B (2014) Maximizing exposure therapy: an inhibitory learning approach. Behav Res Ther 58:10–23

61. Weck F, Neng JMB, Richtberg S, Jakob M, Stangier U (2015) Cognitive therapy versus exposure therapy for hypochondriasis (health anxiety): a randomized controlled trial. J Consult Clin Psychol 83(4):665–676. https://doi.org/10.1037/ccp000001

62. Wolpe J (1969) The practice of behavior therapy. Pergamon, Oxford

63. Hayes SC, Hofmann SG (2018) Process-based CBT: the science and core clinical competencies of cognitive behavioral therapy. New Harbinger Publications, Oakland, CA

Chapter 14

Cognitive Behavioral Therapy with Youth: These Are a Few of Our Favorite Things

Robert D. Friedberg, Brad J. Nakamura, Anusha Kakolu, Sandra Trafalis, Jasmine Thomas, and Marissa Cassar

Abstract

Cognitive behavioral therapy has established itself as a gold standard approach for a wide variety of youth mental health problems over the past several decades. Rooted firmly in both theoretical and empirical foundations, this treatment approach has continued to innovate in a variety of ways, both within and across varying disorders, ranging from spawning new treatment approaches to the creation of modular paradigms. This concluding section provides brief overviews of each chapter, as well as remarks on potentially particularly noteworthy components contained therein. As demonstrated here and throughout the book overall, the reader is exposed to an extensive array of hands on resources for working with youth and families. Spanning all chapters, this compendium helps to underscore several important points about CBT: excellent translation of basic theory and bench science to clinical practice, seamless and bidirectional integration of assessment and treatment approaches, personalized and patient-centered treatment strategies, and a strong focus on action and experiential learning. Taken together, this work makes a substantial contribution to the literature and provides practitioners with an excellent hands-on resource for working with youth and families.

Key words Translation, Integrated assessment, Patient-centered treatment, Experiential learning

---

1 Introduction

> *Raindrops on roses*
> *And whiskers on kittens*
> *Bright copper kettles and warm woolen mittens*
> *Brown paper packages tied up with strings*
> *These are a few of my favorite things*
> Hammerstein and Rodgers (1965)

In the late 1960s and early 1970s, an audacious and even precocious newcomer entered onto the psychotherapy stage. At the time, skeptics abounded. Now nearly 60 years later, CBT with children is considered the gold standard approach. Some clinical scientists claim the approach should be housed within a neuroscience

Robert D. Friedberg and Brad J. Nakamura (eds.), *Cognitive Behavioral Therapy in Youth: Tradition and Innovation*, Neuromethods, vol. 156, https://doi.org/10.1007/978-1-0716-0700-8_14, © Springer Science+Business Media, LLC, part of Springer Nature 2020

department [1]. The paradigm has been investigated by rando-mized clinical trials (RCT), various experimental designs, meta-analyses, systematic reviews, and case reports. So, what is new with this heretofore spunky upstart treatment approach?

Indeed, much is new in CBT. The text is filled with state-of-the-science information pertaining to the clinical care of young patients and their caretakers. The material represents cutting edge ideas from 44 contributors working in 23 institutions. Common yet challenging clinical disorders such depression, anxiety, autism, post-traumatic stress, eating disorders, substance use, and obsessive-compulsive disorders are addressed. Additionally, key contemporary themes such as the role of case conceptualization and diversity issues in CBT are tackled. Finally, emerging treatment models such as mindfulness-based CBT, transdiagnostic approaches, and modular paradigms are described.

In this concluding chapter, we summarize many of the essential parts of the 13 preceding chapters. Like Maria von Trapp portrayed by the iconic Julie Andrews in the classic musical, *The Sound of Music*, we have favorite things. Our favorite things are not rain-drops on roses or whiskers on kitten but key elements embedded in cognitive behavioral therapy with youth. Accordingly, we conclude the chapter with a discussion of our favorite things about CBT.

2   Brief Chapter Overview

McLeod, Fjermestad Liber, and Violante (Chapter 1) lead off the book with a stirring overview of CBT spectrum approaches. The chapter is foundational for elucidating the behavioral and cognitive theories underlying CBT practice. Further, the work also provides a precise review of the various interventions commonly employed in CBT along with focusing on method and delivery formats. The content is user-friendly, comprehensible, and exceedingly suitable to both novice and seasoned practitioners. Compelling examples succinctly fuel the translation of theory to practice. The cultural and developmental consideration sections are especially pivotal in pro-moting clinicians' understanding of marginalized youths' contex-tual experiences. Overall, the chapter is packed with practical, flexible, and jargon-free information.

CBT is the treatment of choice for depressed youth for many years. The approach enjoys solid empirical grounding and also is widely taught in psychology programs across the world. The chap-ter on depression, written by Lawrence, Buffie, Schwartz-Mette, and Nangle (Chapter 2), emphasizes core CBT techniques and associated clinical adaptations necessary for treating youth diag-nosed with depression. Their contribution highlights theoretical foundations, case conceptualization, and intervention strategies for working with young patients. Various handy worksheets, tables, and rich sample dialogues make this material quite portable. More specifically, sample dialogues are provided in a session-wise order with tips for adaptation to facilitate seamless bench to bedside translations. Finally, Lawrence and colleagues list valuable resources that contain information practitioners can use to better their clinical skills.

Anxiety disorders affect nearly 32% of preadolescent and adolescent youth. Badin, Alvarez, and Chu in Chapter 3 provide a thorough overview of CBT strategies and principles that work in a variety of contexts. The chapter succinctly presents the essentials embedded in a CBT approach to anxiety disorders. In particular, robust case conceptualization places the procedures in a broad clinical context. Badin and colleagues utilize an informative case study that illustrates how to best put CBT into practice. They provide a discussion of ways to work with parents during treatment and how to incorporate cultural context factors into the therapeutic process. With worksheet templates and clear figures, the chapter brings practices and principles to life. The chapter provides useful guidance and principles for therapists who would like to incorporate exposure into their therapeutic practice, work effectively with parents, and navigate youth ecosystems. With a focus on real-world settings, this chapter is a comprehensive, hands-on guide that a clinician will find invaluable.

In Chapter 4, Babinski shares her insight and experience treating young patients diagnosed with OCD. She provides a practical and thorough guide to treatment that maintains fidelity while maximizing flexibility and portability. OCD affects about 1–2% of youth, and CBT is the treatment of choice thanks to sturdy clinical outcomes. CBT also meets the preferences of youth and families as a non-medication treatment option. Clinicians should appreciate the numerous examples and practical suggestions in the chapter along with sample dialogue and case examples taking the clinician through each therapeutic component. For example, in the assessment phase, Babinski provides examples of measures that can be used at intake and follow-up points. Daily report cards track progress, subjective distress is monitored at various points during treatment, and symptom hierarchies are constructed. A particularly helpful dialogue includes interchanges between therapist, child, and mother addressing unwitting reinforcement of symptoms. Overall, the dialogues are positive in tone, enhance motivation, and effectively maintain a positive and collaborative approach. For instance, Babinski models how to facilitate a sense of hope, reinforce treatment gains, and establish realistic expectations.

Allen, Riden, and Shenk explain their approach to trauma-focused cognitive behavioral therapy (TF-CBT) in Chapter 5. They provide a thorough overview of TF-CBT which includes historical, theoretical, and empirical perspectives. The chapter opens with a discussion of classical learning theory and Mowrer's two-factor model. Here the reader gains an understanding of the mechanisms by which traumatic responses to environmental cues are established and how subsequent maladaptive behaviors are maintained through reinforcement. Allen and his team provide sound developmentally oriented instruction of how to implement TF-CBT. For example, in a step-by-step, sequential order starting at a patient's intake, they present wise clinical instruction using relevant examples and dialogue for each treatment component including relaxation skills, affect expression and modulation, cognitive coping, narrative construction, in vivo exposure, conjoint sessions, incorporating caregivers, and safety planning. Numerous resources for clinicians wishing to obtain information about further training and standards of care are offered. The authors recommend ways to avoid potential pitfalls that can undermine or unnecessarily lengthen therapy as well as stressing important developmental considerations.

Treating aggressive young patients is often a challenging and perplexing clinical task. Stromeyer, Lochman, and Kassing in Chapter 6 expertly guide readers through the therapeutic process by discussing the elements and implementation of the *Coping Power Program*. The program aims to divert children from aggressive behaviors that may lead to conduct disorders and later possible substance abuse. Child components of the program focus on emotion recognition and regulation, strategy-learning for a successful future, and practice of coping and anger management skills. The program also recognizes how parent and family dynamics may modulate child behavior and implements a parent component, which has strong roots in social learning theory. Throughout the chapter, the authors provide clear and concise explanations of the theoretical framework that shapes each component of the intervention. They also provide many real-world dialogues that may be modeled. The hands-on treatment methods listed are readily application for clinician use. Regardless of parent participation, which is highly encouraged but not necessary, each child-centered exercise is displayed in an encouraging positive. Stromeyer and colleagues' work delivers much needed resources and straight-forward recommendations for treating aggressive youth and their families.

Banneyer, Fein, and Storch in Chapter 7 provide a comprehensive review of CBT for youth with autism spectrum disorder (ASD) and comorbid anxiety or obsessive compulsive disorders. The treatment includes five essentials to be incorporated such as assessment, psychoeducation, goal-setting, exposure/response prevention (ERP), and termination. Each module of the treatment incorporates family participation and practice between sessions to potentiate outcomes. The authors provide realistic sample dialogues that should ring true to readers. Additionally, the individual exercises that Banneyer and colleagues suggest for children and families are exceedingly portable. Overall, the authors describe their treatment approaches and techniques in an actionable and straightforward way that may be applied to a variety of children with these disorders. Working clinicians should profit greatly from their work.

Bobek and colleagues in their chapter (Chapter 8) on CBT with substance using youth do a commendable job in exploring the core elements of CBT in individual and group settings. They provide a wide-ranging overview on adolescent substance use and also highlight theoretical and empirical foundations of CBT for substance use. The authors knowingly describe procedures for family behavioral therapy and culturally alert adaptations. Clinical examples show readers the way to readily apply concepts and practices. There are numerous worksheets provided in the chapter which are simple and easy to use in individual and group settings. The chapter has a logical flow and is written in simple language, making it an easy read for practicing clinicians to replicate these powerful methods in their own work.

Although cognitive behavioral therapy (CBT) is acknowledged as the most successful and robust treatment for adults with eating disorders (EDs), few studies have examined CBT's effectiveness among younger patients. In Chapter 9, Essayli and Vitousek provide an exhaustive and detailed description of how to adapt CBT for young patients with EDs. The text gives readers tips for case conceptualization, structuring sessions, and properly implementing CBT interventions. The sample dialogue artfully illustrates both the principles and practice of CBT interventions. The chapter moves onto explaining the importance of augmenting CBT interventions with family involvement. Clinicians will appreciate the detailed descriptions and practicality of the interventions which can be seamlessly modified to patients' presenting complaints. The brief clinical vignettes interwoven in the text also helps readers understand the process in action. The text is written in a clear, conversational, and engrossing tone. Many practical suggestions and exercise augment the material. Overall, this chapter represents a remarkable advance in the teaching and dissemination of CBT for EDs.

A creative mindfulness-based approach to treating childhood anxiety disorders is presented in the chapter by Madni, Giambrone, and Semple (Chapter 10). The 12-session intervention provides exercises that may be easily implemented by clinicians and readily digestible for children. MBCT-C is broken down into three units that are designed to engage and educate children in the practice of mindfulness and bodily awareness, the use of the senses, and the implementation of mindfulness practices for long-term coping. Exercises are playful and have a camp-like feeling to them. For example, "Mindfully Moooving Slowly" has a captivating title, is likely attractive to young learners, and encourages the development of awareness body movements and sensations. Another compelling activity includes the music and emotion connection exercise in sessions 5 and 6. In these sessions, children are invited to produce musical tracks that reflect their emotional state and are encouraged to produce interpretive and imaginative titles. Exercises involving the senses also engage children with attractive materials, such as Silly Putty. The authors have also thought systematically about the implementation of these exercises; adaptations are provided for children with special learning requirements or even for children with dietary restrictions that may not be able to participate as heavily in taste-centered activities. In general, the sessions are laid out very clearly that may allow maximum understanding for practitioners but also leave plenty of room for clinician creativity and playfulness. The exercises appear to be a practical and fun treatment that may create a space where children do not even feel as though they are in an intervention program. Overall, the approach embodies innovation that may foster a safe and exciting method of combating childhood anxiety.

Without a doubt, the world is becoming a much more diverse place. In Chapter 11, Cardemil, Straubel, and O'Leary provide readers with a rich up-to-date review of the literature pertaining to the care of diverse youth experiencing emotional and behavioral difficulties. Spanning empirical, theoretical, and clinical issues, the chapter gives readers accessible information and expert guidance. Cardemil et al. include a vividly detailed case description replete with replicable recommendations for interventions. This is a can't miss chapter!

The chapter by Boustani, Regan, and Stanick (Chapter 12) provides a clinician-friendly resource for understanding the practical and technical implementations of a modular approach to CBT. The chapter lays out the fundamental principles of what may be regarded as a modular approach; four criteria include the partially decomposable nature of a treatment, serving a specific purpose, the use of standardized interface, and all information must be self-contained. The chapter distinctly clarifies what it means to modulate a treatment and provides specific logic for clinicians to understand how and when treatments may be altered for individual clients. Throughout, detailed examples are provided as well as miniature vignette examples in order to aid understanding of modular CBT implementation. Realistic dialogues that may take place with young clients are carefully supplied to give clinicians a tangible example of how these ideas may be relevant to specific children. SMART goals, a simple and tactile concept, are also mentioned as a helpful strategy for helping clinicians guide treatment decision-making. Additionally, the authors recognize and discuss the possibility that a modular approach may not be the end-all-be-all fix to child and family engagement in CBT treatment. They recognize and acknowledge that there are many psychosocial factors that may drive the context of a child's treatment that may not necessarily be an issue related to the treatment itself. In conclusion, the principles and guidelines behind modularity are clearly and easily defined by the authors and may be considered by clinicians of all skill levels and with a variety of clinical populations.

A transdiagnostic, unified protocol for treating children and adolescents with emotional disorders is described by Halliday and Ehrenreich-May (Chapter 13). Their work provides a comprehensive overview of the treatment of emotional disorders in children and adolescents that will be much appreciated by practicing clinicians. The authors highlight the empirical support and theoretical foundations for the use of this approach. The interventions included in the protocol are flexible and tailored to particular patients' complaints and circumstances. Session-wise agendas are described which enables clinicians' ability transfer the material from the text to their practices. Specific procedures directly target motivation, emotions, behaviors, awareness of physical sensations, flexible thinking, situational emotion exposure and reviewing accomplishments. In particular, Halliday and Ehrenreich-May also include several worksheets that can be seamlessly integrated into practitioners' work with young patients. For example, the *Detective Thinking* worksheet is designed as a game that helps children learn to look for clues to assess their surroundings.

3   Our Favorite Things

3.1   Translate Basic Theory and Bench Science to Clinical Practice

CBT is relevant! For the most part, CBT does a good job of translating basic theory and bench science findings to clinical practice [2]. The CBT methods in this book are diverse and far-ranging. However, all these multiple procedures share common theoretical DNA. Allegiance to learning theory, information processing principles, and self-efficacy theory are all maintained.

CBT is a learning theory based treatment paradigm including classical, operant, and social learning theory models. Simply, maladaptive behaviors are acquired phenomena. The premise is then if one can learn unproductive coping patterns, they can unlearn them and replace the problematic action patterns with more functional ones. The theory and science buttressing these notions date back nearly a century [3, 4]. Indeed, reinforcement is one of the few laws in psychological science [5]. Classical, operant, and social learning principles pervade proper CBT practice. Classical conditioning is the hub of exposure treatments. Operant canons are fundamental to case conceptualization, functional analyses, and behavior management programs.

Pioneers such as Aaron T. Beck and Phil Kendall advanced the understanding of cognitive processes in psychopathology. Ingram and Kendall [6] divided the information processing system into three crucial components: cognitive structures (e.g., schemata), cognitive products (e.g., automatic thoughts), and cognitive processes (e.g., distortions). The content-specificity hypothesis [7–9] proposed that different feelings state are demarcated by unique cognitive products. Accordingly, the CSH yielded an actionable clinical template for determining "hot" or emotionally meaningful cognitions. Finally, recent work by Ehrenreich and her team (Chapter 13, this volume; [8–12]) as well as Chorpita and his colleagues (Chapter 12, this volume; [13–20]) modernized CBT into even more clinician-friendly packages by developing unified transdiagnostic protocols and modular based methods.

The tight nexus connecting bench science to office practice is perhaps one of the most compelling elements of CBT. Clinicians schooled in the basic theory and empirical foundations of CBT do not need to reach far to find direct applications. While some critics of the approach may argue its common sense, straight-forward approach, and readily measurable components appear too simplistic, advocates counter that CBT's accessibility and portability represents weighty significance.

3.2 Seamlessly Integrates Assessment and Treatment

Seamless integration of assessment and treatment is a CBT tradition [21]. Assessment builds a platform that supports case conceptualization, treatment planning, progress monitoring, and outcome evaluation. Each innovation in this text is rooted in this tradition. Reliance upon regular assessment and tracking avoids "seat of the pants" navigation approach to treatment (e.g. "This feels right … We'll get to the goals somehow and sometime.").

Currently, measurement-based care (MBC) is increasingly demanded by payers and implemented by providers. Indeed, there are no shortages of professionals who advocate for this practice [18, 22–26]. Practitioners reading this text are likely well aware that accountability and value-based service are contemporary watchwords. Business leaders have long advocated that if something cannot be measured, it is not important. Now and in the coming decade patient outcomes will be prized and paid at premium. Documenting favorable outcomes is fundamental to sustainability of services and professional stability. MBC facilitates demonstrating value-added benefits of CBT.

The strong and bidirectional links between assessment and treatment help to make CBT a premium front-line approach for treating young patients' psychological difficulties. In all chapters of this book, assessment feeds the treatment process. If metrics indicate treatment is successful, clinicians should stay the course. If the data show treatment is not working, practitioners can recalculate the route. Accordingly, clinical flexibility is served.

While there are a diversity of validated measures employed by the chapter authors and elsewhere, competent functional analysis (FA) lies at the heart of effective CBT. FA is learning theory based and heavily influenced by operant models. Simply, FA plots the cues, consequences, and ambient conditions that initiate, maintain, exacerbate, and attenuate particular behaviors [27]. Behavior is seen as purposeful and goal-oriented. Accordingly, it is the nucleus of self-control oriented interventions. FA is one handy clinical tool! Chapters in this text pay homage to this foundational practice.

The bond joining assessment and treatment is firmly cemented in CBT approaches. Tying metrics to methods contributes to clinical accountability. Clinicians who engage in measurement-based care are rewarded with reliable direction, purpose, and discernible outcomes.

3.3 Treatment Is Collaborative and Patient-Centered

CBT like most forms of psychotherapy is a fundamentally interpersonal venture. Collaboration is a central element of any CBT based intervention package. Far from being prescriptive and clinician-driven, proper implementation of CBT requires a robust patient–therapist partnership. Young patients and their patients are encouraged to take active ownership in their return to emotional health. Collaboration is empowering and ensures genuinely informed consent.

Personalized medicine is in fashion. Individualized treatment plans are necessary. Active patient collaboration propels this personalized approached to behavioral health care. When patients and therapists are genuine partners, building a customized evidence based treatment that suits individual patients' circumstances becomes easier. Each of the previous 13 chapters adopts collaborative, patient-centered approaches. Contextual vicissitudes are explicitly addresses in the diverse clinical applications presented. In this way, various processes and procedures are made more relevant and real to particular patients.

A robust therapeutic partnership enables both patients and clinicians to optimize the clinical encounter. Shared goals and mutual commitment propel therapeutic momentum. True patient–therapist collaboration coupled with theoretical soundness, empirical support, and technical proficiency is a recipe for success.

3.4 Focus on Action and Experiential Learning

In CBT, actions speak louder than words. A.T. Beck [7] emphasized that the "experiential approach exposes the patient to experiences that are in themselves powerful enough to change misconceptions (p. 214)." Nearly all the chapters in this text include a healthy dose of behavioral experimentation and experiential learning. And why not! Silverman and Kurtines [28] considered exposure a common factor across all successful psychotherapies. Experiential learning in CBT with youth helps young patients "show that they can [29]. In this way, CBT is emotionally evocative and catalyzes young peoples' efforts toward genuine self-efficacy [30, 31]. It has been argued that without exposures and behavioral experiments, CBT risks being an intellectual exercise [32]

Throughout this book, the authors of individual chapters emphasize the fundamental notion that enduring change is initiated and sustained by prompting productive persistence in the face of distress or negative emotional arousal. I (RDF) often teach my supervisees that we as therapists are in the change business not the stay the same business. Therapeutic change is difficult and, at times, even painful. Altering heavily reinforced action patterns, firmly held emotional reactions, and rigid appraisals is not easy. Navigating a clinical course cannot be done abstractly. Like learning to swim or ride a bike, coping skills must be acquired directly in the context in which they are applied. One cannot learn to swim without getting in the water. In CBT, patients and therapists must get wet!

4   Conclusion

Improving the emotional well-being of vulnerable children represents a public health imperative. As the book documents, CBT spectrum approaches offer a first-line intervention option. However, there is a lack of practitioners who are adequately trained to deliver these powerful methods [33]. Thus, many youth are left to battle with their distressing emotions with ineffective strategies. Ideally, this compendium of potent psychotherapeutic methods will equip practitioners from various disciplines with user-friendly tools to help vulnerable patients.

References

1. March J (2009) The future of psychotherapy for mentally ill children and adolescents. J Child Psychol Psychiatr 50:170–179

2. Friedberg RD, Thordarson MA (2018) Cognitive behavioral therapy. In: Matson J (ed) Handbook of child psychopathology and developmental disabilities treatment. Springer, New York, NY, pp 43–62

3. Pavlov IP (1927) Conditioned reflexes: an investigation of the physiological activity of the cerebral cortex. Oxford University, London

4. Skinner BF (1938) The behavior of organisms. Appleton-Century-Crofts, New York, NY

5. Thorndike EL (1898) Animal intelligence: an experimental study of the associative process in animals. Psychol Rev Mono Supp 4:2

6. Ingram RE, Kendall PC (1986) Cognitive clinical psychology: Implications of an information processing perspective. In: Ingram RE (ed) Information processing approaches to clinical psychology. Academic, Orlando, FL, pp 3–21

7. Beck AT (1976) Cognitive therapy and the emotional disorders. International University Press, New York, NY

8. Clark DM, Beck AT (1988) Cognitive approaches. In: Last CG, Hersen M (eds) Handbook of anxiety disorders. Pergamon, New York, NY, pp 95–122

9. Laurent J, Stark KD (1993) Testing the cognitive content-specificity hypothesis with anxious and depressed youngsters. J Abn Psychol 102:226–237

10. Ehrenreich-May J, Bilek EL (2012) The development of a transdiagnostic cognitive behavioral group intervention for childhood anxiety disorders and co-occurring depression symptoms. Cogn Behav Prac 19:41–55

11. Kennedy S, Bilek EL, Ehrenreich-May J (2018) A randomized controlled pilot trial of the unified protocol for transdiagnostic treatment of emotional disorders in children. Behav Modif Adv. https://doi.org/10.1177/0145445517753940

12. Ehrenreich-May J, Kennedy SM, Sherman JA et al (2018) Unified protocol for transdiagnostic treatment of emotional disorders in children and adolescents: therapist guide. Oxford University Press, New York, NY

13. Chorpita BF, Weisz JR (2009) Modular approach to therapy for children with anxiety, depression, trauma or conduct problems (MATCH-ADTC). Satellite Beach, FL, Practicewise

14. Chorpita BF, Daleiden EL (2009) Mapping evidence-based treatments for children and adolescents: applications of the distillation and matching model to 615 treatments from 322 randomized trials. J Consult Clin Psychol 77:566–579

15. Chorpita BF, Weisz JR, Daleiden EL et al (2013) Research network on youth mental health. Long term outcomes for Child STEPS randomized effectiveness trial: a comparison of modular and standard treatment designs with usual care. J Consult Clin Psychol 81:999–1009

16. Weisz JR, Chorpita BF (2012) "Mod squad" for youth psychotherapy: restructuring evidence based treatment for clinical practice. In: Kendall PC (ed) Child and adolescent therapy: cognitive-behavioral procedures. Guilford, New York, NY, pp 379–397

17. Becker KD, Lee BR, Daleiden EL, Lindsey M et al (2015) The common elements of engagement in children's mental health services; which elements for which outcomes. J Clin Child Adolesc Psychol 44:30–43

18. Chorpita BF, Daleiden EL, Ebsutani C et al (2011) Evidence-based treatments for children and adolescents: an updated review of indicators of efficiency and effectiveness. Clin Psychol Sci Pract 18:154–172

19. Chorpita BF, Park A, Tsai K et al (2015) Balancing effectiveness with responsiveness: therapist satisfaction across different treatment designs in the Child STEPS randomized effectiveness trials. J Consult Clin Psychol 83:709–718

20. Weisz JR, Chorpita BF, Palinkas LA et al (2012) Testing standard and modular designs for psychotherapy treating depression, anxiety, and conduct problems in youth: a randomized effectiveness trial. Arch Gen Psychiatr 69:274–282

21. Nelson-Gray R (2003) Treatment utility of psychological assessment. Psychol Assess 15:521–531

22. Bickman L (2008) A measurement feedback system is necessary to improve mental health outcomes. J Am Acad Child Adolesc Psychiatr 47:1114–1119

23. Chorpita BF, Daleiden EL, Bernstein A (2016) At the intersection of health information technology, and decision support: Measurement feedback systems…and beyond. Adm Policy Ment Health 43:471–477

24. Jensen-Doss A (2015) Practical evidence-based clinical decision-making: Introduction to the special issue. Cogn Behav Pract 22:1–4

25. McLeod BM, Jensen-Doss A, Ollendick TH (2013) Diagnostic and behavioral assessment in children and adolescents: a clinical guide. Guilford Press, New York, NY

26. Scott K, Lewis CC (2015) Using measurement-based care to enhance any treatment. Cogn Behav Pract 22:49–59

27. Kazdin AE (2001) Behavior modification in applied settings. Wadsworth, Belmont, MA

28. Silverman WK, Kurtines WM (1996) Anxiety and phobic disorders: a pragmatic approach. Plenum, New York, NY

29. Kendall PC, Robin JA, Hedtke KA (2006) Considering CBT with anxious youth? Think exposures. Cogn Behav Pract 12:136–148

30. Bandura A (1977a) Self-efficacy: toward a unifying theory of behavior change. Psychol Rev 84:191–215

31. Bandura A (1977) Social learning theory. Prentice-Hall, Englewood Cliffs, NJ

32. Friedberg RD (2015) Where's the beef: concrete elements in supervision with CBT with youth. J Am Acad Chld Adoles Psychiatr 54:527–531

33. Comer JA, Barlow DC (2014) The occasional case against broad dissemination and implementation: retaining a role for specialty care in the delivery of psychological treatment. Am Psychol 69:1–18

INDEX

Robert D. Friedberg and Brad J. Nakamura (eds.), *Cognitive Behavioral Therapy in Youth: Tradition and Innovation*, Neuromethods,
vol. 156, https://doi.org/10.1007/978-1-0716-0700-8, © Springer Science+Business Media, LLC, part of Springer Nature 2020

Printed in the United States
by Baker & Taylor Publisher Services